Applied Sequential Methodologies

Additional Volumes in Preparation

Applied Sequential Methodologies

Real-World Examples with Data Analysis

edited by

Nitis Mukhopadhyay

University of Connecticut
Storrs, Connecticut, U.S.A.

Sujay Datta

Northern Michigan University
Marquette, Michigan, U.S.A.

Saibal Chattopadhyay

Indian Insitute of Management
Calcutta, India

CRC Press

Taylor & Francis Group
Boca Raton London New York

CRC Press is an imprint of the
Taylor & Francis Group, an **informa** business

CRC Press
Taylor & Francis Group
6000 Broken Sound Parkway NW, Suite 300
Boca Raton, FL 33487-2742

First issued in paperback 2019

ISBN-13: 978-0-8247-5395-5 (hbk)
ISBN-13: 978-0-367-39456-1 (pbk)

Library of Congress Cataloging-in-Publication Data
A catalog record for this book is available from the Library of Congress.

Visit the Taylor & Francis Web site at
http://www.taylorandfrancis.com

and the CRC Press Web site at
http://www.crcpress.com

In celebration of the brilliant career of Professor Anis Mukhopadhyay, my elder brother and teacher, and in recognition of his recent retirement from the Indian Statistical Institute, Calcutta, this volume is presented to him with love and affection.

Nitis Mukhopadhyay

In the memory of my late father, in recognition of my mother's lifelong dedication to the well-being of her son, with deepest gratitude to my ever-inspiring wife, and in loving acknowledgement of many sweet distractions from my two-year old daughter.

Sujay Datta

In loving memories of my late sister and father, in admiration of the untiring effort of my mother toward my upbringing, and in recognition of the support and encouragement of my wife and children.

Saibal Chattopadhyay

APPRECIATION

To those colleagues who most kindly offered to help and freely shared their expertise and vision at various junctures of editing this volume, the Co-Editors express their sincerest gratitude and appreciation.

Many colleagues helped tremendously in the editorial process by diligently sharing the burden of refereeing one or more articles. What a difference each individual has made! The Co-Editors thank each referee for showing unselfish dedication and unmistakable enthusiasm.

The Co-Editors consider it a privilege on their part to mention all referees by name:

Douglas A. Abraham
Makoto Aoshima
Uttam Bandyopadhyay
Tathagata Banerjee
Michael I. Baron
Atanu Biswas
Arup Bose
Benzion Boukai
Probal Chaudhuri
Pinyuen Chen
Sam Efromovich
Joseph Glaz
Robert W. Keener
Subrata Kundu

Wei Liu
Rahul Mukerjee
Adam T. Martinsek
Madhuri Mulekar
Connie Page
William F. Rosenberger
Pranab K. Sen
Sugata Sen Roy
Tumulesh K. S. Solanky
Alexander G. Tartakovsky
Rand R. Wilcox
Peter Willett
Linda J. Young
Shelemyahu Zacks

Preface

Since the publication of Abraham Wald's classic text, *Sequential Analysis*, in 1947, a particularly impressive list of monographs has appeared in this field. These have led to an enormous growth in research methodologies. Some monographs boldly charted the research-track and created an indelible mark of tradition that is so well-known in sequential analysis. It is a fitting tribute to the authors that these volumes continue to serve as flag-bearers and resource guides in this field. We cite some of these influential volumes here:

Bechhofer, R.E., Kiefer, J. and Sobel, M. (1968). *Sequential Identification and Ranking Procedures*. University of Chicago Press: Chicago.

Berry, D.A. and Fristedt, B. (1985). *Bandit Problems*. Chapman & Hall: New York.

Chernoff, H. (1972). *Sequential Analysis and Optimal Design*. CBMS #8. SIAM: Philadelphia.

Chow, Y.S., Robbins, H. and Siegmund, D. (1971). *Great Expectations: The Theory of Optimal Stopping*. Houghton Mifflin: Boston.

Ghosh, B.K. (1970). *Sequential Tests of Statistical Hypotheses*. Addison-Wesley: Reading.

Ghosh, B.K. and Sen, P.K. (1991). *Handbook of Sequential Analysis*, edited volume. Marcel Dekker: New York.

Ghosh, M., Mukhopadhyay, N. and Sen, P.K. (1997). *Sequential Estimation*. Wiley: New York.

Govindarajulu, Z. (1981). *The Sequential Statistical Analysis*. American Sciences Press: Columbus.

Gut, Allan (1988). *Stopped Random Walks: Limit Theorems and Applications*. Springer-Verlag: New York.

Mukhopadhyay, N. and Solanky, T.K.S. (1994). *Multistage Selection and Ranking Procedures*. Marcel Dekker: New York.

Sen, P.K. (1981). *Sequential Nonparametrics*. Wiley: New York.

Sen, P.K. (1985). *Theory and Applications of Sequential Nonparametrics*. CBMS #49. SIAM: Philadelphia.

Shiryaev, A.N. (1978). *Optimal Stopping Rules*. Springer-Verlag: New York.

Wald, A. (1947). *Sequential Analysis*. Wiley: New York.

Wetherill, G.B. (1975). *Sequential Methods in Statistics*, 2nd ed. Chapman & Hall: London.

Woodroofe, M. (1982). *Nonlinear Renewal Theory in Sequential Analysis*. CBMS #39. SIAM: Philadelphia.

While one continues to draw inspiration from these exclusive publications, we m ay a dd that some leading 'non-sequential' books have included important material from this area too. We cite, for example, the following monographs:

> Bechhofer, R.E., Santner, T.J. and Goldsman, D.M. (1995). *Design and Analysis of Experiments for Statistical Selection, Screening, and Multiple Comparisons.* Wiley: New York.
> Gibbons, J.D., Olkin, I. and Sobel, M. (1977). *Selecting and Ordering Populations.* Wiley: New York.
> Gupta, S.S. and Panchapakesan, S. (1979). *Multiple Decision Theory.* Wiley: New York.
> Rao, C.R. (1973). *Linear Statistical Inference,* 2nd ed. Wiley: New York.
> Zacks, S. (1971). *The Theory of Statistical Inference.* Wiley: New York.

Wald's monograph was unique in its style in 1947 and in many ways it still remains unique largely because Wald's elegantly original mathematical and statistical contributions played a fundamental role in solving practical problems of real-life importance at the time. We surmise, however, that over the years these other volumes have pointed more toward theoretical advancements. Directly or indirectly, purely theoretical contributions have received more encouragement from many quarters and hence the theory of sequential analysis has indeed become very rich. Unfortunately, at the same time, real applications have taken serious hits.

We are personally convinced that this field can and should interface with every conceivable applied area of statistics. But since this field has not been accessible to practitioners for widespread real-world applications, we believe that its popularity among statisticians has dwindled. Real-life experimental data are rarely presented or discussed in sequential books and journal articles. This frustrating situation amounting to what may be viewed as a 'death sentence' has developed over many decades and sadly, this otherwise attractive field with such great promise has alienated itself nearly completely from most practitioners in statistical sciences.

One notable exception, in our view, is the area of clinical trials which has continued to be the major beneficiary of some of the basic research in sequential methodologies. Again, we cite some influential volumes in this area:

> Armitage, P. (1975). *Sequential Medical Trials,* 2nd ed. Blackwell Scientific Publications: Oxford.
> Jennison, C. and Turnbull, B.W. (1999). *Group Sequential Methods with Applications in Clinical Trials.* Chapman & Hall: London.

Rosenberger, W.F. and Lachin, J.M. (2002). *Randomization in Clinical Trials: Theory and Practice*. Wiley: New York.

But, a specialized field such as sequential analysis cannot be expected to thrive solely on applications in just one area of statistics. It is time for everyone involved to join an aggressive pursuit of real applications of sequential methodologies in as many contemporary and interesting problems of statistics as possible. We believe that time is quickly running out for purely theoretical researchers in this field to continue building newer levels of ivory towers and living in them!

Together, we all must make sequential analysis accessible to all practitioners in statistics. The idea that it is all right for sequential analysis to remain esoteric since a practitioner can seek assistance from a sequential analyst whenever needed remains as far-fetched as ever. That attitude has not worked in the last fifty years and is certainly not about to work now. We urge sequential analysts to take the initiative to vigorously 'market' their methodologies themselves — someone else can hardly ever be expected to do that for us. The field has survived thus far largely by perpetuating the idea of potential applications in the sense that somebody else may eventually use sequential methodologies somewhere in solving real-life problems some day! But, when we look at the bigger picture today, it becomes abundantly clear that sequential analysis has nearly lost its deserving place in the realm of applied statistics. This field has been ignored by nearly every practicing statistician. This is why we strongly feel that it is incumbent on all researchers in sequential analysis to try to rebuild this field's image and market their products themselves. It may eventually mean the difference between 'life' and 'death' of our wonderful field.

We urge everyone to energetically engage in turning the situation around in a positive way because there is still a great deal of hope out there. We believe that the spectrum of applications of sequential methodologies is much broader than what one finds in some of the so-called mainstream statistical monographs and journals. A variety of interesting and important real applications already exist. We dare to dream that the present volume will help in narrowing the unhealthy gap that has existed far too long between the theory and practice of sequential methodologies in problems and issues of broader interest. For us it will indeed be a dream come true if this volume serves as a catalyst to raise the level of consciousness of all sequential analysts about the current status of the field and to inspire them to fine-tune the focus of their initiatives appropriately from time to time so that sequential analysis becomes more relevant to contemporary statistical applications.

The contributing authors for the present volume of collected papers were earnestly requested to adhere to a set of general guidelines including the following:

> *"Every article should discuss clearly at least one substantive applied problem and the appropriate sequential method(s). Tangential references to potential applications are strongly discouraged. A specific application should remain in focus and guide throughout the development and/or implementation of a methodology. That is, each article should justify the relevance, importance, and usefulness of sequential methodology by highlighting an application and the associated gains with the help of real data. Theoretical developments, specific to a problem on hand, will be most welcome but their practical usefulness should be demonstrated.*
>
> *Real applications are encouraged rather than potential applications. An article will preferably include the data or refer its readers to the source of the data or provide a web-site-address if appropriate. Each article will be anonymously refereed.*
>
> *The exposition should be such that any interested reader may readily appreciate the importance of the practical problem(s) discussed and the conclusions drawn. The idea is that the variety of applied problem(s) considered in this volume will ultimately entice readers to take a look at the methodologies even if they do not consider themselves as sequential analysts. We hope to demonstrate that mathematical sophistication and complexity need not deter enthusiastic practitioners to take a look at this field which has plenty to offer in terms of everyday statistics and as it turns out, sequential methodologies are indeed often essential for solving today's challenging practical problems. It is our belief that with sufficient care, the technical coverage can be judiciously blended with lucidity of presentation so that the volume may remain accessible to many users including graduate students and budding researchers, statisticians or otherwise, looking for exposure to this area. At the same time, some hard-core researchers in sequential analysis would be expected to benefit significantly from seeing real-world applications of our craft."*

We had a modest set of goals. We clearly understood that we simply could not continue doing business as usual. We wanted to present the material in such a way that sequential analysts would get a taste of real-life problem-solving which could, in turn, inspire more methodological work in the near future. At the same time, we wanted to make sure that those scientists who were not thoroughly familiar with sequential analysis would also benefit from this volume by observing sequential methodologies at work in the real world. We thank the authors for trying their very best to address our seemingly unending list of demands like these and others.

In the early stages of planning, we invited a number of leading scientists in many substantive areas of applications including Agricultural Statistics, Animal Abundance, Bayesian Strategies, Biometry, Clinical Trials,

Computer Simulation, Data Mining, Ecology, Engineering, Finance, Fisheries, Genetics, Multiple Comparisons, Multivariate Analysis, Nonparametrics, Psychology, Sonar Detection, Tracking, and Time Series to contribute specially prepared articles. However, on account of tight deadlines set by us or due to other commitments, some invited authors could not participate in this project. We deeply regret this and their contributions are sorely missed. To those who have kindly participated in our crusade to revive our field's relevance and image in today's statistical world, we remain eternally grateful.

To our true delight, we report that this volume includes interesting methodological articles on:

- *passive acoustic detection of marine mammals* (Abraham)
- *selecting the best component* (Aoshima, Aoki, and Kai)
- *randomization tests* (Banerjee and Ghosh)
- *multistate processes* (Barón)
- *adaptive designs for clinical trials with longitudinal responses* (Biswas and Dewanji)
- *data mining* (Chang and Martinsek)
- *approximations for moving sums of discrete random variables* (Chen and Glaz)
- *measurement-error model* (Datta and Chattopadhyay)
- *density estimation of wool fiber diameter* (de Silva and Mukhopadhyay)
- *financial applications of nonparametric curve estimation* (Efromovich)
- *interim and terminal analyses of clinical trials* (Lai)
- *tests for target tracking* (Li and Solanky)
- *multiple comparisons* (Liu)
- *designing computer simulations* (Mukhopadhyay and Cicconetti)
- *estimation in the agricultural sciences* (Mulekar and Young)
- *contrasting group-sequential and time-sequential interim analysis in clinical trials* (Sen)
- *change-point detection in multichannel and distributed systems* (Tartakovsky and Veeravalli)
- *two-stage multiple comparison procedures in Psychology* (Wilcox)
- *testing in the agricultural sciences* (Young), and
- *ordering genes* (Zacks and Rogatko).

We believe that this is quite an impressive list indeed.

More than one colleague refereed each paper anonymously. The authors revised their manuscripts diligently by taking into account all constructive suggestions and criticisms from the referees. What one finds in this volume is a direct result of total commitment as well as unending patience and support from all parties involved. We remain indebted to this enthusiastic group of colleagues.

We admit that the twenty articles included here are not all written at the same level and personally, we view this disparity positively. Seeing this unevenness in some places, the readers will probably come to realize more that routine phrases such as "applied statistics" and "usefulness of a methodology" are also subject to interpretation.

It is our belief that many students, researchers, or practitioners will find in this volume some important and interesting material. The volume can be used both as reference material as well as a solo textbook. One can also use it as a companion with another book while offering a senior undergraduate or graduate level course in sequential methods. An experienced teacher may also discover a number o f h idden or not-so-hidden ideas on conducting hands-on practical experiments to gather real or realistic data that would make a traditional offering of a course in sequential analysis more interesting, lively, and above all, relevant.

Even though this is a substantial volume in itself on applications of sequential methodologies, in no way do we claim that this represents all types of applications. To be truthful, it is far from it. We hope to have other opportunities in the future to be more inclusive and capture a greater diversity of applications. If the present volume makes readers realize that this is a field with a great deal of promise for both intra-disciplinary work in statistics as well as for all types of inter-disciplinary work, then that will be our most gratifying reward.

It has been a real pleasure to work with the editorial and production staff at Marcel Dekker in planning and completing this project. We specially mention Ms. Maria Allegra and Ms. Helen Paisner and thank them both. Without their patience and constant support, this project could not have reached this stage. We are also thankful to several technical experts who helped us at various stages of the compilation process, especially to Dr. Andrew A. Poe from the Department of Mathematics and Computer Science, Northern Michigan University and Professor Uttam Sarkar from the Indian Institute of Management Calcutta. In addition, two of us (Datta and Chattopadhyay) gratefully acknowledge the support received in the form of a faculty grant from Northern Michigan University and a research grant from the Center for Management Development Studies, Indian Institute of Management

Calcutta, respectively. One of us (Mukhopadhyay) gratefully acknowledges the support received through a sabbatical leave from the University of Connecticut in the fall semester 2003 and expresses heartfelt gratitude to Dr. Tom F. Babor, Professor and Head of the Department of Community Medicine, University of Connecticut Health Center, for gracious hospitalities and use of their facilities.

Both collectively and individually, we express indebtedness to our colleagues, students and staff at our home institutions. We must apologize, however, for not mentioning them by name. Mr. Ranjan Mukhopadhyay and Ms. Cathy Brown have rendered invaluable help during the final preparation of a camera-ready copy. Our sincerest thanks go to both Ranjan and Cathy.

Finally, we express our deepest sense of gratitude to our families for their never-ending encouragement and love that gave us the ultimate courage to shoulder this challenging project in the first place.

Nitis Mukhopadhyay
Sujay Datta
Saibal Chattopadhyay

Contents

18. Extension of Hochberg's Two-Stage Multiple Comparison Method   371
Rand R. Wilcox

19. Sequential Testing in the Agricultural Sciences   381
Linda J. Young

Contributing Authors

Douglas A. Abraham

Applied Research Laboratory, Pennsylvania State University, P.O. Box 30, State College, PA 16804, U.S.A. E-mail: daa10@psu.edu

Mitsuru Aoki

Institute of Mathematics, University of Tsukuba, Ibaraki 305-8571, Japan. E-mail: aoki@math.tsukuba.ac.jp

Makoto Aoshima

Institute of Mathematics, University of Tsukuba, Ibaraki 305-8571, Japan. E-mail: aoshima@math.tsukuba.ac.jp

Tathagata Banerjee

Department of Statistics, Calcutta University, 35 Ballygunge Circular Road, Kolkata 700019, India. E-mail: btathaga@yahoo.com

Michael I. Barón

Department of Statistics, University of Texas at Dallas, Richardson, TX 75080, U.S.A. E-mail: mbaron@utdallas.edu

Atanu Biswas

Applied Statistics Unit, Indian Statistical Institute, 203 B. T. Road, Kolkata 700 108, India. E-mail: atanu@isical.ac.in

Yuan-Chin Ivan C. Chang

Institute of Statistical Science, Academia Sinica, 128, Sec. 2, Academia Road, Taipei, Taiwan 11529. E-mail: ycchang@sinica.edu.tw

Saibal Chattopadhyay

Operations Management Group, Indian Institute of Management Calcutta,

Joka, D.H. Road, PO Box 16757, Alipore, Kolkata 700027, INDIA. E-mail: chattopa@hotmail.com

Jie Chen

Department of Computing Services, University of Massachusetts, 100 Morrissey Boulevard, Boston, MA 02125, U.S.A. E-mail: jie.chen @umb.edu

Greg Cicconetti

Department of Mathematical Sciences, Muhlenberg College, Allentown, PA 18104, U.S.A. E- Mail: cicconet@muhlenberg.edu

Sujay Datta

Department of Mathematics, Statistics and Computer Science, Northern Michigan University, 1401 Presque Isle Avenue, Marquette, MI 49855, U.S.A. E-mail: sdatta@euclid.acs.nmu.edu

Basil M. de Silva

Department of Mathematics and Statistics, RMIT University City Campus, GPO Box 2476V, Melbourne, Victoria 3001, Australia. E-Mail: desilva@rmit.edu.au

Anup Dewanji

Applied Statistics Unit, Indian Statistical Institute, 203 B. T. Road, Kolkata 700 108, India. E-mail: dewanjia@isical.ac.in

Sam Efromovich

Department of Mathematics and Statistics, University of New Mexico, Albuquerque, NM 87131-1141, U.S.A. E-mail:efrom@math.unm.edu

Onkar P. Ghosh

Directorate General of Commercial Intelligence and Statistics, Govt. of India, 1 Council House Street, Kolkata 700001, India. E-mail: onkar-ghosh@yahoo.co.uk

Joseph Glaz

Department of Statistics, UBox 4120, University of Connecticut, Storrs, CT 06269-4120, U.S.A. E-Mail: glaz@uconnvm.uconn.edu

Masaki Kai

Insurance Distribution, Fidelity Investments Japan Limited, Tokyo 104-0033, Japan. E-mail: masaki.kai@fid-intl.com

Tze L. Lai

Department of Statistics, Stanford University, Stanford, CA 94305-4065, U.S.A. E-mail: lait@stat.stanford.edu

X. Rong Li

Department of Electrical Engineering, University of New Orleans, New Orleans, LA 70148, U.S.A. E-mail: xli@uno.edu

Wei Liu

Department of Mathematics, University of Southampton, SO17 1BJ, U.K. E-mail: W.Liu@maths.soton.ac.uk

Adam T. Martinsek

Department of Statistics, University of Illinois, 725 South Wright Street, Urbana, IL 61801, U.S.A. E-mail: martins@stat.uiuc.edu

Nitis Mukhopadhyay

Department of Statistics, UBox 4120, University of Connecticut, Storrs, CT 06269-4120, U.S.A. E-Mail: mukhop@uconnvm.uconn.edu

Madhuri Mulekar

Department of Mathematics and Statistics, University of South Alabama, 307 University Boulevard, ILB 325, Mobile, AL 36688-0002, U.S.A. E-mail: mmulekar@jaguar1.usouthal.edu

André Rogatko

Department of Biostatistics, Fox Chase Cancer Center, Philadelphia, PA 19111-2497. E-mail: A_Rogatko@fccc.edu

Pranab K. Sen

Department of Biostatistics, 3105E McGavran-Greenberg, CB #7420, School of Public Health, University of North Carolina, Chapel Hill, NC 27599-7400, U.S.A. E-mail: pksen@bios.unc.edu

Tumulesh K. S. Solanky

Department of Mathematics, University of New Orleans, New Orleans, LA 70148, U.S.A. E-mail: tsolanky@uno.edu

Alexander G. Tartakovsky

Center for Applied Mathematical Sciences, University of Southern California, 1042 Downey Way, DRB-155, Los Angeles, CA 90089-1113, U.S.A. E-mail: tartakov@math.usc.edu

Venugopal V. Veeravalli

Department of Electrical and Computer Engineering, University of Illinois at Urbana-Champaign, 128 Coordinated Science Laboratory, 1308 West Main Street, Urbana, IL 61801, U.S.A. E-mail: vvv@ uiuc.edu

Rand R. Wilcox

Department of Psychology, University of Southern California, Los Angeles, CA 90089-1061, U.S.A. E-mail: rwilcox@usc.edu

Linda J. Young

Department of Biostatistics, College of Medicine, P.O. Box 100212, Gainesville, FL 32610, U.S.A. E-mail: lyoung@biostat.ufl.edu

Shelemyahu Zacks

Department of Mathematical Sciences, Binghamton University, Bingham-ton, NY 13902-6000, U.S.A. E-mail: Shelly@math.Binghamton.edu.

Chapter 1

Passive Acoustic Detection of Marine Mammals Using Page's Test

DOUGLAS A. ABRAHAM
SACLANT Undersea Research Centre, La Spezia, Italy

1.1 INTRODUCTION

Sonar signal processing is a subset of signal processing related to the analysis of acoustic signals measured underwater. Applications of sonar signal processing lie in diverse fields such as oil exploration, marine mammal study and naval warfare. The fundamental objectives in sonar signal processing are the detection, classification, and localization of sounds that are heard under water. Sonar systems can be either active or passive in their use of sound (Burdic (1984), Urick (1983)). Passive sonar systems only process signals that are recorded on underwater microphones called hydrophones. A typical sonar system will use many hydrophones that are held together physically in what is called an array. An active sonar system transmits a signal using an underwater loud speaker and processes the subsequently heard reflections.

Many of the signal processing algorithms that are used in sonar systems were developed using methodologies from statistical decision theory. Detectors and classifiers may be formulated as binary and multiple hypothesis tests. Localization is simply the estimation of the physical location of the sound emitter or reflector. Sequential methodologies have primarily been used for the detection of sounds, but have also seen use in localization procedures such as target tracking (Lerro and Bar-Shalom (1993)). This paper will focus on the application of the cumulative summation type of sequential procedure called Page's test. Page (1954) developed the procedure to determine the time at which a change in the distribution of sequentially obtained data had occurred. It has seen significant applications in areas such as fault detection (Basseville and Nikiforov (1993)), but has also been used for the detection of unknown but finite duration signals (Han et al. (1999)). It is this latter application that is useful in both active and passive sonar signal processing. Page's test, when it utilizes the log-likelihood ratio, is known to be optimal in the sense of minimizing the average delay before detection while constraining the average time between false alarms (Lorden (1971), Moustakides (1986)). Analysis of its performance at detecting finite duration signals is not trivial, though accurate approximations exist for simple configurations (Han et al. (1999)).

As an example of an active sonar signal processing application, Page's test has been used to detect target echoes when they are corrupted by propagation through a shallow water environment (Abraham and Willett (2002), Abraham (1996b)). In shallow water environments, the sound travels from the source to the target and then from the target to the receiving hydrophones through many paths. As the time it takes to traverse each path differs, the received signal appears to be spread over time. Owing to the difficulty in accurately modeling the ocean propagation, Page's test was applied to obtain adequate detection performance over a wide range of environmental conditions.

An example of a passive sonar signal processing application, and the focus of this paper, may be found in the detection of marine mammals by their acoustic emissions. Such non-invasive detection of marine mammals is necessary for the study of their habits as well as to aid in ensuring their absence prior to any potentially harmful activity such as oil exploration or naval testing. In this passive sonar application, the sounds generated by marine mammals are recorded on hydrophone arrays and must be discerned from all background noises. As marine mammals vary considerably in size and vocal characteristics, the detector must be quite flexible and able to detect both short and long duration signals with potentially widely varying frequency content. In the following sections the problem will be described in more detail along with the sequential detector that was implemented and some results from the analysis of real data. A more detailed description of the analysis may be found in the SACLANT Undersea Research Centre report (Abraham (2000)) from which most of this article is derived.

1.2   ACOUSTIC DETECTION OF MARINE MAMMALS

Detection of the presence of marine mammals is crucial in the vicinity of a research vessel carrying out operations involving the projection of acoustic energy into the local ocean environment. The following sections detail an analysis of passive acoustic data garnered from data recorded during the SACLANT Undersea Research Centre's SWAC4 sea-trial in Kyparissiakos Gulf off the coast of Greece during May 1996. Specifically, the automatic detection processing that was implemented will be described and examples of the data processing results presented.

The automatic detection of marine mammal acoustic emissions is a difficult task. The data available are in the form of passive sonar recordings from a hydrophone array that have been "beamformed" to emphasize signals coming from certain directions. This process increases the signal-to-noise ratio and aids in the localization of the sound source by dictating on which bearing it is heard. Unfortunately, it also means that the detection algorithm must be applied to the data from each beam (direction). The data from the SWAC4 sea-trial were beamformed to point to 120 directions spanning from forward to aft. Designing the detection algorithm is difficult owing to the wide variety of frequencies, bandwidths, and time characteristics of marine mammal acoustic emissions from different species. Take, for example, the sperm whale clicks shown in Figure 1.2.1. These data are in the frequency range of 750–1500 Hertz owing to the limitations imposed by the experimental conditions and acquisition systems. It is known that marine mammals emit sound over a much wider frequency band. Nevertheless, even with this limited band, it is seen that the spectrum is not necessarily flat and that the time series exhibits the effects of multipath propagation, spreading the signal energy in time. As other types of waveforms are expected (for example, whistles or sweeps) a detector is desired that is flexible both in the time duration of the waveform and its frequency content.

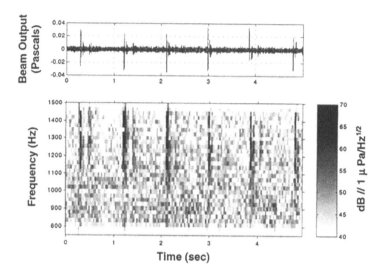

Figure 1.2.1: Time Series and Frequency Spectrum of Sperm Whale Click Train Limited to 750–1500 Hertz Frequency Band.

1.3   METHODOLOGY AND ANALYSIS

In hypothesis testing, there must be some difference between the null and alternative hypotheses in order for the Type-I and Type-II errors to be small. In terms of detector design, this translates into requiring that there be some means for distinguishing the signal to be detected from the background noise and any interferences. As this situation calls for a very general detector, not tuned to any specific characteristics of individual marine mammal emissions, the only distinction from the background or interferences is time duration. Thus, a detector is desired that finds short duration signals that are not similar to the more slowly changing background (for example, ambient ocean noise from the ocean surface or distant shipping) or interferences (for example, near-by surface vessels including the ship towing the hydrophone array). Additionally, the detector should be robust to varying signal duration and frequency content. A detector with these characteristics was proposed by Abraham and Stahl (1996) by combining Page's (1954) test with the power-law processor of Nuttall (1994) for the combination of *discrete Fourier transform* (DFT) bin outputs. A block diagram of this detector structure is shown in the left half of Figure 1.3.1. The time series data are transformed into the frequency domain by overlapping DFTs. The duration of the DFT should be near or less than the duration of the shortest signal that may be encountered. Define the magnitude squared of the DFT bins of interest for the kth DFT as $\{X_{k,1}, X_{k,2}, \ldots, X_{k,m}\}$. For the data presented in this paper the bins of interest are those in the 750–1500 Hertz band. Let the estimated background power in the jth DFT bin at time k be $\lambda_{k,j}$ for $j = 1, \ldots, m$. As will soon be described, the background power levels in each DFT bin need to be estimated from previously observed data. These estimates are used to form normalized DFT bin data,

$$Y_{k,j} = \frac{X_{k,j}}{\lambda_{k,j}}, \qquad (1.3.1)$$

which are combined into a single test statistic by a power-law non-linearity,

$$Z_k = \left(\frac{1}{m} \sum_{j=1}^{m} Y_{k,j}^p \right)^{\frac{1}{p}}. \qquad (1.3.2)$$

The power-law non-linearity provides robustness against varying signal bandwidth or frequency structure and can have the effect of either picking the maximum DFT bin output (high power law) or summing all the DFT bins together (power law equal to one) as in an energy detector. This one statistic from each DFT is then used in Page's test to detect the onset of a signal. The update for Page's test has the form

$$W_{k+1} = \max \{0, W_k + Z_{k+1} - b_0\} \qquad (1.3.3)$$

where b_0 is a bias and a detection is declared when W_k exceeds a threshold. The bias is most easily chosen through Dyson's (1986) method which uses the average value of Z_k under the null and alternative hypotheses. In certain cases it is possible to choose the bias to optimize an asymptotic performance measure (Abraham (1996a)). The threshold may be chosen according to a desired average time between false alarms through the standard Wald or Siegmund approximations or quantization based approximations (Basseville and Nikiforov (1993), Brook and Evans (1972), Abraham and Stahl (1996)). Assuming perfect normalization, the performance of a power-law non-linearity feeding a Page test is examined in terms of the average sample numbers (average time between false alarms and average delay before detection) in Abraham and Stahl (1996). A more relevant measure of detection performance is the probability of detection, which may be obtained from quantization based methods (Han et al. (1999)). Theoretical analyses such as those found in Abraham and Stahl (1996), Han et al. (1999) or Nuttall (1994) typically assume that the normalization is perfect and hence that the DFT bin data are exponentially distributed with unit mean when no signal is present. It is often assumed that the signal manifests itself as either a change in the scale or as a constant additive component to the complex DFT data, yielding a non-central chi-squared distribution for the DFT power data. Though these theoretical analyses help evaluate the performance of various detector structures under ideal conditions, the processing of real data introduces many difficulties ranging from data quality issues (for example, data glitches or drop-outs that can occur with varying frequency) to inadequate modeling (simplifications in the modeling that allow analysis but only approximately represent the real data). Thus, in practice, signal strength based parameters such as the bias in (1.3.3) are usually set for a minimum detectable signal level and thresholds are typically chosen so that the operator and any post-detection processing are not overloaded with false alarms.

In this application, Page's test provides robustness to the unknown duration of the signal compared with the use of a sliding fixed block detector that would be tuned to a single signal duration. Robustness in this sense means that the detector provides adequate, though perhaps sub-optimal, performance over a wide range of signal durations. As estimates of both the start and end times of the signal are desired, the alternating-hypothesis form of Page's test (Streit (1995), Abraham and Willett (2002)) must be implemented. The update structure for this form is similar to (1.3.3) and is described in the following section. In the alternating-hypothesis configuration, the start and end times of a detected event are estimated by the most recent reset of the Page test to its null state as described in Abraham (1997).

Necessary to the implementation of the detector is the estimation of the background noise and interference power at the output of each DFT bin. As these may be considered nuisance parameters (that is, parameters that need to be estimated but are not used to describe the signal), the scheme proposed in Abraham (1996c) which exploits the structure of the Page test to isolate data believed to be signal-free is appropriate. This detector structure uses data prior to the most recent

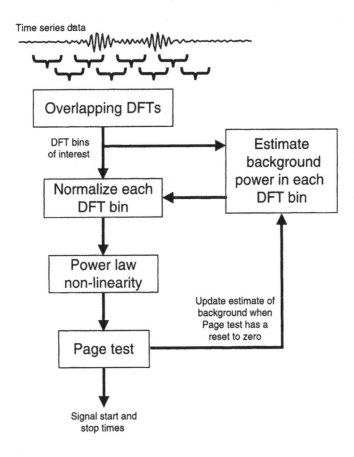

Figure 1.3.1: Block Diagram of Detection Scheme.

reset of the Page test to estimate the nuisance parameters. The form of the background estimator may be of a sliding block or exponentially averaged type. In this case, the latter is chosen because it is (marginally) easier to implement. The exponential averager simply applies a single pole filter to the time series; thus, updating the background estimate results in

$$\lambda_{k+1,j} = \alpha\lambda_{k,j} + (1 - \alpha)\,X_{k,j} \qquad (1.3.4)$$

where $\alpha \in (0,1)$ is usually near one to provide a long averaging time. Each time the Page test has a reset to zero, the background estimate is updated as in (1.3.4) but using only the non-signal data between the previous reset to zero and the current reset to zero (that is, if a signal was detected in between, those data would not be included). The estimates should be initialized with either a fixed block estimate or an exponential averager that has been running for a short time.

1.3.1   Algorithm Specifics

Following are the algorithm (in pseudocode) and definitions of the variables used to implement the detector along with their values when they require setting. The dependence of the variables on DFT snapshot k is suppressed as it is not necessary when implementing the algorithm.

Detection Algorithm

(1) Initialization

- Set $W = 0$, $k = 1$, $i_0 = 1$, $i'_0 = 1$
- Form initial estimate of $\{\lambda_1, \lambda_2, \ldots, \lambda_m\}$

(2) Normalization and power-law

- Form normalized DFT bin outputs
  $Y_j = \frac{X_{k,j}}{\lambda_j}$ for $j = 1, \ldots, m$
- Apply power-law to normalized data
  $$Z = \left[\frac{1}{m} \sum_{j=1}^{m} Y_j^p\right]^{\frac{1}{p}}$$

(3) If $W < h_0$,

- Set $W = \max\{0, W + Z - b_0\}$
- If $W \geq h_0$,
  - The leading edge of a signal has been detected
  - An estimate of the starting time index is i_0
  - Set $W = h_0 + h_1$ and $i_1 = k$
- Else if $W = 0$,
  - A reset to zero has occurred, update background estimate
    for $i = i'_0$ to k and $j = 1, \ldots, m$
    $$\lambda_j = \alpha\lambda_j + (1 - \alpha) X_{i,j}$$
    end
  - Set $i_0 = k$ and $i'_0 = k$

(4) If $W \geq h_0$,

- Set $W = \min\{h_0 + h_1, W + Z - b_1\}$
- If $W \leq h_0$,
  - The lagging edge of a signal has been detected
  - An estimate of the stopping time index is i_1
  - Set $W = 0$, $i_0 = k$, and $i'_0 = i_1$
- If $W = h_0 + h_1$, set $i_1 = k$

(5) Set $k = k + 1$ and go to (2)

Description of variables

p - power law ($p \geq 1$, $p = 1$ was used)

h_0 - threshold for signal onset detection ($h_0 = 12$)

b_0 - Page test bias for signal onset detection ($b_0 = 2.5$)

h_1 - threshold for signal termination detection ($h_1 = 10$)

b_1 - Page test bias for signal termination detection ($b_1 = 5$)

α - time constant for exponential averager ($0 < \alpha < 1$, $\alpha = 0.95$ was used)

N_{fft} - size of DFT block ($N_{fft} = 128$)

N_{off} - offset from one DFT block to next ($N_{off} = 32$)

W - Page test statistic

i_0 - index to most recent reset to zero

i'_0 - index for updating background power estimates

i_1 - index to most recent reset to $h_1 + h_0$ (signal present state)

It should be noted that the indices for the starting and stopping times are in terms of DFT blocks and must be converted to time samples based on the DFT size and amount of overlap. Also, the power-law was kept at unity because of the small bandwidth of the data being processed relative to the potential bandwidth of the marine mammal acoustic emissions. In general, data used for the acoustic detection of marine mammals would have a higher bandwidth and may exhibit tonal emissions. In this more common situation, a higher power law would improve detection performance.

1.3.2   Detection Results

The algorithm of Section 1.3.1 was applied to data from Run 9[1] of the
SWAC4 data set to obtain a series of start and stop times for every
detection. From these start and stop times, the total signal energy was
estimated and tabulated along with the current average noise power
estimate (here it is assumed that the detected signal is an energy signal
and that the background noise is a power signal). Additionally, an
estimate of the noise background after removal of the detected signals
was formulated every 12 seconds. The detection processing results are
then displayed in Figure 1.3.2 as the total energy-to-noise power ratio
(ENR) detected on each beam over 6 second intervals for the nearly
three hour Run 9. This display only shows the results of the detection
processing for short duration signals. Thus, when detections occur over
an extended period, as seen in the figure, then the time domain signal
has persistent non-stationary components as illustrated by the sperm
whale clicks in Figure 1.2.1. As previously mentioned, the beams span
from forward to aft and are spaced equally in the cosine space of bearing
so that there are more beams broadside to the hydrophone array (which
was in the shape of a line) than near the forward or aft directions.

   The detection results are grouped into events by visual association over beam and time, as indicated by the numbers in Figure 1.3.2.
Event time series are then formed by choosing the beam containing the
largest ENR over each 12 second period. These time series were then
submitted to Prof. G. Pavan of the University of Pavia, Pavia, Italy for
classification. Those shown in Figure 1.3.2 were all classified as sperm
whale click trains. The detector also found many signals associated
with surface vessels, particularly those from the Research Vessel Alliance (the ship towing the hydrophone array) in the forward beams,
and a plethora of isolated detections that could be marine mammal,
fish, man-made or false alarms. It is possible to associate some of the
detections with surface vessels by overlaying the detection results on
the estimated background noise, as shown in Figure 1.3.3 for the first
20 minutes of Run 9. In this figure the background power is shown
in gray as indicated by the scale on the right. The ENRs of detected
short time duration events are overlaid in black. The surface vessels
are clearly visible in the background power estimate as slowly moving
lines in the gray scale. Short duration signal detections that overlay

---

[1]As will be seen in Section 1.3.3, Run 9 is unique in that the data allow for
localization of two sperm whales.

the surface vessels in bearing and time are most likely originating there as well (though this is not necessarily so). The sperm whale click trains (events 1, 2 and 3) arrive on quiet beams (that is, there is no surface ship in the background), additionally supporting their classification as marine mammal. Event 21 was eventually classified as acoustic emissions from fish of unknown type. It may also be surmised that, of the two sperm whales detected during the first several minutes, event 1 is nearer than event 2, assuming they both produced approximately the same source levels. As will be discussed in the following section, the towed hydrophone array was completing a turn previously carried out by the RV Alliance so localization of these two events is possible, including resolution of the left/right ambiguity inherent in the array signal processing.

1.3.3   Localization

Passive listening of acoustic emissions inherently only provides bearing information. Additionally, owing to the straight line shape of the hydrophone array, there exists a cone of ambiguity; that is, the sound arriving at the array sounds the same if it arrives from anywhere on a cone axially aligned with the towed line array. All of the runs analyzed were such that the tow ship (RV Alliance) was on a constant bearing. Thus, triangulating detections observed over extended periods of time still results in an ambiguity to the left or right side of the array. However, during the first 10 minutes of Run 9, the array was still completing a turn the tow-ship had made prior to commencing the run. From the array heading information (which is quite noisy) it was possible to localize the two sperm whale click trains detected as shown in Figures 1.3.4 and 1.3.5. Lines along the bearing of the detected events from the position of the RV Alliance are shown for events 1, 4, 10 and 15 of Run 9 in Figure 1.3.4. These events are believed to originate from the same whale, though there is no proof of this other than approximate coincidence in space and time. Each line is 15 nautical miles long and when taken in conjunction with the others form a locus where the sperm whale might have been, effectively localizing the whale in range, bearing and resolving the left/right ambiguity. The lines from event 1 (the lighter ones) are shown assuming that the whale was on either the left or right side of the array. The diverging lines seen to the east of the track illustrate an incorrect localization. Figure 1.3.5 contains the localization of events 2, 3, 5, 6, 8, 12, 13 and 17 of Run 9. The ranging

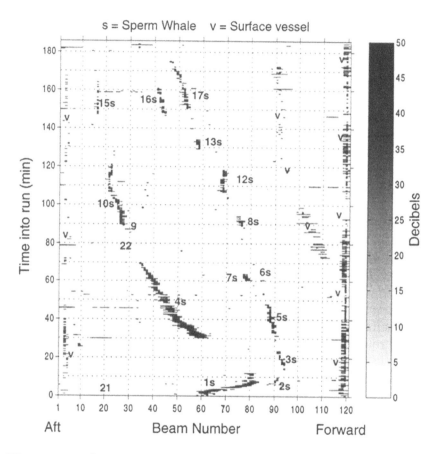

Figure 1.3.2: Signal Energy-to-Noise Power Ratio (ENR) of Detected Short Time Duration Events from Run 9, Combined Over Every 6 Seconds.

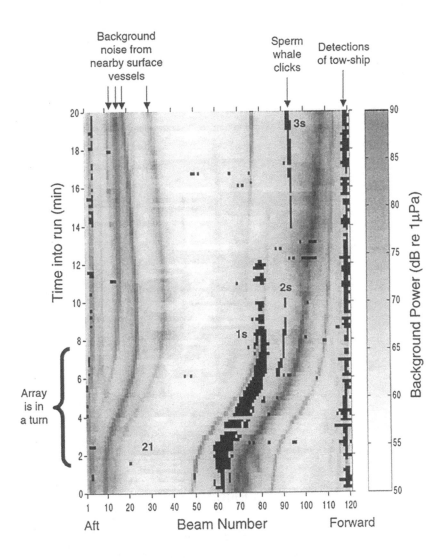

Figure 1.3.3: ENR of Detected Events from First 20 Minutes of Run 9 (Black) Overlaid on Background Noise and Interference Power Estimates (Gray Scale).

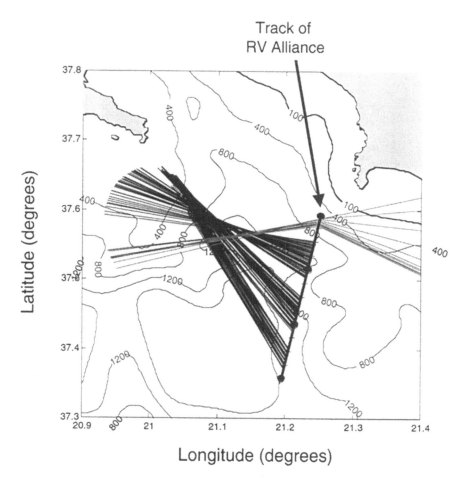

Figure 1.3.4: Lines Along Bearings of Detections for Events 1, 4, 10
and 15 of Run 9. Each Line Is 15 Nautical Miles in Length. The
Lighter Lines Are from Event 1 and Illustrate Localization of a Sperm
Whale to the West Side of the Track of RV Alliance.

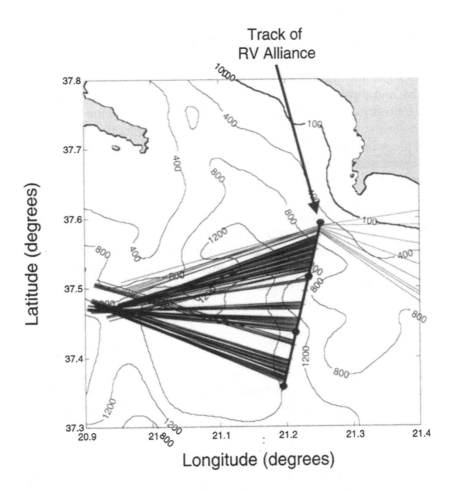

Figure 1.3.5: Lines Along Bearings of Detections for Events 2, 3, 5, 6, 8, 12, 13 and 17 of Run 9. Each Line Is 15 Nautical Miles in Length. The Lighter Lines Are from Event 2 and Illustrate Localization of a Sperm Whale to the West Side of the Track of RV Alliance.

information garnered from Figures 1.3.4 and 1.3.5, that event 2 is farther away than event 1, is corroborated by the ENR levels observed in Figure 1.3.2 where event 2 is weaker than event 1.

1.4 CONCLUDING REMARKS

This paper has illustrated the use of sequential methodologies in both active and passive sonar signal processing. It was shown in detail how Page's test may be applied to the detection of marine mammal acoustic emissions having unknown time and frequency structure. The sequential procedure allows for the detection and segmentation of signals with unknown and potentially widely varying time structure. The algorithm described was effective in estimating the background noise and interference power and then detecting departures from that norm, which in this case were either acoustic emissions from marine mammals or non-stationary acoustic emissions from nearby surface ships. In the data presented, sperm whale sounds were detected, along with numerous other detections not analyzed further owing to limited time and resources. Localization of some of the sperm whale sounds was possible through triangulation.

ACKNOWLEDGMENT

This work was performed while the author was with the SACLANT Undersea Research Centre, La Spezia, Italy. Without the efforts of Walter Zimmer, this work would not have been possible. He designed and implemented a passive sonar system on SACLANT Undersea Research Centre's real-time system, enabling the analysis of large amounts of data and provided the filtered and beamformed data for the automatic detection processing. The author would also like to acknowledge the efforts of Angela D'Amico and Ettore Capriulo who both contributed to the work presented in this paper.

REFERENCES

[1] Abraham, D.A. (1996a). Asymptotically optimal bias for a general non-linearity in Page's test. *IEEE Trans. Aerosp. Elec. Sys.*, **32**, 360–367.

[2] Abraham, D.A. (1996b). A nonparametric Page test applied to active sonar detection. In *Oceans 96 MTS/IEEE Conf. Proc., The Coastal Ocean–Prospects for the 21st Century*, 908–913. IEEE: Piscataway.

[3] Abraham, D.A. (1996c). A Page test with nuisance parameter estimation. *IEEE Trans. Inform. Theory*, **42**, 2242–2252.

[4] Abraham, D.A. (1997). Analysis of a signal starting time estimator based on the Page test statistic. *IEEE Trans. Aerosp. Elec. Sys.*, **33**, 1225–1234.

[5] Abraham, D.A. (2000). Passive acoustic detection of marine mammals. SM-351, SACLANT Undersea Research Centre, La Spezia, Italy.

[6] Abraham, D.A. and Stahl, R.J. (1996). Rapid detection of signals with unknown frequency content using Page's test. In *Proc. 1996 Conf. Inform. Sci. Sys.* (S. Kulkarni and M. Orchard, co-chairs), 809–814. Princeton University: Princeton.

[7] Abraham, D.A. and Willett, P.K. (2002). Active sonar detection in shallow water using the Page test. *IEEE J. Ocean. Eng.*, **27**, 35–46.

[8] Basseville, M. and Nikiforov, I.V. (1993). *Detection of Abrupt Changes: Theory and Applications*. Prentice-Hall: Englewood Cliffs.

[9] Brook, D. and Evans, D.A. (1972). An approach to the probability distribution of cusum run length. *Biometrika*, **59**, 539–549.

[10] Burdic, W.S. (1984). *Underwater Acoustic System Analysis*. Prentice Hall: Englewood Cliffs.

[11] Dyson, T. (1986). *Topics in Nonlinear Filtering and Detection*. Ph.D. thesis, Princeton University, Princeton.

[12] Han, C., Willett, P.K. and Abraham, D.A. (1999). Some methods to evaluate the performance of Page's test as used to detect transient signals. *IEEE Trans. Signal Proc.*, **47**, 2112–2127.

[13] Lerro, D. and Bar-Shalom, Y. (1993). Interacting multiple model

18 Abraham

tracking with target amplitude feature. *IEEE Trans. Aerosp. Elec. Sys.*, **29**, 494–509.

[14] Lorden, G. (1971). Procedures for reacting to a change in distribution. *Ann. Math. Statist.*, **42**, 1897–1908.

[15] Moustakides, G.V. (1986). Optimal stopping times for detecting changes in distributions. *Ann. Statist.*, **14**, 1379–1387.

[16] Nuttall, A.H. (1994). Detection performance of power-law processors for random signals of unknown location, structure, extent, and strength. *Tech. Rep.# 10,751, Naval Undersea Warfare Center*, New London, Connecticut.

[17] Page, E.S. (1954). Continuous inspection schemes. *Biometrika*, **41**, 100–114.

[18] Streit, R.L. (1995). Load modeling in asynchronous data fusion systems using Markov modulated Poisson processes and queues. In *Proc. Signal Processing Workshop*, Washington, D.C., Maryland/District of Columbia Chapter of the IEEE Signal Processing Society.

[19] Urick, R.J. (1983). *Principles of Underwater Sound*. McGraw-Hill: New York.

Address for communication:

DOUGLAS A. ABRAHAM, Applied Research Laboratory, Pennsylvania State University, P.O. Box 30, State College, PA 16804, U.S.A.
E-mail: daa10@psu.edu

Chapter 2

Two-Stage Procedures for Selecting the Best Component of a Multivariate Exponential Distribution

MAKOTO AOSHIMA
University of Tsukuba, Ibaraki, Japan

MITSURU AOKI
University of Tsukuba, Ibaraki, Japan

MASAKI KAI
Fidelity Investments Japan Limited, Tokyo, Japan

2.1 INTRODUCTION

Let $(X_{1r}, X_{2r}, ..., X_{kr})$, $r = 1, 2, ...$ be k-dimensional random sample vectors from a multivariate exponential (MVE) distribution whose survival function is given by

$$P(X_{1r} > x_1, \ X_{2r} > x_2, ..., \ X_{kr} > x_k)$$
$$= \ \exp\left\{-\lambda_1 x_1 - \lambda_2 x_2 - \cdots - \lambda_k x_k - \lambda_0 \max(x_1, x_2, ..., x_k)\right\}$$

where $x_i > 0$, $\lambda_i > 0$ $(i = 1, 2, ..., k)$ and $\lambda_0 \geq 0$. This is a special version of the MVE distribution given by Marshall and Olkin (1967) under the assumption that a failure is caused by $k + 1$ types of Poisson shocks on a system containing k components. The MVE distribution is desirable when the assumption of independence among components is questionable or false. Plenty of literature is available about the MVE distribution because of its applications in life testing, reliability and other fields.

We consider here the problem of selecting the best component with respect to λ_i $(i = 1, 2, ..., k; \; k \geq 2)$ which is viewed as the hazard of the component in lifetime analysis. We define the jth component to be the best component if $\lambda_j = \min(\lambda_1, \lambda_2, ..., \lambda_k)$. Note that

$$E(X_{ir}) = 1/(\lambda_i + \lambda_0);$$
$$Var(X_{ir}) = 1/(\lambda_i + \lambda_0)^2; \qquad\qquad (2.1.1)$$
$$Cov(X_{ir}, X_{jr}) = \lambda_0/(\lambda_i + \lambda_0)(\lambda_j + \lambda_0)(\lambda_i + \lambda_j + \lambda_0)$$

for $1 \leq i, j(\neq i) \leq k$. The best component also corresponds to the largest mean. Throughout this paper, we assume that the 1st component is the best component, that is, $\lambda_1 < \lambda_i$ $(i = 2, 3, ..., k)$ without loss of generality. When $\lambda_0 = 0$, the problem reduces to that of selecting the best one out of k independent univariate exponential populations.

We shall seek a rule R which selects one of the components as the best component. A *correct selection* (CS) occurs when the selected component is indeed the best component. Denote the probability of a correct selection (PCS) with Rule R by $P(CS|R)$. We require that

$$P(CS|R) \geq P^* \quad \text{whenever} \quad \lambda_i/\lambda_1 \geq \delta^* \;\; (i = 2, ..., k), \qquad (2.1.2)$$

where $\delta^*(> 1)$ and $P^* \in (k^{-1}, 1)$ are specified by the experimenter in advance.

Hyakutake (1992) considered the present problem when $k = 2$ and proposed a Stein (1945) type two–stage procedure based on a rule satisfying the requirement (2.1.2). In this paper, we will develop selection procedures for the case where $k \geq 3$. The results obtained here can be easily reduced to similar results for the "$k = 2$" case by modifying design constants accordingly. Techniques given in this paper can also be used to obtain solutions to the problem of detecting the worst (k^{th}) component under the configuration that $\lambda_k/\lambda_i \geq \delta^*$ $(i = 1, ..., k-1)$. In that situation, the MVE distribution is interpreted as a competing-risk model.

Arnold (1968) gave the following result:

$$P(X_{ir} < X_{i'r}, \ i' = 1, ..., i-1, i+1, ..., k) = p_i \qquad (2.1.3)$$

for $i = 1, 2, ..., k$, and

$$P(X_{1r} = X_{2r} = \cdots = X_{kr}) = p_0, \qquad (2.1.4)$$

where $p_i = \lambda_i / \sum_{\ell=0}^{k} \lambda_\ell$ $(i = 0, 1, 2, ..., k)$. Then, selecting the best component with respect to λ_i is equivalent to selecting the best component with respect to p_i. The preference zone, namely, $\lambda_i/\lambda_1 \geq \delta^\star$ $(i = 2, ..., k)$ as in (2.1.2), can be alternatively stated as $p_i/p_1 \geq \delta^\star$ $(i = 2, ..., k)$. We start with this idea to propose a selection rule satisfying the requirement (2.1.2) when $k \geq 3$.

2.2 A SPECIFIC APPLICATION

Let us consider the situation in which a car manufacturer is interested in testing the endurance of a piston in an engine. The engine is run at 13,000 *revolutions per minute* (r.p.m.) and the testing is carried out under a specific heavy-stress condition to cope with some unexpected events such as a "no–oil" situation or "continuous high over-revolution". An engine has pistons and each piston has its rod within a cylinder. All rods are connected to a common crank-shaft. A piston is made of an aluminum alloy with silicon, copper, nickel and magnesium. The mixing proportions (scheme-ratios) of these ingredients mainly determine the endurance of a piston. In order to improve the performance of an engine, the car manufacturer is interested in identifying a scheme-ratio that would result in higher endurance. For cost reduction in running the experiment, it is more likely that the manufacturer will conduct a test on a 4-cylinder engine with four pistons having different scheme-ratios, say A, B, C and D, simultaneously. While testing an engine assembled with four pistons having different scheme-ratios, we are interested in selecting the piston with the best scheme-ratio with respect to the endurance of a piston. The endurance of four pistons in an engine are positively correlated with each other since all pistons are connected by their own rods to a common crank-shaft. We note that if the crank-shaft is damaged during an experiment, the engine dies. We suppose that the survival times of four pistons have a 4-dimensional MVE distribution. Under the heavily stressed condition described above, 4-dimensional data on survival times with an inline 4-cylinder engine are recorded. The 1st piston is made using the scheme-ratio A, the 2nd one is made using the scheme-ratio B, the 3rd one is made using the scheme-ratio C, and the 4th one is made using the scheme-ratio D, respectively. Thus, $k = 4$ and X_i = survival time of the ith piston, $i = 1, ..., 4$, λ_i = failure rate of the ith piston, $i = 1, ..., 4$, and λ_0 = failure rate of the crank-shaft.

Note that p_0 is a nuisance parameter and the frequency of event regarding p_0 in (2.1.4) determines the choice of a methodology for data analysis. For known p_0, we shall consider two selection rules R_1 and R_2. For selecting the component associated with p_1, the *least favorable configuration* (LFC) of $\boldsymbol{\lambda} = (\lambda_0, \lambda_1, ..., \lambda_k)$ is the configuration for which the PCS is minimized subject to the condition that $\lambda_i/\lambda_1 \geq \delta^\star$ ($i = 2, ..., k$). The LFC for the proposed selection rule would be given by

$$\lambda_0, \ \delta^\star \lambda_1 = \lambda_2 = \cdots = \lambda_k \ (\delta^\star > 1). \qquad (2.2.1)$$

We shall obtain a large sample solution for the sample size required by the rule R_ℓ ($\ell = 1, 2$) in order to satisfy the requirement (2.1.2). We evaluate the PCS for R_ℓ under the LFC, which we will denote by $P^\star(CS|R_\ell)$. Under the LFC in (2.2.1), we note that

$$\delta^\star p_1 = p_2 = \cdots = p_k; \ p_1 = (1 - p_0)/\{1 + (k - 1)\delta^\star\}. \qquad (2.2.2)$$

2.2.1   The Rule R_1

Here we describe the first of our two selection rules.

Selection Rule R_1: Take a sample of n observations. Let n_i and n_0 denote the number of observations in the regions $\{x_i < x_{i'}, \ i' = 1, ..., i - 1, \ i + 1, ..., k\}$ and $\{x_1 = \cdots = x_k\}$, respectively. Select the component with the smallest count among $(n_1, ..., n_k)$ as the best component. Use randomization to break ties for the first place.

Hyakutake (1994) showed the equivalence between R_1 and the rule based on Arnold's (1968) estimates. Since $(n_1, n_2, ..., n_k, n_0)$ has a multinomial distribution with $n = \sum_{\ell=0}^{k} n_\ell$ and vector of cell probabilities $(p_1, p_2, ..., p_k, p_0)$, we have:

$$\begin{aligned} P^\star(CS|R_1) &= P(n_1 < n_i, \ i = 2, ..., k) \\ &= \sum_{n_1 < n_2, ..., n_1 < n_k} \frac{n!}{n_1! n_2! \cdots n_k! n_0!} p_1^{n_1} p_2^{n_2} ... p_k^{n_k} p_0^{n_0}. \end{aligned}$$

Now consider (2.2.2) under the configuration given by (2.2.1). Therefore, the PCS becomes

$$P^\star(CS|R_1) = \sum_{n_1 < n_2, \dots, n_1 < n_k} \frac{n!}{n_1! n_2! \cdots n_k! n_0!} \delta^{\star n - n_1 - n_0} p_1^{n - n_0} p_0^{n_0}.$$

It can be easily verified that

$$\begin{aligned}
E(n_i - n_1) &= n p_1 (\delta^\star - 1), \\
Var(n_i - n_1) &= n\{(\delta^\star + 1)p_1 - (\delta^\star - 1)^2 p_1^2\}, \\
Cov(n_i - n_1, n_j - n_1) &= n\{p_1 - (\delta^\star - 1)^2 p_1^2\}
\end{aligned}$$

for all i and j $(i, j = 2, \dots, k;\ i \neq j)$. Now, we shall use the normal approximation to obtain the optimal sample size needed to satisfy the requirement (2.1.2). Define

$$Y_i = \frac{n_i - n_1 - n p_1 (\delta^\star - 1)}{\sqrt{n\{(\delta^\star + 1)p_1 - (\delta^\star - 1)^2 p_1^2\}}}, \quad i = 2, \dots, k.$$

Then, we can write

$$P^\star(CS|R_1) = P(Y_i \leq c_1\sqrt{n},\ i = 2, \dots, k),$$

where

$$c_1 = \left\{ \frac{(1 - p_0)(\delta^\star - 1)^2}{p_0(\delta^\star - 1)^2 + \delta^\star(2 + k - 2\delta^\star + k\delta^\star)} \right\}^{1/2} \equiv c_1(p_0, \delta^\star, k), \text{ say.} \tag{2.2.3}$$

Then, for large n, we use the result that $(Y_2, \dots, Y_k)$ is approximately distributed as $N_{k-1}(0, \Sigma_1)$ where $\Sigma_1 = \rho_1 \mathbf{1}\mathbf{1}' + (1 - \rho_1)I_{(k-1)\times(k-1)}$ and

$$\rho_1 = \frac{p_0(\delta^\star - 1)^2 + (1 + k - \delta^\star)\delta^\star}{p_0(\delta^\star - 1)^2 + \delta^\star(2 + k - 2\delta^\star + k\delta^\star)} \equiv \rho_1(p_0, \delta^\star, k), \text{ say.} \tag{2.2.4}$$

Note that $\rho_1 > 0$ for $\delta^\star < 1 + k$. Thus, for large n and $\delta^\star < 1 + k$, we have

$$P^\star(CS|R_1) \simeq \int_{-\infty}^{\infty} \Phi^{k-1}\left(\frac{x\sqrt{\rho_1} + c_1\sqrt{n}}{\sqrt{1 - \rho_1}}\right) d\Phi(x). \tag{2.2.5}$$

Solving the equation

$$\int_{-\infty}^{\infty} \Phi^{k-1}\left(\frac{x\sqrt{\rho_1} + u_1}{\sqrt{1 - \rho_1}}\right) d\Phi(x) = P^\star \tag{2.2.6}$$

for $u_1 \equiv u_1(p_0, \delta^\star, P^\star, k)$, the optimal fixed sample-size needed to satisfy the requirement (2.1.2) is obtained as

$$n_0^{(1)} = \left\langle \frac{u_1^2}{c_1^2} \right\rangle + 1, \qquad (2.2.7)$$

where $\langle s \rangle$ denotes the greatest integer less than s. However, when p_0 is unknown, Rule R_1 can not meet the requirement (2.1.2) with a finite sample size.

2.2.2 The Rule R_2

This rule is based on the marginal distributions of the MVE distribution. Marginally, X_{ir} has an exponential distribution with parameter $\lambda_i + \lambda_0$. The problem of selecting a component with respect to λ_i is equivalent to that of selecting a component with respect to $\lambda_i + \lambda_0$. Let $\overline{X}_{in} = \sum_{r=1}^n X_{ir}/n$, then $\overline{X}_{in}$ is an unbiased estimator of $1/(\lambda_i + \lambda_0)$. Proschan and Sullo (1976) showed that $\overline{X}_{in}$ is also an intuitive (INT) estimator. Now, we consider the following selection rule.

Selection Rule R_2: Take a sample of n observations. Select the component associated with $\overline{X}_{jn} = \max(\overline{X}_{1n}, ..., \overline{X}_{kn})$ as the best component.

Let us evaluate its PCS under the LFC using the normal approximation. We have:

$$P^\star(CS|R_2) = P(\overline{X}_{1n} > \overline{X}_{in}, \ i = 2, ..., k).$$

It can be easily verified from (2.1.1) that

$$E(\overline{X}_{1n} - \overline{X}_{in}) = \frac{(\delta^\star - 1)\lambda_1}{(\lambda_1 + \lambda_0)(\delta^\star \lambda_1 + \lambda_0)} = \mu, \text{ say,}$$

$$nVar(\overline{X}_{1n} - \overline{X}_{in}) = \frac{1}{(\lambda_1 + \lambda_0)^2} + \frac{1}{(\delta^\star \lambda_1 + \lambda_0)^2}$$
$$- \frac{2\lambda_0}{(\lambda_1 + \lambda_0)(\delta^\star \lambda_1 + \lambda_0)(\lambda_1 + \delta^\star \lambda_1 + \lambda_0)} = \sigma^2, \text{ say,}$$

$$nCov(\overline{X}_{1n} - \overline{X}_{in}, \overline{X}_{1n} - \overline{X}_{jn}) = \frac{1}{(\lambda_1 + \lambda_0)^2} + \frac{\lambda_0}{(\delta^\star \lambda_1 + \lambda_0)^2(2\delta^\star \lambda_1 + \lambda_0)}$$
$$- \frac{2\lambda_0}{(\lambda_1 + \lambda_0)(\delta^\star \lambda_1 + \lambda_0)(\lambda_1 + \delta^\star \lambda_1 + \lambda_0)},$$

for all i and j $(i, j = 2, ..., k; i \neq j)$. Define $Z_i = \sqrt{n}(\overline{X}_{1n} - \overline{X}_{in} - \mu)/\sigma$, $i = 2, ..., k$. Then, we can write

$$P^\star(CS|R_2) = P(Z_i \leq c_2\sqrt{n}, \ i = 2, ..., k),$$

where $c_2 \equiv c_2(p_0, \delta^\star, k) =$

$$\left\{ \frac{(1-p_0)(\delta^\star - 1)^2(1 + \delta^\star - 2p_0\delta^\star + kp_0\delta^\star)}{(1 + p_0 + \delta^\star - 3p_0\delta^\star + 2kp_0\delta^\star)(1 + kp_0\delta^\star + \delta^{\star 2} - 2p_0\delta^{\star 2} + kp_0\delta^{\star 2})} \right\}^{1/2}$$
(2.2.8)

Now, for large n, $(Z_2, ..., Z_k)$ is approximately distributed as $N_{k-1}(\mathbf{0}, \Sigma_2)$ where $\Sigma_2 = \rho_2 \mathbf{11}' + (1 - \rho_2)\mathbf{I}_{(k-1) \times (k-1)}$ with

$$\begin{aligned}
\rho_2 &\equiv \rho_2(p_0, \delta^\star, k) \\
&= \big\{ 2\delta^{\star 3}(1 + \delta^\star) + p_0(1 + k\delta^\star + k\delta^{\star 2} + k\delta^{\star 3} + (7k - 13)\delta^{\star 4}) \\
&\quad + p_0^2(1 + (4k - 6)\delta^\star + (8 - 8k + 3k^2)\delta^{\star 2} + 8(k - 2)\delta^{\star 3} \\
&\quad + (25 - 24k + 5k^2)\delta^{\star 4}) \\
&\quad + p_0^3(k\delta^\star + (3k^2 - 5k - 2)\delta^{\star 2} + (12 - 3k - 6k^2 + 2k^3)\delta^{\star 3} \\
&\quad + (-5k^2 + 17k - 14)\delta^{\star 4}) \big\} \big\{ (p_0 + 2\delta^\star - 3p_0\delta^\star \\
&\quad + kp_0\delta^\star)(1 + p_0 + \delta^\star - 3p_0\delta^\star + 2kp_0\delta^\star) \\
&\quad \times (1 + kp_0\delta^\star + \delta^{\star 2} - 2p_0\delta^{\star 2} + kp_0\delta^{\star 2}) \big\}^{-1}.
\end{aligned}$$

Note that $\rho_2 > 0$. Thus, for large n, we have

$$P^\star(CS|R_2) \simeq \int_{-\infty}^{\infty} \Phi^{k-1}\left(\frac{x\sqrt{\rho_2} + c_2\sqrt{n}}{\sqrt{1 - \rho_2}} \right) d\Phi(x). \qquad (2.2.9)$$

Solving the equation

$$\int_{-\infty}^{\infty} \Phi^{k-1}\left(\frac{x\sqrt{\rho_2} + u_2}{\sqrt{1 - \rho_2}} \right) d\Phi(x) = P^\star \qquad (2.2.10)$$

for $u_2 \equiv u_2(p_0, \delta^\star, P^\star, k)$, the optimal fixed sample-size needed to satisfy the requirement (2.1.2) is obtained as

$$n_0^{(2)} = \left\langle \frac{u_2^2}{c_2^2} \right\rangle + 1. \qquad (2.2.11)$$

Again, when p_0 is unknown, Rule R_2 also can not meet requirement (2.1.2) with a finite sample size.

2.3 METHODOLOGIES AND ANALYSIS

If p_0 is unknown, we will consider Stein (1945) type two–stage procedures. In the first step, take a pilot sample to estimate p_0. If $k = 2$, the design constants u_1 and u_2 in (2.2.7) and (2.2.11) reduce to $u_1 = u_2 = \Phi^{-1}(P^\star)$ and then the two–stage procedures developed by Hyakutake (1992) and Aoshima and Chen (1999) can be used. In this section, let us consider the case where $k \geq 3$. We seek lower bounds for the integrals in (2.2.5) and (2.2.9) which will lead to approximations for u_1 and u_2 that do not depend on the knowledge of p_0. Note that we will keep c_1 and c_2 as functions of p_0.

2.3.1 Two–Stage Procedure for R_1

Because $\rho_1 \equiv \rho_1(p_0, \delta^\star, k)$ is increasing in p_0 for any fixed $\delta^\star > 1$, we get

$$\rho_1 = \rho_1(p_0, \delta^\star, k) > \rho_1(0, \delta^\star, k) = \frac{1 + k - \delta^\star}{2 + k + \delta^\star(k - 2)} = \rho_1^\star, \text{ say.}$$

Then, by the well–known Slepian (1962) inequality, the integral in (2.2.5) is bounded below by

$$\int_{-\infty}^{\infty} \Phi^{k-1}\left(\frac{x\sqrt{\rho_1^\star} + c_1\sqrt{n}}{\sqrt{1 - \rho_1^\star}}\right) d\Phi(x).$$

Solving the equation

$$\int_{-\infty}^{\infty} \Phi^{k-1}\left(\frac{x\sqrt{\rho_1^\star} + u_1}{\sqrt{1 - \rho_1^\star}}\right) d\Phi(x) = P^\star \qquad (2.3.1)$$

for u_1, we get $u_1 \equiv u_1^\star$, which does not depend on p_0. This gives the sample size

$$n_1^{(1)} = \left\langle \frac{u_1^{\star 2}}{c_1^2} \right\rangle + 1. \qquad (2.3.2)$$

A second lower bound for the integral in (2.2.5) can be obtained by noting that

$$\frac{1}{\delta^\star} \geq \frac{\rho_1}{1 - \rho_1} \geq \frac{1 + k - \delta^\star}{1 + (k-1)\delta^\star}. \qquad (2.3.3)$$

The above inequalities are obtained by writing the expression for $\rho_1/(1-\rho_1)$ using (2.2.4) [which is increasing in p_0] and letting $p_0 = 1$ and $p_0 = 0$, respectively. Using (2.3.3), the lower bound is given by

$$\int_{-\infty}^0 \Phi^{k-1}\left(x\sqrt{\frac{1}{\delta^\star}} + \frac{c_1\sqrt{n}}{\sqrt{1-\rho_1}}\right) d\Phi(x)$$

$$+ \int_0^\infty \Phi^{k-1}\left(x\sqrt{\frac{1+k-\delta^\star}{1+(k-1)\delta^\star}} + \frac{c_1\sqrt{n}}{\sqrt{1-\rho_1}}\right) d\Phi(x).$$

Solving the equation

$$\int_{-\infty}^0 \Phi^{k-1}\left(x\sqrt{\frac{1}{\delta^\star}} + u_1\right) d\Phi(x)$$

$$+ \int_0^\infty \Phi^{k-1}\left(x\sqrt{\frac{1+k-\delta^\star}{1+(k-1)\delta^\star}} + u_1\right) d\Phi(x) = P^\star \quad (2.3.4)$$

for u_1, we get the solution $u_1 \equiv \tilde{u}_1$, which does not depend on p_0. This gives the sample size

$$n_2^{(1)} = \left\langle \frac{(1-\rho_1)\tilde{u}_1^2}{c_1^2} \right\rangle + 1. \quad (2.3.5)$$

We investigated the relative merits of $n_1^{(1)}$, $n_2^{(1)}$, and $n_3^{(1)}$ (which is derived by letting $u_1 = c_1\sqrt{n/\rho_1}$ and using the inequality (2.3.3) in (2.2.5)), for obtaining an approximation to $n_0^{(1)}$. By comparing those values with $n_0^{(1)}$ for $p_0 = 0.1(0.1)0.9$, $k = 3(1)10$, $\delta^\star = 1.5(0.5)3.0$ and $P^\star = 0.9$, we observed that in most cases $n_1^{(1)}$ is the best approximation to $n_0^{(1)}$. Only when p_0 is extremely high, $n_2^{(1)}$ gives the best approximation to $n_0^{(1)}$. We omit the details for brevity. The following tables give values of $u_1^\star$ for $n_1^{(1)}$ and $\tilde{u}_1$ for $n_2^{(1)}$, respectively, when $k = 3(1)10$, $\delta^\star = 1.5(0.5)3.0$ and $P^\star = 0.9$.

Table 2.3.1: Values of $u_1^\star$ When $P^\star = 0.9$

$\delta^\star \setminus k$	3	4	5	6	7	8	9	10
1.5	1.596	1.762	1.873	1.956	2.022	2.076	2.122	2.161
2.0	1.609	1.780	1.895	1.980	2.048	2.103	2.151	2.192
2.5	1.617	1.792	1.909	1.996	2.064	2.121	2.169	2.211
3.0	1.624	1.801	1.919	2.007	2.076	2.133	2.182	2.224

Table 2.3.2: Values of $\tilde{u}_1$ When $P^* = 0.9$

$\delta^* \setminus k$	3	4	5	6	7	8	9	10
1.5	2.058	2.272	2.416	2.522	2.607	2.677	2.737	2.788
2.0	1.968	2.175	2.315	2.419	2.502	2.571	2.629	2.679
2.5	1.915	2.116	2.252	2.354	2.436	2.503	2.560	2.610
3.0	1.884	2.077	2.210	2.310	2.390	2.456	2.512	2.561

We now propose the following two–stage procedure based on $n_1^{(1)}$ in (2.3.2) for $\delta^* < 1 + k$.

Procedure S_1: First, take a moderately large sample of size m. Let m_0 denote the count of $\{x_1 = \cdots = x_k\}$. Compute $\hat{c}_1 = c_1(\hat{p}_0, \delta^*, k)$ with $\hat{p}_0 = m_0/m$ in (2.2.3). Compute

$$N^{(1)} = \max\left\{m, \left\langle \frac{u_1^{*2}}{\hat{c}_1^2} \right\rangle + 1\right\}. \qquad (2.3.6)$$

Take an additional sample of size $N^{(1)} - m$. From the total sample of size $N^{(1)}$, let $N_i^{(1)}$ be the count of $\{x_i < x_{i'}, \ i' = 1, ..., i-1, \ i+1, ..., k\}$. Select the component with the smallest count among $(N_1^{(1)}, ..., N_k^{(1)})$ as the best component. Randomize to break ties for the first place.

Procedure S_1 satisfies the requirement (2.1.2). Since m is assumed to be moderately large, the event $\{N^{(1)} = n\}$ and the random variables $\{n_1, ..., n_k\}$ are asymptotically independent, and $\hat{c}_1 \simeq c_1$. So, from (2.2.5), we get

$$P^*(CS \mid S_1) \simeq E\left\{\int_{-\infty}^{\infty} \Phi^{k-1}\left(\frac{x\sqrt{\rho_1} + \hat{c}_1\sqrt{N^{(1)}}}{\sqrt{1 - \rho_1}}\right) d\Phi(x)\right\}$$

$$\geq \int_{-\infty}^{\infty} \Phi^{k-1}\left(\frac{x\sqrt{\rho_1} + u_1^*}{\sqrt{1 - \rho_1}}\right) d\Phi(x) \qquad (2.3.7)$$

$$\geq P^*.$$

The first inequality in (2.3.7) follows from (2.3.6) whereas the second follows from Slepian's inequality and (2.3.1).

Remark 2.3.1 If $\hat{p}_0 = m_0/m$ is extremely high, say $\hat{p}_0 \geq 0.9$, then the following modification is suggested: In the first stage of Procedure S_1, compute $\hat{\rho}_1 = \rho_1(\hat{p}_0, \delta^\star, k)$ as given by (2.2.4). Then compute

$$N^{(1)} = \max\left\{m, \left\langle \frac{(1-\hat{\rho}_1)\tilde{u}_1^2}{\hat{c}_1^2} \right\rangle + 1\right\} \qquad (2.3.8)$$

based on $n_2^{(1)}$ given by (2.3.5) instead of (2.3.6) and proceed to the second stage. However, when p_0 is extremely high, the experiment itself should be reexamined for the purpose of selecting the best component. For example, in the application described in Section 2.2, the car manufacturer should perhaps reduce the stress-level of the test to around 12,000 r.p.m.

2.3.2 Two–Stage Procedure for R_2

Note that $\rho_2 = \rho_2(p_0, \delta^\star, k)$ is decreasing in p_0 for any fixed $\delta^\star > 1$. So, we get

$$\rho_2 = \rho_2(p_0, \delta^\star, k) > \rho_2(1, \delta^\star, k) = \frac{1}{1+\delta^\star} = \rho_2^\star, \text{ say.}$$

Using Slepian's inequality, the integral in (2.2.9) is bounded below by

$$\int_{-\infty}^{\infty} \Phi^{k-1}\left(\frac{x\sqrt{\rho_2^\star} + c_2\sqrt{n}}{\sqrt{1-\rho_2^\star}}\right) d\Phi(x).$$

Solve the equation

$$\int_{-\infty}^{\infty} \Phi^{k-1}\left(\frac{x\sqrt{\rho_2^\star} + u_2}{\sqrt{1-\rho_2^\star}}\right) d\Phi(x) = P^\star \qquad (2.3.9)$$

for u_2. The solution $u_2 \equiv u_2^\star$ does not depend on p_0 and gives the sample size

$$n_1^{(2)} = \left\langle \frac{u_2^{\star 2}}{c_2^2} \right\rangle + 1. \qquad (2.3.10)$$

We compared $n_1^{(2)}$ with the other two approximations $n_2^{(2)}$ and $n_3^{(2)}$ derived similarly to those described in Section 3.1. We observed numerically that $n_1^{(2)}$ is the best approximation to $n_0^{(2)}$ in every case where we considered $p_0 = 0.1(0.1)0.9$ when $k = 3(1)10$, $\delta^\star = 1.5(0.5)3.0$ and $P^\star = 0.9$. The following table gives values of $u_2^\star$.

Table 2.3.3: Values of $u_2^\star$ When $P^\star = 0.9$

$\delta^\star \setminus k$	3	4	5	6	7	8	9	10
1.5	1.594	1.759	1.871	1.954	2.019	2.074	2.120	2.160
2.0	1.603	1.774	1.889	1.974	2.042	2.098	2.146	2.187
2.5	1.609	1.783	1.900	1.987	2.056	2.114	2.162	2.204
3.0	1.613	1.789	1.907	1.996	2.066	2.124	2.173	2.216

We now propose the following two–stage procedure based on $n_1^{(2)}$ given by (2.3.10).

Procedure S_2: Take a moderately large sample of size m. Let m_0 denote the count of $\{x_1 = \cdots = x_k\}$. Compute $\hat{c}_2 = c_2(\hat{p}_0, \delta^\star, k)$ with $\hat{p}_0 = m_0/m$ in (2.2.8). Compute

$$N^{(2)} = \max\left\{m, \left\langle \frac{u_2^{\star 2}}{\hat{c}_2^2} \right\rangle + 1\right\}. \qquad (2.3.11)$$

Take an additional sample of size $N^{(2)} - m$. On the basis of the total sample of size $N^{(2)}$, select the component associated with the largest sample mean among $(\overline{X}_{1N^{(2)}}, ..., \overline{X}_{kN^{(2)}})$ as the best component.

It can be shown that Procedure S_2 also satisfies the requirement (2.1.2) in a way similar to Procedure S_1.

2.4 CONCLUDING REMARKS

Before analyzing the data for our application in Section 2.2, let us compare the procedures S_1 and S_2 in terms of the required sample size and the PCS. For large initial sample size m, we evaluated the relative efficiency of S_1 and S_2 by computing the ratio $n_1^{(1)}/n_1^{(2)}$ of the sample sizes defined by (2.3.2) and (2.3.10) for $p_0 \in (0,1)$ when $k = 3(1)10$, $\delta^\star = 1.5(0.5)3.0$ and $P^\star = 0.9$. We observed that $n_1^{(1)}/n_1^{(2)} > 1$ when $p_0 \leq 0.4$ and $0.4 < p_0 < 0.5$ for most values of $(k, \delta^\star)$, and also that $n_1^{(1)}/n_1^{(2)} < 1$ when $p_0 \geq 0.5$. In addition, we observed that the ratio increases in k and decreases in $\delta^\star$ and p_0.

We also investigated efficiencies of S_1 and S_2 when m is moderate. We estimated the PCS and the expected sample size by simulating $R = 10000$ trials for each procedure. Theorem 2.1 given by Proschan and Sullo (1976) was used to generate k-variate exponential random vectors. Setting $k = 4$ and $\lambda_1 = 3$, under the LFC, the parameters are written as $(\lambda_0, \lambda_1, \lambda_2, \lambda_3, \lambda_4) = \left(\frac{3p_0(1+3\delta^\star)}{1-p_0}, 3, 3\delta^\star, 3\delta^\star, 3\delta^\star \right)$. We used $\delta^\star = 1.5$, 2.0 and $P^\star = 0.9$. In each case, we picked the values of $u_1^\star$ and $u_2^\star$ from Table 2.3.1 and Table 2.3.3 respectively. A pilot sample of size $m = [\ 0.8 \min\{n_0^{(1)}, n_0^{(2)}\}\ ] + 1$ with $p_0 = 0.1(0.2)0.9$, where $n_0^{(1)}$ and $n_0^{(2)}$ are the optimal sample sizes defined by (2.2.7) and (2.2.11) respectively, was used to start off S_1 and S_2.

For each procedure, let us write $n^\star$ instead of n_0. Let n_r be the observed value of N and $p_r = 1(0)$ according as a correct selection occurs(does not occur). Denote $\bar{n} = \sum_{r=1}^{R} n_r / R$, $s^2(\bar{n}) = \sum_{r=1}^{R} (n_r - \bar{n})^2/(R^2 - R)$, $\bar{p} = \sum_{r=1}^{R} p_r / R$ and $s^2(\bar{p}) = \bar{p}(1 - \bar{p})/R$. The quantities $\bar{n}$ and $\bar{p}$ respectively estimate $E(N)$ and $P(CS)$, while $s(\bar{n})$ and $s(\bar{p})$ stand for their corresponding estimated standard errors. The values of p_0, $n^\star$, m, $\bar{p}$, $s(\bar{p})$, $\bar{n}$, $s(\bar{n})$ and $(\bar{n} - n^\star)/n^\star$ for each $\delta^\star$ are reported in Table 2.4.1. For each p_0, the upper(lower) line gives those values for $S_1(S_2)$.

We observe from Table 2.4.1 that, as expected, the proposed two-stage procedures S_1 and S_2 work well. The same conclusion about the efficiencies of S_1 and S_2 seems to hold for moderate values of m as well.

We consequently recommend that for both large and moderate m, after taking a pilot sample, examine the value of $\hat{p}_0$: If $\hat{p}_0 \geq 0.5$, proceed with S_1, otherwise, use S_2. When $\hat{p}_0$ is extremely high, say $\hat{p}_0 \geq 0.9$, refer to Remark 2.3.1.

For the application provided in Section 2.2, survival times (x_1, x_2, x_3, x_4) are recorded in seconds for each piston. From the experimental side, the difference was set as $\delta^\star = 2$ and the confidence was set as $P^\star = 0.9$. We took a pilot sample of size $m = 30$ (see Table 2.4.2).

The situation that a crank-shaft is broken occurs $m_0 = 7$ times when $x_1 = x_2 = x_3 = x_4$. Then, we have $\hat{p}_0 = 7/30 = 0.2333$. Hence, from the discussion given earlier in this section, we adopt the Procedure S_2. From Table 2.3.3, we find $u_2^\star = 1.774$. Then, using (2.3.11), the value of $N^{(2)}$ is

$$n = \max \left\{ 30, \left\langle \frac{5.566 \times 8.733}{(1 - 0.2333) \times 3.933} (1.774)^2 \right\rangle + 1 \right\} = 51.$$

We needed to run 21 additional tests. The additional data collected is given in Table 2.4.3.

Table 2.4.1: Estimated $P(CS)$ and $E(N)$
for S_1 (Upper Entry) and S_2 (Lower Entry) with 10,000 Trials
When $P^\star = 0.9$ and $k = 4$

$$\delta^\star = 1.5$$

p_0	$n^\star$	m	$\bar{p}$	$s(\bar{p})$	$\bar{n}$	$s(\bar{n})$	$(\bar{n} - n^\star)/n^\star$
0.1	186.6	53	0.9151	0.0028	187.5	0.0898	0.0050
	65.48		0.9130	0.0028	72.22	0.1507	0.1030
0.3	240.6	129	0.9145	0.0028	242.3	0.1445	0.0070
	161.2		0.9061	0.0029	170.6	0.2712	0.0579
0.5	337.9	271	0.9173	0.0028	340.3	0.2122	0.0070
	347.1		0.9051	0.0029	358.6	0.4172	0.0333
0.7	565.0	452	0.9171	0.0028	570.4	0.4165	0.0097
	794.6		0.9031	0.0030	811.6	0.8343	0.0215
0.9	1700	1360	0.9192	0.0027	1717	1.4130	0.0098
	3060		0.9081	0.0029	3089	2.8170	0.0097

Table 2.4.1 Continued

$$\delta^\star = 2.0$$

p_0	$n^\star$	m	$\bar{p}$	$s(\bar{p})$	$\bar{n}$	$s(\bar{n})$	$(\bar{n} - n^\star)/n^\star$
0.1	70.72	20	0.9227	0.0027	71.61	0.0603	0.0127
	24.33		0.9246	0.0026	29.36	0.1059	0.2070
0.3	91.70	50	0.9269	0.0026	93.32	0.0934	0.0176
	61.64		0.9195	0.0027	68.70	0.1743	0.1146
0.5	129.5	104	0.9296	0.0026	131.8	0.1387	0.0185
	134.0		0.9171	0.0028	142.1	0.2666	0.0603
0.7	217.6	175	0.9297	0.0026	222.2	0.2651	0.0212
	307.9		0.9153	0.0028	320.4	0.5266	0.0407
0.9	658.0	527	0.9265	0.0026	674.9	0.9269	0.0257
	1187		0.9127	0.0028	1216	1.8310	0.0246

By combining the data from both stages, we get the overall sample mean for survival times of 4 pistons as $(78.0, 93.8, 95.7, 133.0)$. From these mean survival times we conclude that scheme-ratio D is the best with confidence $P^\star = 0.9$.

ACKNOWLEDGMENT

The authors would like to thank the referees for their careful reading of this manuscript. Suggestions made by one of the referees especially led to significant improvement in the overall presentation.

Table 2.4.2: Survival Data of Four Pistons (First Stage)

$(85.6, 163.5, 109.1, 52.2)$	$(97.3, 198.5, 155.0, 43.7)$
$(120.5, 118.7, 120.5, 120.5)$	$(50.3, 144.4, 69.0, 122.7)$
$(135.2, 150.0, 65.1, 406.9)$	$(36.6, 36.6, 36.6, 36.6)$
$(7.7, 184.2, 260.6, 432.5)$	$(118.4, 134.0, 274.9, 305.4)$
$(33.6, 130.8, 130.8, 110.8)$	$(24.5, 2.0, 161.2, 161.2)$
$(9.2, 9.2, 9.2, 9.2)$	$(46.7, 46.7, 38.2, 46.7)$
$(0.9, 108.2, 39.4, 50.0)$	$(45.7, 45.7, 45.7, 45.7)$
$(275.6, 156.0, 58.0, 98.2)$	$(134.1, 165.5, 29.2, 128.5)$
$(175.7, 175.7, 175.7, 175.7)$	$(6.8, 6.8, 6.8, 6.8)$
$(178.7, 174.9, 133.1, 475.3)$	$(44.4, 44.4, 44.4, 42.6)$
$(285.7, 131.1, 282.8, 100.9)$	$(85.8, 17.4, 85.8, 85.8)$
$(51.7, 51.7, 51.7, 51.7)$	$(196.2, 27.2, 128.2, 213.6)$
$(93.6, 172.2, 162.6, 296.0)$	$(58.4, 58.4, 58.4, 58.4)$
$(14.6, 64.7, 64.7, 41.0)$	$(83.3, 75.0, 272.9, 301.7)$
$(158.1, 138.6, 34.9, 306.6)$	$(155.9, 94.4, 64.8, 194.7)$

Table 2.4.3: Suvival Data of Four Pistons (Second Stage)

$(68.8, 68.8, 68.8, 68.8)$	$(75.4, 99.0, 73.4, 257.5)$
$(1.1, 1.1, 1.1, 1.1)$	$(29.9, 166.9, 67.2, 166.9)$
$(31.1, 146.9, 137.9, 146.9)$	$(16.6, 67.1, 95.8, 181.1)$
$(4.4, 21.1, 36.6, 36.6)$	$(5.0, 5.0, 5.0, 5.0)$
$(15.0, 15.0, 15.0, 15.0)$	$(283.3, 327.5, 196.8, 500.6)$
$(126.4, 164.8, 164.8, 60.8)$	$(128.7, 106.4, 2.3, 63.6)$
$(166.4, 149.3, 58.7, 170.9)$	$(14.9, 62.1, 334.0, 142.1)$
$(42.6, 117.7, 109.8, 200.1)$	$(2.3, 59.5, 31.4, 59.5)$
$(16.2, 29.8, 63.1, 63.1)$	$(47.6, 57.4, 156.9, 28.2)$
$(42.1, 42.1, 42.1, 42.1)$	$(22.8, 22.8, 22.8, 22.8)$
$(28.3, 28.3, 28.3, 28.3)$	

REFERENCES

[1] Aoshima, M. and Chen, P. (1999). A two–stage procedure for selecting the largest multinomial cell probability when nuisance cell is present. *Sequential Anal.*, **18**, 143–155.

[2] Arnold, B.C. (1968). Parameter estimation for a multivariate exponential distribution. *J. Amer. Statist. Assoc.*, **63**, 848–852.

[3] Hyakutake, H. (1992). Selecting the better component of a bivariate exponential distribution. *Statist. Decisions*, **10**, 153–162.

[4] Hyakutake, H. (1994). The best component selection procedures for a multivariate exponential distribution. *Kumamoto J. Math.*, **7**, 61–66.

[5] Marshall, A.W. and Olkin, I. (1967). A multivariate exponential distribution. *J. Amer. Statist. Assoc.*, **62**, 30–44.

[6] Proschan, F. and Sullo, P. (1976). Estimating the parameters of a multivariate exponential distribution. *J. Amer. Statist. Assoc.*, **71**, 465–472.

[7] Slepian, D. (1962). On the one–sided barrier problem for gaussian noise. *Bell Sys. Tech. J.*, **41**, 463–501.

[8] Stein, C. (1945). A two–sample test for a linear hypothesis whose power is independent of the variance. *Ann. Math. Statist.*, **16**, 243–258.

Addresses for communication:

MAKOTO AOSHIMA, Institute of Mathematics, University of Tsukuba, Ibaraki 305-8571, Japan. E-mail: aoshima@math.tsukuba.ac.jp
MITSURU AOKI, Institute of Mathematics, University of Tsukuba, Ibaraki 305-8571, Japan. E-mail: aoki@math.tsukuba.ac.jp
MASAKI KAI, Insurance Distribution, Fidelity Investments Japan Limited, Tokyo 104-0033, Japan. E-mail: masaki.kai@jp.fid-intl.com

Chapter 3

Sequential Randomization Tests

TATHAGATA BANERJEE
Calcutta University, Kolkata, India

ONKAR PROSAD GHOSH
Directorate General of Commercial Intelligence and Statistics, Kolkata, India

3.1 MOTIVATING BACKGROUND

Many hypotheses of interest in science can be regarded as alternatives to a null hypothesis of randomness. In other words, the hypothesis under investigation suggests that the data will exhibit a certain kind of pattern, whereas the null hypothesis says that if such a pattern is present then this is purely due to chance of observations in a random order. For example, in a two-sample location and scale problem, it is assumed that under the null hypothesis every permutation of observations in the pooled sample is equally likely. In more complicated experiments such as randomized block designs, under the null hypothesis of no treatment effect, it is assumed that every permutation of observations within a block is equally likely. To carry out randomization tests for a specific problem, an appropriate statistic T is chosen to detect the departure from a null hypothesis. The statistic T should be so chosen as to capture the pattern specified under the alternatives. The observed value t of T is then compared with the randomization distribution of T obtained from the values of T for all permutations of the data, which are assumed equally likely under the null hypothesis. If the observed value t is found to be unlikely under the null hypothesis (that is, under the randomization distribution of T) but it is found to be likely under the alternatives, the null hypothesis is discredited to some extent. As a consequence, the alternative hypothesis is considered more reasonable. A measure of evidence is provided by the *observed significance level* (also known as a p-value). If, under the alternatives, T tends to take large values, the p-value is defined as

$$\hat{\alpha} = P_R\{T \geq t\},$$

where P_R denotes the probability measure corresponding to the randomization distribution of T. If T tends to take either small or large values under the alternatives, the p-value is defined by

$$\hat{\alpha} = 2\min\{P_R\{T \geq t\}, \ P_R\{T \leq t\}\}.$$

Smaller values of $\hat{\alpha}$ indicate stronger evidence against the null hypothesis.

Randomization tests, introduced by Fisher (1935) in the context of designs of experiments, are favored by practitioners for the following reasons:

(i) The tests are valid even if the sample is not randomly chosen from a population.

(ii) It is relatively easy to use a non-standard statistic for testing purposes.

There are plenty of examples of randomization tests being used in different branches of science. To name a few, it has been used in anthropology (Fisher (1936)), agriculture (Kempthorne (1952)), archaeology (Berry and Mielke (1985)), atmospheric science (Tukey et al. (1978)), ecology (Manly (1983)), education (Manly (1988)), epidemiology (Glass et al. (1971)), genetics (Karlin and Wiliams (1984)), geography (Royaltey et al. (1975)), geology (Clark (1989)), molecular biology (Barker and Dayhoff (1972), Karlin et al. (1983)), and palaeontology (Marcus (1969)).

Bradley (1969) termed randomization tests as "stunningly efficient tests that are dismally impractical". Indeed the enumeration of the values of T for all permutations of the data to generate the randomization distribution is a difficult task, if not impossible, even for moderate sample sizes. For instance, in a two-sample problem with sample sizes $m = n = 20$ there are $^{40}C_{20}$ permutations to examine. Even if one considers the modern-day computing facilities, this does not seem to be an easy task. To circumvent this problem, Dwass (1957) proposed a modified randomization test. The modification he proposed is based on a simple and obvious solution. Instead of enumerating the exact randomization distribution of the test statistic, he proposed to carry out the test by examining a "random sample" of permutations and to make the decision of acceptance or rejection of the null hypothesis on the basis of those sampled permutations only. Lock (1986,1991) pointed out that if a test is carried out by repeated sampling from the permutation distribution, it should almost always be possible to decide the outcome before sampling is completed. He then adopted *sequential probability ratio tests* (SPRT) to show that it is possible to achieve considerable savings in terms of computational efforts and still obtain tests that have similar levels and powers. However, Lock confined his discussion to permutation tests only. The same method, in principle, is valid for randomization tests too.

Both Dwass (1957) and Lock (1986,1991) were concerned with testing at a pre-fixed level of significance. However, many statisticians (for example, Kempthorne (1975,1979)) feel that there is nothing sacred about maintaining a 5% or 10% level and instead they prefer reporting the exact p-value. This lets the experts in specific subject-areas arrive at their own conclusions based on "losses" they may associate with each type of error. In such cases, sequential estimation of a p-value is of interest rather than sequential testing of a p-value. Sequential estimation of a p-value may also lead to considerable savings in terms of computing time if investigation of power properties of a randomization test is of prime concern. Oden (1991) considered this problem from a different point of view. In Section 3.2 we formally introduce the basic problem that we have dealt with here. Also, we investigate the application of SPRT in testing the significance of the observed value of a test statistic. A simple sequential procedure is proposed in Section 3.3 to estimate a p-value. The performances of the proposed methodologies are studied in Section 3.4 using extensive computer simulations. These studies show that considerable savings can be achieved in computing time by adopting these procedures. Concluding remarks are given in Section 3.5.

3.2   FORMULATION AND REVIEW OF EXISTING PROCEDURES

Suppose that x represents the data and T, the test statistic to be used for testing a null hypothesis of randomness (H_0) against the alternatives (H_1). The alternatives represent departure from the null hypothesis of a specific pattern. Let G represent a set of permutations considered equally likely under H_0. Corresponding to each g ∈ G, gx represents a permutation of the data x. Suppose that the cardinality of G is denoted by N and t is the value of T at x. The p-value is then defined as

$$\hat{\alpha} = P_R\{T \geq t\} = \frac{\text{Number of } g \text{ belonging to } G \text{ such that } T(gx) \geq t}{N}$$

$$(3.2.1)$$

assuming that large values of T tend to reject H_0. If both large and small values of T indicate the falsity of H_0, a p-value will then be defined as

$$\hat{\alpha} = 2\min\{P_R(T \geq t), \ P_R(T \leq t)\}. \qquad (3.2.2)$$

where $P_R(T \geq t)$ for t varying in the real line represents the randomization distribution of T.

  If the randomization distribution of T is known, $\hat{\alpha}$ can be evaluated and consequently a decision regarding acceptance or rejection of H can be immediately taken. However, to obtain $\hat{\alpha}$, one needs to evaluate T(gx) for all g ∈ G. As stated in the introduction, even for problems with moderate sample sizes, N may become forbiddingly large and hence the enumeration of randomization distribution becomes unmanageable. In order to implement a randomization test one needs to circumvent this problem.

  There are two distinct aspects to this problem. The result may be reported in terms of acceptance or rejection of H at a preassigned level of significance, such as $\alpha = 0.05$ or 0.10. On the other hand, one may choose to report a p-value instead. The former is a testing problem while the latter is an estimation problem. In the following, we describe two procedures for deciding whether to accept or reject H_0. The first one is a modified test procedure based on Monte Carlo simulations proposed by Dwass (1957). It is arguably the most popular method among practitioners for the implementation of randomization tests. The second one is a sequential version of a randomization test based on Wald's SPRT, similar to what Lock (1986,1991) discussed in the context of permutation tests.

A Modified Test Based on Monte Carlo Simulations: This is a method proposed by Dwass (1957) and later studied by Marriott (1979) in the context of Monte Carlo tests. The argument goes as follows. Draw M permutations of x at random and with replacement out of N equally likely permutations. Assuming that under the alternatives T tends to take large values, the null hypothesis is rejected if the observed value t of T is one of the largest $m = \alpha(M + 1)$ values of T (including t) which are ordered from the smallest to the largest. The probability of rejecting the null hypothesis in this situation is given by

$$P = \sum_{r=0}^{m-1} {}^{M}C_r \hat{\alpha}^r (1 - \hat{\alpha})^{M-r}$$

where $\hat{\alpha}$ is defined by (3.2.1).

The following table shows values of P for different values of M and $\hat{\alpha}$. It is evident that when $\alpha = 0.05(0.01)$, M = 1,000(5,000) is almost certain to detect significant results except in some of the borderline cases.

Table 3.2.1: Probabilities (P) of a Significant Result for Tests
Corresponding to Different Values of M, α and $\hat{\alpha}$

$\alpha = 0.05$	$\hat{\alpha}$				
M+1	0.1	0.075	0.05	0.025	0.01
20	0.134	0.228	0.375	0.616	0.824
40	0.087	0.197	0.414	0.744	0.940
100	0.023	0.129	0.447	0.895	0.999
200	0.003	0.067	0.468	0.969	1.000
1000	0.000	0.002	0.473	0.999	1.000

$\alpha = 0.01$	$\hat{\alpha}$				
M+1	0.02	0.015	0.01	0.005	0.001
100	0.134	0.227	0.372	0.606	0.903
200	0.087	0.197	0.406	0.740	0.981
500	0.026	0.129	0.444	0.891	0.999
1000	0.003	0.067	0.457	0.968	1.000
5000	0.000	0.002	0.473	0.999	1.000

A Modified Randomization Test Using SPRT: A sequential version of the randomization test can be formulated based on Wald's SPRT. If the randomization test is to be carried out at a pre-fixed level α, then the problem we need to solve is of testing H_0: $\hat{\alpha} \le \alpha$ against H_1: $\hat{\alpha} > \alpha$. This test can be carried out sequentially by drawing permutations of x one by one, with replacement, from the set of N permutations and deciding at each stage whether to stop or continue sampling. The problem is equivalent to testing sequentially whether a binomial proportion $\hat{\alpha}$ is less than or equal to α against it being larger than α. To apply the SPRT, we need to consider a simple null hypothesis such as "$\hat{\alpha} = \alpha$" against a simple alternative such as "$\hat{\alpha} = \alpha_1$" where α_1 is a pre-fixed number larger than α. Let $\beta_1(0 < \beta_1 < 1)$ and $\beta_2(0 < \beta_2 < 1)$ denote the error-probabilities of rejecting the null hypothesis when it is true and accepting the null hypothesis when it is false. Naturally the choices of α_1, β_1 and β_2 depend on the closeness of approximation desired. If the SPRT has to approximate the original randomization test quite well, α_1 has to be chosen very close to α and β_1, β_2 have to be close to 0. In that case, the ASN will be large.

Let S_n denote the number of times the simple null hypothesis is rejected up to and including the nth drawing of permutations. At the nth stage, a decision regarding whether to stop or continue is based on the following rule: stop and accept the null hypothesis if $S_n \le a_n$. Stop and reject the null hypothesis if $S_n \ge r_n$. Continue otherwise. The constants a_n and r_n are given by

$$a_n = \frac{\log(\frac{\beta_2}{1-\beta_1})}{\log(\frac{\alpha_1}{\alpha}) - \log(\frac{1-\alpha_1}{1-\alpha})} + n\frac{\log(\frac{1-\alpha}{1-\alpha_1})}{\log(\frac{\alpha_1}{\alpha}) - \log(\frac{1-\alpha_1}{1-\alpha})}$$

and

$$r_n = \frac{\log(\frac{1-\beta_2}{\beta_1})}{\log(\frac{\alpha_1}{\alpha}) - \log(\frac{1-\alpha_1}{1-\alpha})} + n\frac{\log(\frac{1-\alpha}{1-\alpha_1})}{\log(\frac{\alpha_1}{\alpha}) - \log(\frac{1-\alpha_1}{1-\alpha})}.$$

Table 3.2.2 furnishes the results of a simulation study. The last two columns of the table respectively show the standard deviation (s_m) of simulated sample sizes and the proportion of rejections of the hypothesis of randomness corresponding to different choices of $(\hat{\alpha}, \beta_1, \beta_2, \alpha, \alpha_1)$ based on 10,000 repetitions of an experiment. From these results, it is apparent that an application of SPRT leads to the rejection of the null hypothesis with certainty if $\hat{\alpha} \le \alpha$. Also, the ASN column (s_m column) clearly shows that an application of SPRT results in substantial savings of computing time.

Table 3.2.2: Testing the Hypothesis of Randomness Via the SPRT
$$\hat{\alpha} = 0.05$$

β_1	β_2	α	α_1	ASN	s_m	prop
		0.01	0.05	123	89.69	0.0106
			0.10	192	84.08	0.2888
0.01	0.01	0.05	0.10	272	152.51	0.9920
			0.20	48	24.36	0.9935
		0.01	0.05	79	64.05	0.0511
			0.10	47	38.17	0.3959
0.05	0.05	0.05	0.10	163	106.43	0.9601
			0.20	30	17.13	0.9695
		0.01	0.05	55	46.90	0.0986
			0.10	29	22.41	0.4339
0.10	0.10	0.05	0.10	111	74.66	0.9192
			0.20	20	12.19	0.9341

We also carried out simulation studies with $\hat{\alpha} = 0.01, 0.10$ but with the same values of β_1, β_2, α and α_1. Those results are not presented here for the sake of brevity. But the savings in terms of sample sizes were considerable. As expected, for $\hat{\alpha} = 0.01(0.10)$, the ASN values were larger (smaller) than the values of ASN given in Table 3.2.2.

3.3 SEQUENTIAL ESTIMATION OF A p-VALUE

The modified randomization test using SPRT described above is an efficient method to carry out the test as long as interest is centered on merely the acceptance or rejection of the null hypothesis at a pre-fixed level of significance rather than on reporting the exact p-value. In the latter case, estimation of a p-value with pre-specified accuracy is of prime concern. The following table shows the confidence limits of standard large sample confidence intervals of $\hat{\alpha}$ with confidence coefficient 0.99 for different values of M.

From Table 3.3.1, it is evident that for the estimation of $\hat{\alpha} = 0.05$, even M = 5000 does not produce too much accuracy. Comparing these

Table 3.3.1: Confidence Interval for $\hat{\alpha}$ with Confidence Coefficients
0.99 Corresponding to Different Values of M, $\hat{\alpha}$

$\hat{\alpha} = 0.05$			$\hat{\alpha} = 0.01$		
M	Lower Limit	Upper Limit	M	Lower Limit	Upper Limit
200	0.010	0.090	600	0.000	0.020
500	0.025	0.075	1000	0.002	0.018
1000	0.032	0.068	5000	0.006	0.014
5000	0.042	0.058	10000	0.007	0.013

results with those given in Table 3.2.1, we find that for obtaining a
reasonably good estimate of a p-value, one needs much larger sample
sizes than is necessary for simply accepting or rejecting a null hypothesis. Also, the smaller the values of $\hat{\alpha}$, the larger the sample sizes one
would need to achieve similar accuracy. Thus, $\hat{\alpha}$ being unknown, the
application of a sequential estimation procedure is expected to achieve
considerable savings in computing time, compared to a fixed sample-
size estimation procedure. In the following, we implement a simple
sequential estimation procedure and investigate its efficiency compared
to a fixed-sample-size procedure.

The sequential estimation procedure we adopt here is a slight modification of the procedure proposed by Anscombe (1953). The logic goes
like this. Suppose that it is required to estimate $\hat{\alpha}$ with a standard error
that is within a fraction c of $\hat{\alpha}$ itself. Anscombe (1953) had shown that,
to a reasonable degree of approximation, the sampling should continue
until the estimated standard error is within the fraction c of the es-
timated $\hat{\alpha}$. This means that if at the nth stage, the estimate of $\hat{\alpha}$ is
l/n (l being a non-negative integer), then one should continue sampling
until $(l/n)(1-l/n)/n \leq (cl/n)^2$ i.e., $l \leq (c^2 + n^{-1})^{-1}$. We find that an
application of Anscombe's original procedure results in large standard
errors and less accurate estimates. However, a simple modification of
the procedure aimed at allowing the estimate to become stable (stated
below in Step # 1 and Step # 3) results in better estimates. The simulation studies furnished below confirm it. The improvement is found
to be appreciable in case $\hat{\alpha}$ is 0.2 or larger. It has also been observed
that increasing L (defined below in Step # 1) to a number more than
10 does not affect the final results significantly. A second modification
that we have made is given in Step # 3 below. This is to truncate the
sampling procedure at some pre-specified stage, say, at the uth stage.
This is to avoid unduly large sample sizes, especially when $\hat{\alpha}$ and c are
small. The modified procedure runs as follows.

Step # 1: Continue sampling from the randomization distribution until a prefixed number, say, L rejections of the null hypothesis are obtained. Suppose that the Lth rejection occurs at the mth stage.

Step # 2: At the mth stage, if $L \geq (c^2 + m^{-1})^{-1}$ where c is a prefixed small positive number, then stop and estimate $\hat{\alpha}$ by L/m.

Step # 3: At the mth stage, if $L < (c^2 + m^{-1})^{-1}$, then continue sampling. At each successive stage, say, at the nth stage, we check a condition similar to that in Step # 2 with L replaced by the number of rejections obtained up to and including the nth stage, and m replaced by n. However, we may continue in this manner only up to a prefixed number of stages (say, u) and then stop. Finally, we estimate $\hat{\alpha}$ by the proportion of rejections.

The results obtained from an extensive simulation study have been reported in Tables 3.3.2-3.3.4. These tables show the values of ASN, the standard deviation of the simulated sample sizes (s_m), $\hat{\alpha}$ and the standard deviation of the simulated values of $\hat{\alpha}$ (s_α) for the following choices of $\hat{\alpha} = 0.01, 0.05$; $L = 2, 10, 20$; $u = 5000, 10000, 15000, 20000$; and $c = 0.01, 0.05, 0.10, 0.15, 0.20$. The patterns that emerge from this study are summarized below.

(i) For $\hat{\alpha} = 0.01$, the choice of L does not affect the performances. As expected, the larger the value of u and the smaller the value of c, the better the estimate of $\hat{\alpha}$. A comparison of the entries from Table 3.3.1 corresponding to $\hat{\alpha} = 0.01$, M = 5000(10000) with those in Table 3.3.2 corresponding to c = 0.15, u = 5000(10000) clearly reveal that similar estimate of $\hat{\alpha}$ can be obtained with considerably smaller sample sizes. The respective ASN's are 4354 and 4447 if one adopts the above sequential estimation procedure.

(ii) For $\hat{\alpha} = 0.05$, the effect of the choice of L on the estimate of $\hat{\alpha}$ with similar accuracy is clearly evident from Tables 3.3.3 - 3.3.4. From the s_α column, it is evident that the choice L = 2 produces a poor estimate of $\hat{\alpha}$ compared to the choice L = 10 or 20. Also it has been observed that choices of L in excess of 10 do not affect the estimate noticeably. A comparison of Table 3.3.1 with Table 3.3.4 clearly shows that the sequential procedure produces similar estimates of $\hat{\alpha}$ using considerably smaller sample sizes.

Table 3.3.2: Simulation Results for the Modified Anscombe's
Procedure ($\hat{\alpha} = 0.01$)

L	u	c	ASN	s_m	$\hat{\alpha}$	s_α
2 or 10 or 20	5000	0.01	5000	–	0.010000	0.001393
		0.05	5000	–	0.010000	0.001393
		0.10	5000	–	0.010000	0.001393
		0.15	4354	555.44	0.010199	0.001607
		0.20	2493	493.61	0.010439	0.002167
	10000	0.01	10000	–	0.010000	0.000997
		0.05	10000	–	0.010000	0.000997
		0.10	9553	502.70	0.010060	0.001064
		0.15	4447	700.72	0.010242	0.001561
		0.20	2493	493.61	0.010439	0.002167
	15000	0.01	15000	–	0.010010	0.000816
		0.05	15000	–	0.010010	0.000816
		0.10	9947	1034.69	0.010105	0.001018
		0.15	4447	700.72	0.010242	0.001561
		0.20	2493	493.61	0.010439	0.002167
	20000	0.01	20000	–	0.010011	0.000706
		0.05	20000	–	0.010011	0.000706
		0.10	9947	1034.69	0.010105	0.001018
		0.15	4447	700.72	0.010242	0.001561
		0.20	2493	493.61	0.010439	0.002167

It is to be mentioned here that we have carried out similar simulation
experiments with $\hat{\alpha} = 0.1$, 0.2 while keeping the values of the other
parameters L, u and c fixed as before. The patterns that emerged were
similar to what we have discussed so far.

3.4   AN ILLUSTRATIVE EXAMPLE

To illustrate the methodology developed in Section 3.3, we consider the
following example:

Fisher (1935) utilized the data collected by Charles Darwin on the
plant 'Zea mays' to introduce randomization tests. Darwin had 15 pairs
of plants. The plants in each pair had exactly the same age, grew under
the same conditions, and descended from the same parents. From each
pair, one plant was selected at random and was cross-fertilized, while
the other one was self-fertilized. The heights of offsprings were then
measured to the nearest eighth of an inch. Table 3.4.1 gives the results
in eighths of an inch over 12 inches. The question of interest to Darwin

Table 3.3.3: Simulation Results for the Modified Anscombe's
Procedure ($\hat{\alpha} = \mathbf{0.05}$)

L	u	c	ASN	s_m	$\hat{\alpha}$	s_α
		0.01	4999	1.80	0.050239	0.013783
		0.05	4999	1.80	0.050239	0.013783
	5000	0.10	1907	201.35	0.050744	0.014353
		0.15	852	134.64	0.051547	0.182180
		0.20	481	100.32	0.052394	0.019658
		0.01	9996	6.27	0.050427	0.019118
		0.05	7598	402.69	0.050284	0.009819
		0.10	1907	201.35	0.050744	0.014353
2	10000	0.15	852	134.64	0.051547	0.182180
		0.20	481	100.32	0.052394	0.019658
		0.01	14996	4.83	0.050319	0.016548
		0.05	7598	402.69	0.050284	0.009819
	15000	0.10	1907	201.35	0.050744	0.014353
		0.15	852	134.64	0.051547	0.182180
		0.20	481	100.32	0.052394	0.019658
		0.01	19994	8.67	0.050322	0.016522
		0.05	7598	402.69	0.050284	0.009819
	20000	0.10	1907	201.35	0.050744	0.014353
		0.15	852	134.64	0.051547	0.182180
		0.20	481	100.32	0.052394	0.019658

was whether these confirm a general belief that the offsprings from
crossed plants are superior to those from either parent. Because of
pairing, it is natural to take pairwise differences in Table 3.4.1, as they
will indicate the *superiority* of "crossing" over "self fertilization" under
similar conditions. Fisher (1935) argued that if the pair of observations
corresponding to cross-fertilized and self-fertilized seeds in a pair were
exchangeable (which is true under the null hypothesis), then the 15
differences were equally likely to be positive or negative. He considered
T, the sum of these differences, and the randomization distribution of
T generated by $2^{15}(= 32768)$ equally likely realizations of this sum. In
this example, t = 314 and $\hat{\alpha} = \frac{863}{32768} = 0.026$. Table 3.4.2 shows that at
1% level, use of SPRT does not yield good results as the true value of
$\hat{\alpha}$ (=0.026) lies midway between the values specified under the null and
the alternative. However, at 5% level, the correct decision is made with
practical certainty. A close look at Table 3.4.3 clearly reveals that an
application of a sequential estimation procedure results in very accurate
estimates. Also, we achieve considerable savings in terms of computing
time.

Table 3.3.4: Simulation Results for the Modified Anscombe's
Procedure ($\hat{\alpha} = 0.05$)

L	u	C	ASN	s_m	$\hat{\alpha}$	s_α
		0.01	5000	–	0.050050	0.003083
		0.05	5000	–	0.050050	0.003083
	5000	0.10	1907	200.07	0.050558	0.005104
		0.15	853	133.64	0.051260	0.007856
		0.20	481	99.88	0.052107	0.010806
		0.01	10000	–	0.050046	0.002158
		0.05	7599	395.42	0.050189	0.002489
		0.10	1907	200.07	0.050558	0.005104
10 or 20	10000	0.15	853	133.64	0.051260	0.007856
		0.20	481	99.88	0.052107	0.010806
		0.01	15000	–	0.050034	0.000177
		0.05	7599	395.42	0.050189	0.002489
	15000	0.10	1907	200.07	0.050558	0.005104
		0.15	853	133.64	0.051260	0.078560
		0.20	481	99.88	0.052107	0.010806
		0.01	20000	–	0.050037	0.001519
		0.05	7599	395.42	0.050189	0.002489
	20000	0.10	1907	200.07	0.050558	0.005104
		0.15	853	133.64	0.051260	0.007856
		0.20	481	99.88	0.052107	0.108060

Table 3.4.1: The Heights of Offsprings of 'Zea mays' As Reported by
Charles Darwin in Eighths of an Inch Over 12 Inches

Pair	1	2	3	4	5	6	7	8	9	10	11	12	13	14	15
Cross-fertilized	92	0	72	80	57	76	81	67	50	77	90	72	81	88	0
Self-fertilized	43	67	64	64	51	53	53	26	36	48	34	48	6	28	48
Difference	49	-67	8	16	6	23	28	41	14	29	56	24	75	60	-48

Table 3.4.2: Results from the Application of SPRT

β_1	β_2	α	α_1	ASN	s_m	prop
		0.01	0.05	352	274.814210	0.4298
0.01	0.01	0.05	0.10	135	40.736258	1.0000
		0.01	0.05	154	118.970954	0.4761
0.05	0.05	0.05	0.10	87	33.737235	1.0000

Table 3.4.3: Application of a Sequential Procedure to Estimate a
p-value

L	u	c	ASN	s_m	$\hat{\alpha}$	s_α
		0.01	5000	–	0.026641	0.002439
		0.05	4999	–	0.026739	0.010034
2	5000	0.10	3673	558.334954	0.026909	0.002668
		0.15	1648	443.605940	0.027222	0.003975
		0.01	5000	–	0.026641	0.002440
		0.05	5000	–	0.026643	0.002433
10	5000	0.10	3673	558.957937	0.026908	0.002668
		0.15	1648	443.485336	0.027222	0.003974
		0.01	5000	–	0.026641	0.002435
		0.05	5000	–	0.026642	0.002439
20	5000	0.10	3673	558.958080	0.026908	0.002668
		0.15	1648	443.686500	0.027221	0.003974

3.5 CONCLUDING REMARKS

In this article, we confine ourselves to a sequential version of a randomization test. There are non-sequential methods available for reducing the computing time. They basically involve search algorithms developed by exploiting the inherent structure of a problem. Green (1977) suggested a "branch and bound" method for use in two-sample tests. Good (1976), Gail and Mantel (1977), Pagano and Halvorsen (1981), and Patfield (1982), among others, joined in the search for a speedy method to evaluate tail probabilities for Fisher's exact test. A quantum leap in this direction took place with the "network approach" of Mehta and Patel (1980). Of course, all these methods are directed toward evaluating the tail probabilities of a randomization test. However, if we are interested in studying the power of a randomization test, then a computer code must involve two main loops. In the outer loop having "O" iterations, data are generated under a specific alternative hypothesis. For every outer iteration, there may be "I" inner iterations in each of which, data are permuted and the test statistic computed. The total number of iterations is, therefore, "O" times "I". The sequential estimation approach for a p-value that we have proposed here is aimed at reducing the number of inner iterations "I". Consequently, this leads to substantial savings in computing time. There exists an extensive literature on the sequential estimation of a binomial success probability p (Robbins and Siegmund (1974), Cabilio and Robbins (1975) and Cabilio (1977)). For an up-to-date review, we refer to Ghosh et al. (1997). However, the method we adopt here is novel, simple, and easy to implement. At the same time, the extensive simulation results provided here clearly indicate that our method performs very well compared with fixed sample procedures.

ACKNOWLEDGMENT

We thank two anonymous reviewers for their constructive suggestions which led to a considerable improvement in our presentation.

REFERENCES

[1] Anscombe, F.J. (1953). Sequential estimation. *J. Roy. Statist. Soc., Ser. B*, **15**, 1-29.

[2] Barker, W.C. and Dayhoff, M.V. (1972). Atlas of protein sequences and structure. National Biomedical Research Foundation, Washington, D.C.

[3] Berry, K.J. and Mielke, Jr., P.W. (1985). Computation of exact and approximate probability values for a matched pair permutation test. *Commun. Statist., Ser. B*, **14**, 229-248.

[4] Bradley, J.V. (1969). *Distribution-Free Statistical Test.* Prentice Hall: Englewood Cliffs.

[5] Cabilio, P. (1977). Sequential estimation in Bernoulli trials. *Ann. Statist.*, **5**, 342-356.

[6] Cabilio P. and Robbins, H. (1975). Sequential estimation of p with squared relative error loss. *Proc. Nat. Acad. Sci., U.S.A.*, **72**, 191-193.

[7] Clark, R.M. (1989). A randomization test for the comparison of ordered sequences. *Math. Geol.*, **11**, 429-442.

[8] Dwass, M. (1957). Modified randomization tests for nonparametric

hypotheses. *Ann. Math. Statist.*, **28**, 181-187.

[9] Fisher, R.A. (1935). *Design of Experiments.* Oliver and Boyd: Edinburg.

[10] Fisher, R.A. (1936). Coefficient of racial likeness and the future of craniometry. *J. Roy. Anthrop. Soc.*, **66**, 57-63.

[11] Gail, M. and Mantel, N. (1977). Counting the number of r × c contingency tables with fixed marginals. *J. Amer. Statist. Assoc.*, **72**, 859-862.

[12] Ghosh, M., Mukhopadhyay, N. and Sen, P.K. (1997). *Sequential Estimation.* John Wiley: New York.

[13] Glass, A.G., Mantel, N., Gunz, F.W. and Spears, G.F.S. (1971). Time-space clustering of childhood leukemia in New Zealand. *J. Nat. Cancer Inst.*, **47**, 329-336.

[14] Good, I.J.(1976). On the analysis of symmetric Dirichlet distributions and their mixtures to contingency tables. *Ann. Statist.*, **4**, 1159-1189.

[15] Green, B.F. (1977). A practical interactive program for randomization tests of location. *Amer. Statist.*, **31**, 37-39.

[16] Karlin, S., Ghandour, G., Ost, T. S., and Korph, K. (1983). New approaches for computer analysis of DNA sequences. *Proc. Nat. Acad. Sci.*, U.S.A., **80**, 5660-5664.

[17] Karlin, S. and Williams, P.T. (1984). Permutation methods for the structured exploratory data analysis (SEDA) of familial trait values. *Amer. J. Human Genetics*, **36**, 873-898.

[18] Kempthorne, O. (1952). *Design and Analysis of Experiments.* Wiley: New York.

[19] Kempthorne, O. (1975). Inferences from experiments and randomisation. In *A Survey of Statistical Design and Linear Models* (J.N. Srivastava, ed.), 303-332. North Holland: Amsterdam.

[20] Kempthorne, O. (1979). In dispraise for the exact test; reactions. *J. Statist. Plan. Inf.*, **3**, 199-213.

[21] Lock, R.H. (1986). Using the computer to approximate permutation tests. In *Proc. Statist. Comp. Sec., Amer. Statist. Assoc. Conf.*, 349-352.

[22] Lock, R.H. (1991). A sequential approximation to a permutation test. *Commun. Statist.-Simul.*, **20**, 341-363.

[23] Manly, B.F.J. (1983). Analysis of polymorphic variation in different types of habitat. *Biometrics*, **39**, 13-27.

[24] Manly, B.F.J. (1988). The comparison and scaling of student assessment marks in several subjects. *Appl. Statist.*, **37**, 385-395.

[25] Marcus, L.F. (1969). Measurement of selection using distance statistics in prehistoric Orang-utan pongo pygamous palaeosumativens. *Evolution*, **23**, 301-307.

[26] Mariott, F.H.C. (1979). Barnard's Monte Carlo tests: how many simulations? *Appl. Statist.*, **28**, 75-77.

[27] Mehta, C.R. and Patel, N.R. (1980). A network algorithm for the exact treatment of $2 \times k$ contingency table. *Commun. Statist., Ser. B*, **9**, 649-664.

[28] Oden, N.L. (1991). Allocation of effort in Monte Carlo simulation for power of permutation tests. *J. Amer. Statist. Assoc.*, **86**, 1074-1076.

[29] Pagano, M. and Halvorsen, K. (1981). An algorithm for finding the exact significance levels of $r \times c$ contingency tables. *J. Amer. Statist. Assoc.*, **76**, 931-934.

[30] Patfield, W.M. (1982). Exact tests for trends in ordered contingency tables. *Appl. Statist.*, **31**, 32-43.

[31] Robbins, H. and Siegmund, D. (1974). Sequential estimation of p in

Bernoulli trials. In *Studies in Probability and Statistics* (E.J. Williams, ed.), 103-107. Academic Press: Jerusalem.

[32] Royaltey, H.H., Astrachen, E. and Sokal, R.R. (1975). Tests for patterns in geographic variations. *Geog. Anal.*, **6**, 369-395.

[33] Tukey, J.W., Brillinger, D.R. and Jones, L.V. (1978). The management of weather resources II: The role of statistics in weather resource management. U.S. Government Printing Office, Washington, D.C.

Addresses for Communication:

TATHAGATA BANERJEE, Department of Statistics, Calcutta University, 35 Ballygunge Circular Road, Kolkata 700019, India. E-mail: btathaga@yahoo.com
ONKAR PROSAD GHOSH, Directorate General of Commercial Intelligence and Statistics, Govt. of India, 1 Council House Street, Kolkata 700001, India. E-mail: onkar-ghosh@yahoo.co.uk

Chapter 4

Sequential Methods for Multistate Processes

MICHAEL I. BARÓN
University of Texas at Dallas, Richardson, U.S.A.
and IBM Research Division, Yorktown Heights, U.S.A.

4.1 INTRODUCTION AND THE GENERAL SCHEME

This paper justifies the use of sequential methods for non-sequential problems that arise in the analysis of multistate processes.

In a variety of applications, the observed dynamical system switches between different modes. Thus, the observed process is non-stationary, but it consists of homogeneous segments separated by mode switch times, or *change-points*. Each mode corresponds to a particular distribution, however, the change-points and the modes are usually unknown.

Examples of such models are found in quality control (in-control and out-of-control modes: Lai (1995), Montgomery (1997)), economics (growth and recession), energy finance (one regular state and many spike states: Baron et al. (2001), Ethier and Dorris (1999), Rosenberg et al. (2002)), climatology (global cooling and warming trends: Calvin (2002)), developmental psychology (different phases during development, learning and problem solving: Baron and Granott (2003), Granott and Parziale (2002), Piaget (1970)), epidemiology (regular, pre-epidemic, epidemic and post-epidemic periods: Baron (2002), Bridges et al. (2000)).

We will refer to the general class of such stochastic processes as *multistate processes*. A subclass of multistate processes with the additional assumption of a Markov chain that governs the mode switching is called *hidden Markov chains*. The latter are widely used in speech recognition, biological sequences alignment, biochemistry, genetics, and signal processing (Cloth and Backofen (2000), Durbin et al. (1998), Rabiner (1989)).

In general, a multistate process generates a sequence $\mathbf{X} = \{\mathbf{X}_j\}_{j=1}^{\lambda}$ with

$$
\begin{cases}
\mathbf{X}_1 = (X_{\nu_0}, \ldots, X_{\nu_1}) & \sim f_1 \\
\mathbf{X}_2 = (X_{\nu_1+1}, \ldots, X_{\nu_2}) & \sim f_2 \\
\cdots\cdots\cdots\cdots\cdots\cdots \\
\mathbf{X}_\lambda = (X_{\nu_{\lambda-1}+1}, \ldots, X_{\nu_\lambda}) & \sim f_\lambda
\end{cases}
$$

where $\mathbf{X}_j$'s are single-mode segments, $f_1, \ldots, f_\lambda$ are the corrseponding known or unknown densities, ν_i, $i = 1, \ldots, \lambda - 1$, are change-points (with a convention that $\nu_\lambda = N$ is the total sample size and $\nu_0 = 1$), and λ is the unknown number of switches. The modes may be repeated, but any two successive segments are assumed to belong to different modes, that is, there are no fake change-points. In the simplest situation, the process oscillates between two possible modes only. As another extreme, all segments may be generated from different densities, where each density is selected from a family according to some prior distribution.

In different applications, one may be interested in

(a) segmenting the observed sequence into a number of homogeneous subsequences by estimating all the change-points and all the modes (this is known as a *decoding problem*),

(b) estimating the first change point ν_1 only, as the time when the process went out of control, or

(c) estimating ν_1 sequentially, in order to detect the disorder "as soon as possible".

Whereas the last problem is extensively studied in the literature (see Basseville and Nikiforov (1993), Lai (1995), and Zacks (1983) for the survey of existing algorithms), only a few methods have been proposed for problems (a) and (b).

In applications to climatology and energy finance, described in this paper, there are infinitely many possible states. In this case, none of the proposed *off-line* algorithms seems to estimate change-points adequately.

The overall maximum likelihood (ML) procedure (Fu and Curnow (1990)) and the related *Viterbi algorithm* (Rabiner (1989)) are likely to detect false change-points. Indeed, even in the simplest case of Bernoulli variables, adding a change-point forces the likelihood function to increase if and only if the sample contains at least two different observations. For the same reason, a restricted ML procedure with conditions on the number of change-points or the maximum distance between them (Lee (1996)) is likely to report the maximum allowed number of change-points.

A conceptually different *binary segmentation* procedure (Vostrikova (1981)) initially assumes exactly one change-point in the dataset and estimates its location. It divides the observed sequence into two parts. If it is found significant, the procedure is applied to each part, and so on. The deficiency of the binary segmentation scheme is that during each step, it assumes only one change-point and disregards all other possible change-points in the same data. Thus, a point separating two dissimilar patterns may not divide the whole dataset into two dissimilar parts. This happens, for instance, in rather practical ABABA patterns, where all A-segments come from close distributions, and so do all B-segments.

Clearly, the difficulties in applying off-line estimation procedures to heterogeneous datasets are caused by the need for artificial assumptions about the number of change-points.

These problems can be resolved by a suitable *sequential scheme*, that can resample the data sequentially and handle one homogeneous segment at a time. At each step, a sequential change-point detection tool is applied to detect the occurrence of the next change-point. If detected, the location of the change-point is estimated, and after that, the algorithm is applied to the post-change data. The initially obtained set of change-point estimates can be refined by re-estimating each change-point based on the preceeding and succeeding segments only. Significance of each change-point is naturally verified by a two-sample test comparing the two adjacent segments. Segments are merged if the change-point separating them is found insignificant.

This scheme does not require any restriction on the possible number of segments. Depending on the stopping boundaries at each sequential step, it may result in frequent change-points. However, a change-point that separates two short segments generated by close distributions will not pass the test of significance and will be deleted. Thus, the scheme is able to detect a change either when sufficiently large segments provide enough evidence for a change, or when a change is dramatic. In the latter case, it will be detected even from short segments (like the spikes in electricity price data, as shown in Figure 4.2.1). There is also a possibility of reporting no change-point in the observed data.

After homogeneous segments of a multistate process are identified, they can be compared in terms of similarities. Similar segments can then be clustered into one class for further analysis and more detailed calibration. For example, the processes of learning and development typically go through a sequence of stages that are rarely repeated, whereas the prices of electricity between the spikes always return to the same "inter-spike" or regular mode.

The described approach can also be used to solve problem (b) of estimating the first change-point that may be followed by more complicated patterns and additional change-points.

We elaborate the proposed scheme for the analysis and modeling of electricity prices in the North Atlantic region of the US. We also apply it to detect global climate changes and main economic patterns from historical data.

4.2   ANALYSIS AND PREDICTION OF ELECTRICITY PRICES

Deregulation and restructuring of American electricity markets caused major changes in the U.S. energy industry. Because of the recent developments and particular features of "power economy" (lack of storage, transmission constraints, dependence on economic and climatic factors, occasional shutouts and surges in prices), the industry needs delicate statistical analysis that would result in

(a) a working stochastic model for the process of electricity prices that can be used for the Monte Carlo simulation study leading to proper valuation of energy derivatives, contracts, and physical assets;

(b) a forecast in the form of a predictive distribution for electricity prices on any given day.

Electricity prices contain a clear trend that includes seasonal (weekly and daily) variations. Also, analysis of detrended hourly electricity prices reveals nonstationary patterns (Figure 4.2.1). In an extreme situation, during a season of peak demand (such as a period of consistently high temperatures, an abrupt increase in the demand of energy, and/or the closure or maintenance of a power plant), electricity prices experience a sudden and very significant increase that forms *a spike*. Spikes may last from a few hours to several days, but the price of electricity can increase tenfold during this short period. This makes the application of traditional analysis of energy finance particularly difficult.

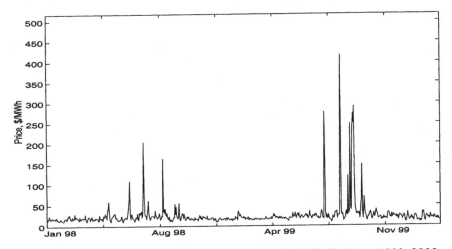

Figure 4.2.1: Electricity Prices in the PA–NJ–MD Region, 1998–2000.

Recently proposed models (Cleslow and Strickland (2000), Ethier and Dorris (1999)) attempted to model the two-stage behavior of electricity markets by adding a Poissonian jump term to the typically used mean reverting process for electricity prices. This term is intended to reflect abrupt spikes. Such models have been criticized for their insensitivity to weather, market conditions, and other factors. For example, spikes can only occur during a peak season, and only during a day, whereas according to these models, they may occur at any time.

In fact, due to differences in supply and demand curves during spikes and inter-spike periods, the prices behave differently during these two phases of the process. Thus, they can not be effectively described by a single equation. A more realistic model is a *multistate process* where spikes and inter-spike periods form homogeneous segments, separated by *change-points*. An analysis of such a model requires a sequential scheme, such as the one described in Section 4.1, to identify the homogeneous segments that can then be treated separately.

Inter-spike (regular) mode: Analysis of data from different energy markets in the U.S.A. supports the following model for electricity prices

P_t between the spikes:

$$\log(P_t) = \alpha + \beta t + \gamma(m_t) + \delta(w_t) + X_t, \qquad (4.2.1)$$

where t is the day, $t = 1, 2, \ldots$, $m(t)$ is the month corresponding to t, $w(t)$ is the day of the week, γ and δ are their respective effects $(\sum_1^{12} \gamma(m) = \sum_1^7 \delta(w) = 0)$. The residuals X_t form an autoregressive process

$$X_t = \phi X_{t-1} + \sigma \varepsilon_t \qquad (4.2.2)$$

with respect to a normalized white noise sequence ε_t. Parameters ϕ, σ, α, β, $\gamma(m)$ and $\delta(w)$, $m = 1, \ldots, 12$, $w = 1, \ldots, 7$, may vary from one region to another (although the slope β usually remains the same). They should be estimated separately for each market.

Spike modes: Although the spikes are short, electricity prices during spikes can differ very significantly (Figure 4.2.1). It is clear that different spikes are generated by different underlying distributions. Thus, the underlying multistate process has one regular mode and many spike modes, where detrended prices X_t form a sample from a lognormal distribution

$$f(x \mid \mu, \tau) = \frac{1}{\sqrt{2\pi} x \tau} \exp\left\{-(\log x - \mu)^2 / 2\tau^2\right\}, x > 0, \mu > 0, \tau > 0.$$

The parameter μ, and also possibly the parameter τ, are different for different spikes. Hence, a Bayesian model with suitable prior distributions on μ and τ seems appropriate. X_t's are conditionally independent and identically distributed, given μ and τ. Unconditionally, they form a multistate process, and they are dependent within each spike.

This agrees with our intuition. For example, if the price reaches \$200/MWh during a spike, then prices will stay around this mark until the end of the spike. However, they will not affect the next spike.

Further, the prior mean of the distribution of μ varies in time,

$$\theta = \boldsymbol{E}(\mu) = \theta(t),$$

and it is high during the peak season, reducing gradually during the off-peak months.

Transitions: A two-phase Markov model (Baron et al. (2001), Rosenberg et al. (2002)) differentiates between the spike and inter-spike phases, realizes the change of phases according to a non-stationary Markov chain, and associates parameters of distributions with different factors. The change of phases occurs according to a Markov chain with the transition probability matrix

$$\Pi = \begin{pmatrix} 1 - p(t) & p(t) \\ q & 1 - q \end{pmatrix}, \qquad (4.2.3)$$

where $p(t)$ is the probability for a spike to start during day t and q is the probability for a spike to end on the next day given that it has already started. The frequency of spikes increases during a peak season, and the possibility of a spike is negligible during weekends. Therefore, the transition probability $p(t)$ varies depending on a day. At the same time, durations of spikes are barely predictable based on t, m_t, and w_t, and q can be assumed constant.

In the next section, we apply the proposed sequential scheme to divide the sequence of electricity prices into spikes and inter-spike segments, from which all model parameters can be estimated.

4.3 METHODOLOGY AND IMPLEMENTATION

As proposed in Section 4.1, we construct a sequential change-point detection algorithm and apply it repeatedly to detect change-points and isolate all the spikes. Before this algorithm is applied, the data are detrended, and possible information about the states is collected.

Detrending: Ordinary least squares estimates are inappropriate because they are significantly influenced by the presence of spikes. We recommend the weighted least squares method with significantly reduced weight for the most influential observations. For example, a small weight of 0.01 assigned to all observations with the studentized residual of $\log(P_t)$ exceeding 2.0 practically eliminates the effect of spikes on the estimates. The resulting parameter estimates in (4.2.1) for the electricity prices in the North Atlantic region are $\hat{\alpha} = 2.95$, $\hat{\beta} = 2.29 \times 10^{-4}$, and the month effects are as given below:

m	1	2	3	4	5	6
$\hat{\gamma}(m)$	(0.09)	(0.18)	(0.05)	(0.02)	0.05	0.05
m	7	8	9	10	11	12
$\hat{\gamma}(m)$	0.41	0.29	0.02	(0.07)	(0.22)	(0.19)

with $R^2 = 0.29$. All the interaction terms are insignificanti with p-values between 0.39 and 0.99. Also, we found no significant differences between prices on Saturday and Sunday, and during different weekdays, so that $\hat{\delta}(w) = 0.056$ for the weekdays and $\hat{\delta}(w) = -0.139$ for the weekends.

Estimation of parameters in "regular" mode: In practical change-point problems, the pre-change "in-control" distribution is typically known whereas the after-change "out-of-control" distribution is often unknown. Not having or not using the information about the "in-control" distribution leads to a loss of power of the sequential change-point detection procedure (Baron (2000), Section 4). For a preliminary estimation, one can use the off-season (detrended) electricity prices, because no spikes occurred during an off-peak season in 1998-2000. Fitting an AR(1) model (4.2.2) to the residuals leads to the following estimates: $\hat{\phi} = 0.464$, $\hat{\sigma} = 0.178$.

Sequential segmentation: To simplify notation, at each step, we re-enumerate the data so that X_1 is the first observation after the most recently detected and estimated change-point. Thus, every time, we detect the next change-point based on $X_1, X_2, \ldots$.

A CUSUM algorithm is particularly popular for sequential detection of change-points. Various optimal features of the CUSUM scheme are discussed in Baron (2001), Basseville and Nikiforov (1993), and Ritov (1990). Our situation is featured by (a) nuisance parameters of the distribution during spikes, and (b) autocorrelation of data during a regular mode. Thus, a modified CUSUM scheme is needed. Different mechanisms are used to detect transitions from a regular mode to spikes and vice versa, that is, to detect the beginning and end of a spike.

Detecting the beginning of a spike: In a regular mode, for each potential change-point k, we use the joint distribution of detrended (pre-change) log-prices $x_{1:k} = (x_1, \ldots, x_k)$,

$$f(x_{1:k}) = (2\pi \det \Sigma)^{-k/2} \exp\left\{ -x'_{1:k}\Sigma^{-1}x_{1:k}/2 \right\},$$

with $\Sigma_{ij} = \hat{\sigma}^2 \hat{\phi}^{|i-j|}/(1 - \hat{\phi}^2)$, according to the AR(1) model (4.2.2). The distribution of detrended (after-change) log-prices during spikes is $N(\mu, \tau^2)$. The mean μ differs from one spike to another, and it is assumed to have a conjugate normal(θ, η) distribution, while τ may be assumed constant for all spikes. Under these conditions, the unconditional distribution of X_t during spikes is normal with nuisance parameters θ and $\eta + \tau^2$. As in Baron (2000), the nuisance parameters are estimated for each potential change-point k. Then, the CUSUM stopping rule for detecting the beginning of a spike is:

$$T = \inf \left\{ n : \max_{k<n} f(\boldsymbol{x}_{1:k}) \prod_{j=k+1}^{n} g_{kn}(x_j)/f(\boldsymbol{x}_{1:n}) > h \right\},$$

where

$$g_{kn}(x) = (2\pi s_{kn}^2)^{-1/2} \exp \left\{ -(x - \bar{x}_{kn})^2/2s_{kn}^2 \right\}, \qquad (4.3.1)$$

$\bar{x}_{kn} = \frac{1}{n-k} \sum_{k+1}^{n} x_j$, and $s_{kn}^2 = \frac{1}{n-k} \sum_{k+1}^{n} (x_j - \bar{x}_{kn})^2$. The threshold h is usually chosen from the desired Type-I and Type-II error probabilities p_I and p_{II}. For example, one may choose $h = (1 - p_{II})/p_I$, as in Govindarajulu (1987, Section 2.2). In our setting, p_I is the probability of a false alarm, and p_{II} is the probability of failing to detect the beginning of a spike.

Detecting the end of a spike: Detecting a change from a spike mode to a regular mode is quite different because all spikes have different distributions, so that there is no good estimate for the pre-change (spike) parameters. In this case, the modified CUSUM scheme has a considerable chance of failing to detect a change-point (Baron (2000)). Thus, it is necessary to use the geometric(q) prior distribution of spike durations that follows from (4.2.3) and forces the corresponding Bayes rule to detect a change-point before long.

One possibility is to construct a stopping rule as a result of a sequence of Bayes tests (Baron (2001), Section 2),

$$T' = \inf \left\{ n : \pi_n(\mathbf{X}) > 1 - p_I \right\}, \qquad (4.3.2)$$

where

$$\pi_n(\boldsymbol{x}) = \boldsymbol{P} \left\{ \nu \leq n | \boldsymbol{x}_{1:n} \right\} = \frac{\sum_{k=1}^{n} (1-q)^k f(\boldsymbol{x}_{k+1:n}) \prod_1^k g_{0k}(x_j)}{\sum_{k=1}^{\infty} (1-q)^k f(\boldsymbol{x}_{k+1:n}) \prod_1^k g_{0k}(x_j)}$$

is the posterior probability of a change-point at $t = n$. If exact parameters of the (pre-change) spike distribution g are used in (4.3.1) instead of estimates $\bar{x}_{0k}$ and s_{0k}^2, then (4.3.2) becomes the Bayes sequential rule under the risk function $R(T, \nu) = \lambda \boldsymbol{E}(T - \nu)^+ + \boldsymbol{P} \left\{ T < \nu \right\}$ (Shiryaev (1978), Section 4.3, Baron (2001), Section 3).

A similar rule can be used to detect the beginning of spikes too. But an expression for the transition probability function $p(t)$ will have to

Table 4.3.1: Estimated Parameters of the Model

	Total	1998	1999
Mean spike duration $(1/q)$	2.3636	2	2.8
Mean interspike period (during the peak season)	18.7778	23	13.5
Mean spike effect on log-prices θ	1.5473	1.1611	2.0107
Within-spike variance τ^2	0.3730	0.1313	0.2693
Between-spike variance η	0.0765	0.0648	0.1079
Transition probabilities			
$P_{peak}\{spike \to control\}$	0.4231	0.5000	0.3571
$P_{peak}\{control \to spike\}$	0.0533	0.0435	0.0740
$P_{off-peak}\{control \to control\}$	1	1	1

be obtained first. Conversely, estimation of q is rather simple, namely, $\hat{q} = $ (average spike length)$^{-1}$

Refinement and final parameter estimation: During a sequential scheme, we used a minimal amount of data to detect each change-point. After the first set of estimated change-points $(\hat{\nu}_1, \ldots, \hat{\nu}_{\hat{\lambda}-1})$ is obtained, each change-point ν_k can be re-estimated based on $x_{\hat{\nu}_{k-1}+1:\hat{\nu}_{k+1}}$. This subsample contains only one change-point, and the maximum likelihood estimator is typically used here. However, when it follows sequential detection, the maximum likelihood estimator is *not distribution consistent*, that is, it fails to estimate the change-point with no error when the change is very significant (Baron and Granott (2003)), such as the change from a regular mode to a spike. A *distribution consistent* estimator here is the minimizer of the p-value from the likelihood ratio test, comparing subsamples $x_{\hat{\nu}_{k-1}+1:j}$ and $x_{j+1:\hat{\nu}_{k+1}}$ for $\hat{\nu}_{k-1} < j < \hat{\nu}_{k+1}$. Naturally, this scheme is repeated until the set of estimated change-points stabilizes.

Results: Applied to the electricity prices in Figure 4.2.1, the described scheme detected 11 spikes during 700 days. Spikes occurred on 05/19-21, 06/25-27, 07/21-23, 08/25, 09/15, and 09/22 in 1998, and on 06/08-09, 07/06-07, 07/24, 07/27-08/01, and 08/12-14 in 1999. The estimated parameters and other characteristics are summarized in Table 4.3.1.

Application of results: The proposed model with estimated parameters can be used to forecast the price of electricity on any given day. Examples of predictive densities are depicted in Figure 4.3.1.

Results of our analysis were also used in a Monte Carlo study that generated sequences of future electricity prices in order to estimate values of derivatives and financial deals.

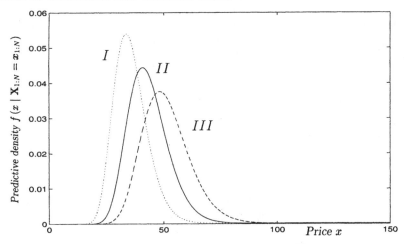

Figure 4.3.1: Predictive Densities for a Weekend (I) and a Weekday (II) Price three Years Ahead, and for a Weekday Price five Years Ahead (III).

4.4 OTHER APPLICATIONS

This section briefly mentions some applications of the described general scheme in economics and climatology. Depending on the situation, the details from each step are modified. In economics and climatology, usually there are no spikes. On the other hand, all segments are rather long, and the patterns are rarely repeated.

It is interesting to determine different phases of economy based on the proportional daily changes in the Dow Jones index. One would expect positive changes during a period of growth and negative changes during a recession.

The proposed algorithm can also be used to detect global climate changes. Analysis of the average temperatures in U.S.A. from 1895 till 2001 uncovers three change-points, significant at the 0.002 level. The dataset and the estimated change-points are shown in Figure 4.4.1. Although the differences among four segments are not obvious from Figure 4.4.1, the summary statistics (Table 4.4.1) show that each transition was marked by warming.

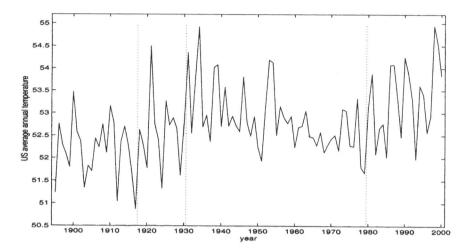

Figure 4.4.1: Mean Annual Temperatures in U.S.A. and Three
Significant Change-Points.

Table 4.4.1: Summary Statistics of Temperatures for the Four Periods

Years	Temperatures		
	mean	median	standard deviation
1895–1917	52.1696	52.2900	0.6619
1918–1930	52.5869	52.6600	0.7963
1931–1979	52.8649	52.7100	0.7009
1980–2001	53.3214	53.3400	0.8466

The datasets analyzed in this article are available from PJM Interconnection, L.L.C. (http://www.pjm.com/), National Climatic Data Center in Asheville, North Carolina (http://lwf.ncdc.noaa.gov/), and Yahoo! Finance (http://finance.yahoo.com/).

4.5 CONCLUDING REMARKS

A sequential scheme is proposed for an effective analysis of multistate processes. The described scheme detects one change-point at a time and divides a heterogeneous sequence into homogeneous segments that can then be analyzed separately. The general idea is to use a sequential change-point detection tool at each step, followed by re-estimation and refinement of the detected set of change-points. Details of each step, distributions of states and transition patterns are elaborated for each particular application. Estimating parameters and understanding the mechanism that governs the change of phases shed light on the behavior of a complicated multistate process and allow its modeling and forecasting.

REFERENCES

[1] Baron, M. (2000). Nonparametric adaptive change-point estimation and on-line detection. *Sequential Anal.*, **19**, 1–23.

[2] Baron, M. (2001). Bayes stopping rules in a change-point model with a random hazard rate. *Sequential Anal.*, **20**, 147–163.

[3] Baron, M. (2002). Bayes and asymptotically pointwise optimal stopping rules for the detection of influenza epidemics. In *Case Studies of Bayesian Statistics* (C. Gatsonis et al., eds.), **6**, 153-163. Springer-Verlag: New York.

[4] Baron, M. and Granott, N. (2003). Consistent estimation of early and frequent change points. In *Foundations of Statistical Inference, Proc. Shoresh Conf.* (Y. Haitovsky et al., eds.), 181-1922. Springer-Verlag: New York.

[5] Baron, M., Rosenberg, M. and Sidorenko, N. (2001). Automatic spike detection for the modeling and prediction of electricity pricing. *Energy & Power Risk Manage.*, **October**, 36-39.

[6] Basseville, M. and Nikiforov, I.V. (1993). *Detection of Abrupt Changes: Theory and Application.* Prentice-Hall: Englewood Cliffs.

[7] Bridges, C.B., Winquist, A.G., Fukuda, K., Cox, N.J., Singleton, J.A. and Strikas, R.A. (2000). Prevention and control of influenza. *Morbidity and Mortality Weekly Reports, Centers for Disease Control and Prevention*, **49**, 1–38.

[8] Calvin, W.H. (2002). *A Brain for All Seasons: Human Evolution*

and Abrupt Climate Change. University of Chicago Press: Chicago.

[9] Cleslow, L. and Strickland, C. (2000). *Energy Derivatives: Pricing and Risk Management.* Lacima Publications: London.

[10] Cloth, P. and Backofen, R. (2000). *Computational Molecular Biology.* Wiley: Chichester.

[11] Durbin, S., Eddy, S., Krogh, A. and Mitchison, G. (1998). *Biological Sequence Analysis. Probabilistic Models of Proteins and Nucleic Acids.* Cambridge University Press: Cambridge.

[12] Ethier, R. and Dorris, G. (1999). Don't ignore the spikes. *Energy & Power Risk Manage.*, **July/August**, 31-33.

[13] Fu, Y.-X. and Curnow, R.N. (1990). Maximum likelihood estimation of multiple change points. *Biometrika*, **77**, 563–573.

[14] Govindarajulu, Z. (1987). *The Sequential Statistical Analysis of Hypothesis Testing, Point and Interval Estimation, and Decision Theory.* American Sciences Press: Columbus.

[15] Granott, N. and Parziale, J. (2002). Microdevelopment: A process-oriented perspective for studying development and learning. In *Microdevelopment: Transition processes in development and learning* (N. Granott and J. Parziale, eds.), 1-28. Cambridge University Press: Cambridge.

[16] Lai, T.L. (1995). Sequential changepoint detection in quality control and dynamical systems. *J. Roy. Statist. Soc., Ser. B*, **57**, 613–658.

[17] Lee, C.-B. (1996). Nonparametric multiple change-point estimators. *Statist. Probab. Lett.*, **27**, 295–304.

[18] Montgomery, D.C. (1997). *Introduction to Statistical Quality Control*, 3rd ed. Wiley: New York.

[19] Piaget, J. (1970). Piaget's theory. In *Carmichael's Manual of Child Psychology* (P.H. Mussen, ed.), 703-732. Wiley: New York.

[20] Rabiner, L.R. (1989). A tutorial on hidden markov models and selected applications in speech recognition. *IEEE Proc.*, **77**, 257–285.

[21] Ritov, Y. (1990). Decision theoretic optimality of CUSUM procedure. *Ann. Statist.*, **18**, 1464–1469.

[22] Rosenberg, M., Bryngelson, J.D., Sidorenko, N. and Baron, M. (2002). Price spikes and real options: transmission valuation. In *Real Options and Energy Management* (E.I. Ronn, ed.), 323-370. Risk Books: London.

[23] Shiryaev, A.N. (1978). *Optimal Stopping Rules*. Springer-Verlag: New York.

[24] Vostrikova, L. Ju. (1981). Detecting "disorder" in multidimensional random processes. *Sov. Math. Dokl.*, **24**, 55–59.

[25] Zacks, S. (1983). Survey of classical and Bayesian approaches to the change-point problem: Fixed sample and sequential procedures in testing and estimation. In *Recent Advances in Statistics*, 245-269. Academic Press: New York.

Address for communication:

MICHAEL BARÓN, Department of Statistics, University of Texas at Dallas, Richardson, TX 75080, U.S.A. E-mail: mbaron@utdallas.edu

Chapter 5

Sequential Adaptive Designs for Clinical Trials with Longitudinal Responses

ATANU BISWAS
Indian Statistical Institute, Kolkata, India

ANUP DEWANJI
Indian Statistical Institute, Kolkata, India

5.1 INTRODUCTION

In many phase III clinical trials, patients enter the study sequentially. Response-adaptive allocation designs may be useful in such cases, where the idea is to randomize entering patients to one of two treatments in such a way that a larger number of patients are eventually treated by the more favorable treatment resulting in ethical gain. This can be done by skewing the allocation probability of a patient in favor of the treatment doing better. To achieve this, one needs to incorporate the allocation and response history up to the entry time of a patient, instead of adopting a balanced allocation scheme.

Most of the literature on adaptive allocation designs in the context of clinical trials is concerned with single binary (success/failure) responses. The *randomized play-the-winner* (RPW) rule (see Wei and Durham (1978)), a randomized version of Zelen's (1969) *play-the-winner*

(PW) rule, is perhaps the most well-known. According to the PW rule, the treatment to be given to the first patient is decided by tossing a fair coin. Subsequently, if the response of a patient is a "success", the next patient is given the same treatment. On the other hand, if the response of a patient is a "failure", the next patient is treated by the other treatment. As a modification if this PW rule, Wei and Durham (1978) introduced the RPW rule as a randomized version which can be interpreted in terms of an urn model quite easily. Here, we briefly discuss widely used format of the RPW rule. We start with an urn having two types of balls, A and B, representing the two treatments. Initially the urn contains 2α balls, α balls of each type. An entering patient is given one treatment or the other according to the draw of a ball from the urn. The patient is given the treatment represented by the ball drawn, and the ball is put back into the urn immediately. The urn is updated with every observed response. If the response of a patient is a "success", we add β balls of the same kind as was given to the patient. On the other hand, if the response is a "failure", we add β balls of the other kind to the urn. Here, one implicit assumption is that the responses are immediate. Note that, the number of balls in the urn before the entry of the $(i+1)^{th}$ patient is $2\alpha + i\beta$. The choices of α and β are to be made by the experimenters.

Adaptive designs have occasionally been used in real-life applications. Iglewicz (1983) reports the usefulness of data-dependent allocation in an unpublished application by M. Zelen to a certain lung cancer trial. The RPW rule has been used in clinical trials including the Michigan *Extracorporeal Membrane Oxygenation* (ECMO) trial (Bartlett et al. (1985)), and two anti-depression trials of fluoxetine hydrochloride sponsored by Eli Lilly (Tamura et al. (1994)). Also, Ware (1989) described the Boston ECMO trial and Rout et al. (1993) discussed an application of the PW rule.

As far as we know, no adaptive allocation design has been implemented or theoretically studied for longitudinal responses, although such responses are quite common. In this paper, we develop some adaptive allocation designs when the responses are longitudinal. These designs incorporate both allocation and response history accumulated up to the current allocation and the most recent entry-time, so as to make the allocation probability lean toward the potentially favorable treatment group. We discuss several response patterns, including binary, ordinal categorical, multivariate ordinal categorical and continuous, for each of the longitudinal responses. We also discuss the possibility of incorporating prognostic factors in a design.

The need for developing such designs arose from the prospect of a trial with *Pulsed Electro-Magnetic Field* (PEMF) therapy being conducted in the Biometry Unit of the Indian Statistical Institute, Kolkata, involving *Rheumatoid Arthritic* (RA) patients. A pilot study (Ganguly et al. (1993)) of PEMF therapy on RA patients indicates some beneficial effect of the therapy, at least in terms of short-term symptomatic improvement. Therefore, this doubly-blinded controlled trial with placebo and therapy groups has been planned to investigate the beneficial effect of the therapy with a statistically sound design. Each patient, after entering the trial, is monitored once a week for nearly 16 weeks leading to longitudinal (weekly) responses. Although the primary response from each week is multivariate ordinal categorical on some parameters (for example, pain, tenderness, swelling, joint Stiffness), there are also continuous measurements on physiological parameters (for example, uric acid, hemoglobin). See Section 5.7 and Biswas and Dewanji (2001) for further details regarding this trial. Allocation probability of an entering patient to either the therapy or the placebo group should ideally depend on all previous allocations and responses. Considering the complicated response pattern, it becomes necessary to summarize weekly responses in some way (for example, "dichotomized" leading to binary response, "categorized" leading to ordinal response, and so forth) and then work with the summarized responses longitudinally. Additionally, information on some prognostic factors are available which are likely to influence the response. For example, the sero-positive/sero-negative status of a patient is important in this context, a fact that became apparent from the past study of Ganguly et al. (1993).

In the present paper, we restrict ourselves only to developing designs for different response summaries and studying the associated allocation probabilities. The proposed design for binary longitudinal responses (Section 5.2) is applied in the PEMF trial described above involving 22 RA patients. This is possibly the first adaptive clinical trial designed for longitudinal responses. The application is reported in Section 5.7. Sections 5.3 and 5.4 present designs for ordinal categorical responses and multivariate ordinal categorical responses respectively. In Section 5.5, we discuss inclusion of prognostic factors while Section 5.6 presents a design for continuous responses. Section 5.8 ends with some concluding remarks.

5.2 LONGITUDINAL BINARY RESPONSES

Suppose that we have two treatments A and B under consideration, whose effects are being compared in a phase III clinical trial. Usually, one of these is a standard treatment and the other one is under development. We assume repeated or longitudinal responses for every single patient in each of the two groups. Suppose we have a fixed number (n) of patients under study and the ith patient provides k_i longitudinal binary responses at the corresponding monitoring times $t_{i1} < \cdots < t_{ik_i}$ (assumed non-stochastic and not necessarily equispaced). Also k_i's are assumed non-stochastic and not dependent on any particular treatment applied to the ith patient. The patients make staggered entries into the study and there may be some entries in batches. Let x_i be the entry time and y_i be the exit time of the ith entering patient. Assume that x_i and y_i are non-stochastic as well.

The first $2m$ patients $(2m < n)$ are to be randomly assigned in such a manner that exactly m patients will be assigned to A and the remaining m patients by B. That is, all $\binom{2m}{m}$ combinations are equiprobable. This is done to ensure having at least m patients in each treatment group, since subsequently all the remaining patients may be treated exclusively by one or the other treatment due to randomization. These initial $2m$ allocations may be viewed as a compromise between fully adaptive and fully balanced allocations. We then develop an urn model for the subsequent allocations. We start with an urn having two types of balls, A and B. It has α balls of each kind, so that the total number of balls is 2α. At each monitoring time point, for every existing patient, we observe whether a "success" (S) or a "failure" (F), properly defined at the outset, occurs. If a "success" occurs, we add β balls of the same kind as was given to the patient. On the other hand, a "failure" results in an addition of β balls of the opposite kind. Thus, at every monitoring time, the number of balls being added is β times the number of patients who are monitored at that time. For an entering patient, we draw a ball from the urn with replacement, and assign the newly entered patient to the treatment indicated by the ball drawn. Clearly, the allocation probability is skewed in favor of the treatment that seems potentially better at the moment. This design may be viewed as a longitudinal version of the randomized play-the-winner rule, and will be denoted by $RLPW(\alpha, \beta)$.

For the ith patient, the observation is $\{\delta_i; Z_{ij}, j = 1, \cdots, k_i\}$. Here δ_i indicates the allocated group, taking values 1 or 0 for treatment A or B, respectively. Z_{ij}'s are the binary responses taking values 1 or 0 for a "failure" or a "success", respectively, at k_i monitoring times. Then, for $i \geq 2m + 1$, the number of failures and the number of monitoring for treatment A up to the entry of the ith patient, that is, at time x_i-, are

$$R_{Ax_i} = \sum_{l=1}^{i-1} \delta_l \sum_{j=1}^{k_l} Z_{ij} I_{\{t_{lj} < x_i\}}, \quad N_{Ax_i} = \sum_{l=1}^{i-1} \delta_l \sum_{j=1}^{k_l} I_{\{t_{lj} < x_i\}},$$

respectively, where $I_{\{.\}}$ is an indicator function. Similarly, R_{Bx_i} and N_{Bx_i} can be obtained by replacing δ_l by $(1 - \delta_l)$ in the expressions of R_{Ax_i} and N_{Ax_i}. Then, the conditional probability that the ith entering patient will be allocated to treatment A, given the allocation and response history up to time x_i-, is

$$P(\delta_i = 1 | R_{Ax_i}, N_{Ax_i}, R_{Bx_i}, N_{Bx_i}) = \frac{\alpha + \beta\{(N_{Ax_i} - R_{Ax_i}) + R_{Bx_i}\}}{2\alpha + \beta(N_{Ax_i} + N_{Bx_i})},$$

$$(5.2.1)$$

which is the proportion of type A balls in the urn at time x_i. Clearly, the denominator of (5.2.1) is non-random.

Now, we write

$$P(\delta_i = 1) = r_i \text{ for } i \geq 2m + 1,$$

the unconditional probability that the ith patient is allocated to treatment A. Note that $r_i = 0.5$ for $i = 1, \cdots, 2m$. We also write

$$P(Z_{ij} = 1 | \delta_i) = \pi_{Aj} \delta_i + \pi_{Bj}(1 - \delta_i),$$

the probability of a failure at the jth monitoring time for the ith patient. Note that, besides j, it also depends on the corresponding treatment group.

The idea behind adopting such an allocation rule is to make the allocation pattern favor the potentially better treatment and eventually have a larger proportion of patients allocated to this treatment. Thus, a meaningful performance characteristic of the design is the proportion of patients assigned to treatment A, say. We denote it by $A_{prop} = n^{-1} \sum_{i=1}^{n} \delta_i$. Clearly,

$$E(A_{prop}) = \frac{1}{n} \sum_{i=1}^{n} r_i = \bar{r},$$

where r_i's can be obtained by taking expectations on both sides of (5.2.1), which provides the recursive relation:

$$r_i = \frac{\alpha + \beta \sum_{l=1}^{i-1} \sum_{j:t_{lj}<x_i}[(1-\pi_{Aj})r_l + \pi_{Bj}(1-r_l)]}{2\alpha + \beta(N_{Ax_i} + N_{Bx_i})}, \quad (5.2.2)$$

for $i \geq 2m+1$.

A closely related performance characteristic is the proportion of type A balls in the urn, denoted by A_{u-prop}. Clearly, A_{u-prop} is given by the right hand side of (5.2.1), evaluated at $x_{n+1}(\geq y_n)$. It may be interesting to study the nature of the plots of $A_{prop}(t) = \frac{1}{n_t} \sum_{i=1}^{n_t} \delta_i$ and $A_{u-prop}(t) = \frac{\alpha + \beta\{(N_{At}-R_{At})+R_{Bt}\}}{2\alpha + \beta(N_{At}+N_{Bt})}$ for different t's, where n_t is the number of entries up to time $t-$. It can be easily checked that the expected allocation proportion is:

$$E[A_{prop}(t)] = \frac{1}{n_t}\left[m + \sum_{i=2m+1}^{n_t} r_i\right] = E_1(t), \text{ say,} \quad (5.2.3)$$

and the expected urn proportion is

$$E(A_{u-prop}(t)) = \frac{\alpha + \beta \sum_{l=1}^{n_t} \sum_{j:t_{lj}<t}[(1-\pi_{Aj})r_l + \pi_{Bj}(1-r_l)]}{2\alpha + \beta(N_{At} + N_{Bt})}. \quad (5.2.4)$$

Let us denote the expression in (5.2.4) by $E_2(t)$. If all the k_i's are the same ($= k$), then both $\{E_1(t)\}$ and $\{E_2(t)\}$ converge to the limiting value

$$\pi_0 = \frac{\sum_{j=1}^{k} \pi_{Bj}}{\sum_{j=1}^{k}(\pi_{Aj} + \pi_{Bj})}.$$

Also, A_{prop} converges in probability to this same limiting value. The proofs are given in Biswas and Dewanji (2001).

Recall that both (5.2.3) and (5.2.4) depend solely on the marginal probabilities π_{Aj}'s and π_{Bj}'s. However, since specification of these entities seems somewhat artificial, we consider the following simple model for illustration. Let us write $\pi_{A1} = q_A$ and $\pi_{B1} = q_B$. Assume that $P(Z_{ij} = 1|Z_{i1} = z_{i1}, \cdots, Z_{i,j-1} = z_{i,j-1}, \delta_i)$ depends on δ_i and only on the time since last recurrence (that is, on $j - l_0$, where l_0 is the maximum of $l < j$ such that $z_{il} = 1$). For example, if $z_{il} = 1$ and $z_{i,l+1} = \cdots = z_{i,j-1} = 0$, the above probability depends only on d_i and $j - l$ and it is denoted by $q_{A,j-l}$ or $q_{B,j-l}$ according as $\delta_i = 1$ or 0, respectively. In particular, if $z_{i1} = \cdots = z_{i,j-1} = 0$, the above probabi-

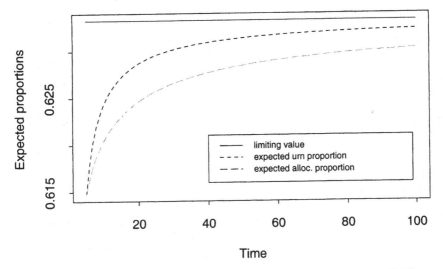

Figure 5.2.1: Expected Urn and Allocation Proportions at Different Time Points with $(q_A, q_B) = (0.2, 0.5)$ and the Limiting Value π_0. Here $k = 10$ and Patients Enter the Study One at a Time with a Gap of 12 Monitoring Times.

lity is q_{Aj} or q_{Bj}. This modeling is similar to that of Bonney (1987) for correlated binary data with some "natural ordering" in the observations. We now consider further modeling of the q_{uj}'s. Let

$$q_{uj} = 1 - (1 - q_u)^j \text{ for } u = A \text{ and } B,$$

so that q_{uj} increases with j. Between Z_{il} and $Z_{i,l+j}$, this model is found to have increasing odds-ratio and decreasing correlation with j, the features which seem so desirable in this context.

We assume equi-spaced monitoring, and also an equal number ($k = 10$) of monitoring for each of the $n = 100$ patients. Patients enter the study one at a time with a gap of 12 monitoring times. We fix $m = 2$, $\alpha = 2$, and $\beta = 1$. Figures 5.2.1 and 5.2.2 provide plots of $E_1(t)$ and $E_2(t)$ against t for $(q_A, q_B) = (0.2, 0.5)$ and $(0.5, 0.8)$, respectively. The figures also show the horizontal line π_0 to which both $E_1(t)$ and $E_2(t)$ converge. Note that the plots are given for $t \geq x_{2m+1}$. In Figure 5.2.1, both the curves are monotonically increasing with $\pi_0 \geq E_2(t) \geq E_1(t)$, while the reverse occurs in Figure 5.2.2. This monotonicity and order-specificity are generally the nature of the convergence shown by $E_1(t)$ and $E_2(t)$, depending on the choice of (q_A, q_B). This feature has been observed when $t \geq x_{2m+1}$, the time it takes to move from the initial

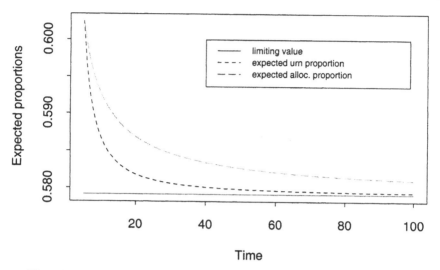

Figure 5.2.2: Expected Urn and Allocation Proportions at Different
Time Points with $(q_A, q_B) = (0.5, 0.8)$ and the Limiting Value π_0.
Here $k = 10$ and Patients Enter the Study One at a Time with a
Gap of 12 Monitoring Times.

balanced position. It is also necessary that at the entry time of the
i^{th} patient (for $i \geq 2m + 1$), all responses from the previous $(i - 1)$
patients are available. For this reason, we have fixed the length of the
gap between successive entries to be 12 monitoring times. However, for
smaller lengths of this gap $(< k)$, this monotonicity is observed only
after the first few time points. Extensive numerical work indicates that,
even for unequi-spaced and unequal monitoring times and with uneven
gaps between successive entries, similar monotone convergence of $E_1(t)$
and $E_2(t)$ to π_0 occurs after the first few time points.

5.3   LONGITUDINAL ORDINAL RESPONSES

In some situations, the responses may be recorded using an ordinal
categorical scale. For example, in the PEMF study as mentioned in
Section 5.1, some of the RA-parameters (for example, pain) are measured with a four-point ordinal scale (namely, severe, moderate, mild
and nil). In this situation, the responses on all RA-parameters can
be taken together to form an overall ordinal scale. Suppose Z_{ij}, the

j^{th} response of the ith patient, is ordinal categorical with categories $0, 1, \cdots, K$. Without loss of generality, we assume that higher values of Z_{ij} indicate worse health conditions. We can easily extend the design from Section 5.2 to fit this setup by adding $(K - Z_{ij})\beta$ balls of the same kind along with $Z_{ij}\beta$ balls of the opposite kind in the urn corresponding to a response. In this case, the expression of R_{Ax_i} will remain the same as in Section 5.2, but the N_{Ax_i} there will be replaced by

$$N_{Ax_i} = K \sum_{l=1}^{i-1} \delta_l \sum_{j=1}^{k_l} I_{\{t_{lj} < x_i\}}.$$

The conditional probability of allocation will have the same expression as in (5.2.1). If we write $E(Z_{ij}|\delta_i) = e_{Aj}\delta_i + e_{Bj}(1 - \delta_i)$, then the recursive relation (5.2.2) for r_i will be modified to:

$$r_i = \frac{\alpha + \beta \sum_{l=1}^{i-1} \sum_{j:t_{lj}<x_i}[(K - e_{Aj})r_l + e_{Bj}(1 - r_l)]}{2\alpha + \beta(N_{Ax_i} + N_{Bx_i})}.$$

Consequently, we can obtain $E_1(t)$ and $E_2(t)$ as before.

5.4 LONGITUDINAL MULTIVARIATE ORDINAL RESPONSES

We now propose to incorporate all ordinal categorical responses at each monitoring time individually in the design. So, we consider multivariate ordinal categorical response at each monitoring time. Suppose that for each pair (i, j), we have an M-component response vector $Z_{ij} = (Z_{ij1}, \cdots, Z_{ijM})'$, where each Z_{ijs}, $s = 1, \cdots, M$, can take values $0, 1, \cdots, K$ (K may be allowed to vary with s). Without loss of generality, we assume that higher values of Z_{ijs} indicate worse health conditions. In this case one can write,

$$R_{Ax_is} = \sum_{l=1}^{i-1} \delta_l \sum_{j=1}^{k_l} Z_{ljs} I_{\{t_{lj}<x_i\}} \quad \text{and} \quad N_{Ax_i} = K \sum_{l=1}^{i-1} \delta_l \sum_{j=1}^{k_l} I_{\{t_{lj}<x_i\}},$$

for $s = 1, \cdots, M$, and similarly express R_{Bx_is} and N_{Bx_i}. The number of balls added to the urn for the sth component of the response vector up to time x_i- for patients in the treatment group A is $\beta_s N_{Ax_i}$, of which, $\beta_s R_{Ax_is}$ are balls of the type B. Thus, using the accumulated data up to time x_i-, the conditional probability that the ith patient is allocated to treatment A can be set as

$$P(\delta_i = 1|\text{ accumulated data })$$
$$= \frac{\alpha + \sum_{s=1}^{M} \beta_s \{(N_{Ax_i} - R_{Ax_is}) + R_{Bx_is}\}}{2\alpha + (N_{Ax_i} + N_{Bx_i}) \sum_{s=1}^{M} \beta_s}.$$

Writing $E(Z_{ijs}|\delta_i) = e_{Ajs}\delta_i + e_{Bjs}(1-\delta_i)$, one can obtain r_i's recursively as

$$r_i = \frac{\alpha + \sum_{s=1}^{M} \beta_s \sum_{l=1}^{i-1} \sum_{j:t_{lj}<x_i}[(K - e_{Ajs})r_l + e_{Bjs}(1 - r_l)]}{2\alpha + \beta(N_{Ax_i} + N_{Bx_i})},$$

and then find $E_1(t)$ and $E_2(t)$, as before.

5.5 INCORPORATING PROGNOSTIC FACTORS

In Sections 5.2-5.4, the group of patients was assumed homogeneous with respect to response probabilities. Often, in clinical trials, the prognostic factors are important and they influence the responses to a great extent. Inclusion of prognostic factors in the allocation design is, of course, a difficult but desirable task. Some work has been done for a single-response adaptive designs (not longitudinal). The treatment allocation problem in the presence of prognostic factors was considered by Begg and Iglewicz (1980). They used optimal design theory to suggest a deterministic design criterion which was subsequently modified for computational convenience. Bandyopadhyay and Biswas (1999) incorporated prognostic factors for allocation through an urn design. In that article, they considered polychotomous prognostic factors and the responses were binary. Bandyopadhyay and Biswas (2001) incorporated prognostic factors for allocation with continuous responses.

Here, we consider only the case of binary longitudinal responses from Section 5.2. We deal with a simple case where there is only one binary prognostic factor w (for example, the sero-group indicator). We consider the value $w = 1$ to be an indicator of the more favorable condition. Thus, along the line of Bandyopadhyay and Biswas (1999), for any γ $(< \beta)$, a "success" from either treatment at any monitoring time with $w = 1$ leads to the addition of γ balls of the same kind, along with $(\beta - \gamma)$ balls of the opposite kind (instead of simply adding β balls of the same kind). A "failure" when $w = 1$ will lead to the addition of β balls of the opposite kind. On the other hand, when $w = 0$, a "success" will result in the addition of β balls of the same kind, and a "failure" results in the addition of $(\beta - \gamma)$ balls of the same kind along with γ balls of the opposite kind. We can obtain $R_{ux_i}^w$ and $N_{ux_i}^w$, $u = A, B$, and $w = 1$ and 0 as:

$$R_{Ax_i}^1 = \sum_{l=1}^{i-1} \delta_l w_l \sum_{j=1}^{k_l} Z_{ij} I_{\{t_{lj} < x_i\}}, \quad N_{Ax_i}^1 = \sum_{l=1}^{i-1} \delta_l w_l \sum_{j=1}^{k_l} I_{\{t_{lj} < x_i\}},$$

where w_l is the value of w for the lth patient, and $R_{Ax_i}^0$ and $N_{Ax_i}^0$ having similar forms with w_l replaced by $(1 - w_l)$. The conditional probability of allocation for the ith entering patient will be

$$P(\delta_i = 1| \text{ Accumulated data }) = \{\alpha + \gamma(N_{Ax_i}^1 - R_{Ax_i}^1 + R_{Bx_i}^0)$$
$$+ (\beta - \gamma)(N_{Bx_i}^1 - R_{Bx_i}^1 + R_{Ax_i}^0) + \beta(R_{Bx_i}^1 + N_{Ax_i}^0 - R_{Ax_i}^0)\}$$
$$/\{2\alpha + \beta(N_{Ax_i} + N_{Bx_i})\},$$

where $N_{ux_i} = N_{ux_i}^1 + N_{ux_i}^0$, $u = A, B$. The r_i's can be recursively obtained, as in Section 5.2, using the additional modeling

$$P(Z_{ij} = 1|\delta_i, w_i) = (\pi_{Ai}\delta_i + \pi_{Bi}(1 - \delta_i))a^{w_i},$$

where $a \in (0, 1)$.

5.6 LONGITUDINAL CONTINUOUS RESPONSES

As mentioned in Section 5.1, the longitudinal responses may be continuous. Let us also allow presence of prognostic factor(s), denoted by w, as in the previous section, with w_{ij} being the vector of such factors for the ith patient at the jth monitoring time. Extending the idea of Bandyopadhyay and Biswas (2001), we can represent the model for the response y_{ij} at the jth monitoring time for the ith patient as

$$y_{ij} = \mu_A \delta_i + \mu_B(1 - \delta_i) + w_{ij}'\theta + e_{ij},$$

where $e_i = (e_{i1}, \cdots, e_{ik_i})' \sim N_{k_i}(0, \Sigma_{k_i})$, the k_i-variate normal distribution, μ_A and μ_B denote the mean baseline (for $w_{ij} = 0$) response in the treatment groups A and B, respectively, and θ is the vector of parameters associated with w_{ij}. To allocate the ith entering patient, we obtain $\hat{\mu}_A(x_i-) - \hat{\mu}_B(x_i-)$, the estimate of $\mu_A - \mu_B$ based on the accumulated data up to time x_i-, eliminating the effects of w_{ij}. The estimation procedure of $\mu_A - \mu_B$ is not relevant in this context and, therefore, we do not discuss the estimation problem here.

Now, the conditional allocation probability can be expressed as

$$P(\delta_i = 1 | \text{ Accumulated data }) = G\left[\hat{\mu}_A(x_i-) - \hat{\mu}_B(x_i-)\right],$$

where G is a suitably chosen distribution function of a symmetric (about 0) random variable. One is referred to Bandyopadhyay and Biswas (2001) for details.

5.7 AN APPLICATION

The trial of PEMF therapy for RA patients was conducted at the Indian Statistical Institute, Kolkata, during the period January (1999)–March (2000), to study the effect of PEMF therapy. After an initial adjustment period of 4 weeks, the patients were put into the trial and treated either by the PEMF therapy (treatment A) or by a placebo (treatment B) using an adaptive allocation design. Each patient was treated for nearly 16 weeks, three times a week. The conditions of the patients were monitored once a week during this trial.

As mentioned in Section 5.1, each longitudinal response was multivariate ordinal categorical in nature with some continuous components. There were also some prognostic factors. Thus, the situation was ideal for applying the methodologies from Sections 5.2-5.6 by using different response summaries. But since this was the very first trial of this nature, we decided to conduct it with a simpler protocol. The complicated response structure was suitably dichotomized by consulting with the medical expert involved in this trial. For allocation purposes, the corresponding binary response was recorded as non-recurrence (S) or recurrence (F). Moreover, we ignored the prognostic factors and applied the methodology from Section 5.2.

The first 4 patients were randomly assigned between the two treatments, that is, we have $m = 2$. In the urn design we considered $\alpha = 2$ and $\beta = 1$. While implementing the design, we made one practical modification in our methodology. When a patient experienced recurrence (F), he/she was given a pain-killer and the possibility of recurrence in the next few monitoring times (depending on the dose of the pain-killer) became negligible. Hence, we did not monitor that patient any more after that.

There were 18 subsequent patients making a total of 22 patients. These patients were treated by one treatment or the other, depending on the allocation probabilities indicated by the urn proportion at their entry times. Table 5.7.1 provides the summarized data. Note that the entry times are reported in therapy days (three times a week). The monitoring in the case of the last five patients (who entered the study on the 149th therapy day) became fewer in number since the randomized study terminated soon after.

Table 5.7.1: Summarized Data from the PEMF Trial

Entry times	No. of entries	Treatment applied	Number of observations	Number of recurrences
0	4	T, T, P, P	61, 62, 34, 48	0, 0, 5, 5
23	1	T	45	0
51	3	T, T, P	49, 47, 48	0, 1, 1
62	5	T, T, P, T, T	46, 46, 47, 45, 41	0, 0, 0, 0, 1
109	4	T, T, P, T	40, 36, 35, 21	0, 1, 1, 0
149	5	T, P, T, T, T	10, 10, 7, 10, 10	0, 0, 1, 0, 0

From the data, we observe 16 recurrences (out of 798 monitoring) for the 22 patients among which, 4 were in the PEMF group and 12 were in the placebo group. Figure 5.7.1 shows the graph for the observed urn and allocation proportions during the study. We observe that both proportions are increasing apparently to an asymptote (as in Figure 5.2.1). The convergence stage seems to have been reached even with such a small number of patients. As the apparent asymptote in Figure 5.7.1 is almost 0.73 (much larger than 0.5), one may conclude that the PEMF therapy is superior to placebo.

5.8 CONCLUDING REMARKS

It would be relevant to carry out some suitable inference procedure following the implementation of the design. But it is not our intention to discuss the inference issue in this paper. Biswas and Dewanji (2001) carried out various test procedures following the PEMF trial. They carried out an intuitive test procedure based on the test statistic $prop_A - prop_B$. The distribution of this statistic is obtained by means

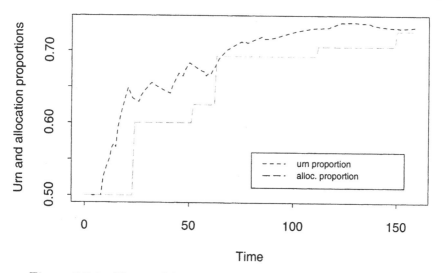

Figure 5.7.1: Observed Urn and Allocation Proportions at Different
Time Points in the PEMF Trial.

of simulation accompanied by some power calculation. This simulation
technique incorporates the design into the analysis. A likelihood ratio
test and a score test had also been carried out. These tests rejected the
equivalence of the PEMF therapy and the placebo with overwhelming
evidence in favor of the therapy.

Once we successfully implemented the methodology from Section
5.2 in this trial, we had enough confidence to apply the more complicated methodologies from Sections 5.3-5.6, in practice. Currently, we
are planning another trial which will combine the methodologies from
Sections 5.4 and 5.5.

Note that the allocation probabilities of Sections 5.3 and 5.4 (using
ordinal response history) depend on the labels of the categories. This
dependence is through the R_{ux_i}'s and N_{ux_i}'s, for $u = A, B$ which change
with the labels. However, it is our belief that the designs will not change
drastically because of this. Note that the expected urn and allocation
proportions also depend on such labels through the e_{Aj}'s and e_{Bj}'s.

Although we have worked on a variety of response patterns in Sections 5.2-5.6, there are still a few left which can be relevant and interesting. For example, we assumed that the responses were instantaneous in
the sense that the response on a particular monitoring day is available
before the next monitoring day. This, however, may not be realistic in
some cases. In those situations, one can update the urn status on each

monitoring day based on the available data up to that day as in Wei (1988), Bandyopadhyay and Biswas (1996), and Bai et al. (2002). It is also important to extend the designs for trials with multiple arms.

ACKNOWLEDGMENT

The authors thank two referees for their valuable suggestions which led to a significant improvement over an earlier version of the manuscript. The authors also thank the Biometry Unit of the Indian Statistical Institute, Kolkata, for providing the data from their PEMF trial.

REFERENCES

[1] Bai, Z.D., Hu, F., and Rosenberger, W.F. (2002). Asymptotic properties of adaptive designs for clinical trials with delayed response. *Ann. Statist.*, **30**, 122-139.

[2] Bandyopadhyay, U. and Biswas, A. (1996). Delayed response in randomized play-the-winner rule: a decision theoretic outlook. *Cal. Statist. Assoc. Bul.*, **46** 69-88.

[3] Bandyopadhyay, U. and Biswas, A. (1999). Allocation by randomized play-the-winner rule in the presence of prognostic factors. *Sankhya, Ser. B*, **61**, 397-412.

[4] Bandyopadhyay, U. and Biswas, A. (2001). Adaptive designs for normal responses with prognostics factors. *Biometrika*, **88**, 409-419.

[5] Bartlett, R.H., Roloff, D.W., Cornell, R.G., Andrews, A.F., Dillon, P.W. and Zwischenberger, J.B. (1985). Extracorporeal circulation in neonatal respiratory failure: A prospective randomized trial. *Pediatrics*, **76**, 479-487.

[6] Begg, C.B. and Iglewicz, B. (1980). A treatment allocation procedure for sequential clinical trials. *Biometrics*, **36**, 81-90.

[7] Biswas, A. and Dewanji, A. (2001). An adaptive clinical trial with repeated monitoring for the treatment of rheumatoid arthritis. *Tech. Rep. # ASD/2001/13*, Indian Statistical Institute.

[8] Bonney G.E. (1987). Logistic regression for dependent binary observations. *Biometrics*, **43**, 951-973.

[9] Ganguly, K.S., Sarkar, A.K., Datta, A.K. and Rakshit, A. (1993). PEMF therapy in rheumatoid arthritis: A case study. *J. West Bengal Orthop. Assoc.*, **10**, 11-31.

[10] Iglewicz, B. (1983). Alternative designs: sequential, multi-stage, decision theory and adaptive designs. In *Cancer Clinical Trials: Methods and Practice* (M.E. Buyse et al., eds.), 312-334. Oxford University Press: Oxford.

[11] Rout, C.C., Rocke, D.A., Levin, J., Gouws, E. and Reddy, D. (1993). A reevaluation of the role of crystalloid preload in the prevention of hypotension associated with spinal anesthesia for elective cesarean section. *Anesthesiology*, **79**, 262-269.

[12] Tamura, R.N., Faries, D.E., Andersen, J.S. and Heiligenstein, J.H. (1994). A case study of an adaptive clinical trials in the treatment of out-patients with depressive disorder. *J. Amer. Statist. Assoc.*, **89**, 768-776.

[13] Ware, J.H. (1989). Investigating therapies of potentially great benefit: ECMO. *Statist. Sci.*, **4**, 298-340.

[14] Wei, L.J. (1988). Exact two-sample permutation tests based on the randomized play-the-winner rule. *Biometrika*, **75**, 603-606.

[15] Wei, L.J. and Durham, S. (1978). The randomized play-the-winner rule in medical trials. *J. Amer. Statist. Assoc.*, **73** , 838-843.

[16] Zelen, M. (1969). Play-the-winner rule and the controlled clinical trial. *J. Amer. Statist. Assoc.*, **64** , 131-146.

Addresses for communication:

ATANU BISWAS, Applied Statistics Unit, Indian Statistical Institute, 203 B.T. Road, Kolkata 700 108, India. E-mail: atanu@isical.ac.in
ANUP DEWANJI, Applied Statistics Unit, Indian Statistical Institute, 203 B.T. Road, Kolkata 700 108, India. E-mail: dewanjia@isical.ac.in

Chapter 6

Sequential Approaches to Data Mining

YUAN-CHIN IVAN CHANG
Academia Sinica, Taipei, Taiwan

ADAM T. MARTINSEK
University of Illinois, Urbana, U.S.A.

6.1 INTRODUCTION

Recently, there is a growing interest in *data mining* (DM) and *knowledge discovery in databases* (KDD) among researchers from many areas such as artificial intelligence, statistics, and data visualization. There are many kinds of data mining tasks (Witten and Frank (2000), Han and Kamber (2000)). These tasks often require different statistical tools and points of view, even for the same data set. For example, information for detecting telephone fraud and for locating potential target customers for new telephone services can both be obtained by analyzing the same dataset.

Modern computing and communication technologies can make data collection procedures very efficient. But our ability to analyze and extract information from large datasets is hard-pressed to keep up with our capacity for data collection. Automatic data collection designs can be found in scientific projects as well as in marketing surveys. For example, environmental data may be collected automatically by a meteorological satellite. In applications to biology, the subjects are monitored by electronic equipment, and the observed data are transferred into computers that may be connected to the equipment directly or through the Internet.

Usually, these kinds of datasets will be stored in a database. Although database technologies provide efficient tools for managing and maintaining datasets, analyzing them remains a difficult problem. Sophisticated statistical methods and data analysis tools are required to analyze such large datasets.

For data mining processes, introduced in popular textbooks such as Han and Kamber (2000) or Witten and Frank (2000), it is clear that the statistical ideas and techniques are in the kernel of the information extraction processes. In particular, clustering and classification analysis are the most commonly used tools in many DM/KDD processes. Many algorithms from the existing literature have been implemented in either commercial or non-commercial packages, such as CART (Breiman et al. (1984)), C4.5 (Quinlan (1993)), SPSS, and IBM's Enterprise Miner.

It is believed in some circles that with enough computing power it should be possible to analyze datasets of any size by using standard statistical tools. Ideally, this would be the case. However, in practice, the amount of data is growing much faster than the increasing speed of computers. A number of statisticians and information scientists have faced this issue (Fayyad et al. (1996) and Hastie et al. (2001)). Unless we are willing to wait for the next generation's computers to solve the problems of our generation, efficient and precise analysis of massive datasets will still be an interesting and challenging problem.

In addition to sheer size, a massive dataset usually has a complicated structure and contains different types of variables (for example, mixture of categorical, ordinal, and interval variables). Traditional clustering and classification algorithms, such as *k-nearest neighborhood* (kNN) or *principal component analysis* (PCA), are not really designed for analyzing such large and complicated datasets. Thus, they are usually not very efficient when the number of variables becomes very large and the data structure quite complicated.

The dataset on which we will focus contains more than 2 million in-patient medical charge records for 1172 hospitals, and there are 75 variables for each record. These 75 variables are a mixture of types: the dataset contains interval, ordinal and categorical variables. The goal of our analysis is to find out if there are any similarities or dissimilarities among hospitals based on the data. Especially, from the administrator's point of view, one would like to see if there are any differences in the medical charges among various hospitals.

The methods of clustering (tree) analysis will be used, and sequential methods will be employed for variable selection. Within each leaf-node, we use PCA to find the "dissimilar" hospitals based on sequential estimates of the medical charges of hospitals. That is, instead of applying cluster analysis to the whole dataset, we will use only a "sub-sample," which is extracted from the original dataset based on some sequential sampling criteria. It will be seen that sequential sampling approaches are more efficient than complete sample analysis in terms of CPU time.

In the rest of this Chapter, we will first briefly describe the dataset, and the analysis procedures. We then summarize the results in Section 6.2. In Section 6.3, only the highlights of the theoretical results of the sequential estimation used here are summarized. A general discussion is given in Section 6.4.

6.2 MEDICAL INSURANCE PROCEEDS

The dataset that we analyze contains in-patient records of 1172 hospitals during a calendar year. There are more than 2 million records and 75 variables. These 75 variables include personal identification, gender, medical administration related variables, and disease related variables. Unfortunately, the patient's medical records are not available. The most important disease related variables within these 75 variables come from the ICD 9 code, which is an international disease classification code (*International Classification of Disease*, 9th revision) originally published by the World Health Organization (WHO). There are 17 major disease groups in ICD 9, ranging from infectious and parasitic diseases (001-139) and neoplasms (140-239) to injury and poisoning (800-999), plus two supplementary classes. These include supplementary classification of factors influencing health status and contact with health services (V01-V82) and supplementary classification of external causes of injury and poisoning (E800-E999).

Each group has been further classified into many sub-disease groups. A brief explanation of the ICD 9 codes can be found from the following URL: http://www.mcis.duke.edu/standards/termcode/icd9. There are only 18 groups in our dataset and the last one, E800-E999, is not included. It is obvious that if all the ICD 9 sub-categories are used then the "data cube" will be very sparse. Moreover, for each record, there are also some itemized medical proceeds, which are the claims of the hospitals to the insurance company for each patient who has been served in their hospitals during the same calendar year that the data were collected.

The goal of the analysis is to find similarities/dissimilarities of medical charges among the 1172 hospitals using this in-patient dataset. In particular, from the health insurance administrator's point of view, it is interesting to know whether there are any differences in medical charges among various hospitals for the same or similar diseases. The similarity of diseases can be defined according to ICD 9. Below, we use two analysis approaches. One is based on clustering tree. The second one is based on using the ICD 9 code to stratify patients into different disease groups within each hospital. Both the entropy and the Gini index are used to measure homogeneity (Breiman et al. (1984)). In the clustering tree approach, we will construct two trees for the hospitals based on both entropy and Gini index, followed by a principal component analysis (Mardia et al. (1979)) for each leaf-node. In the second approach, the whole dataset will be stratified into 17 groups according to the major disease groups classification of the ICD 9 code. Then, the strata are analyzed one by one using PCA. The entropy and Gini (Gini-Simpson) index have been used in many areas, for example, genetics, bio-diversity, economics, and quality of life. For details, we refer to Chakraborty and Rao (1991), Mukherjee (2000), Rao (1982,1983,1984), Nayak (1986), Nayak and Gastwirth (1989), Sen (1999), and Chatterjee and Sen (2000).

6.2.1   Clustering Tree

Here we will first ignore the variables for medical charges, and use the rest of the variables to construct a clustering tree analysis for the hospitals. The monothetic divisive method is used for tree construction (Williams and Lambert (1960), Jongman et al. (1995)). That is, at each node (splitting), we will choose only one variable to split the data. The criterion of variable selection is based on sequential estimates of Shannon's entropy or Gini index. Next, we define these measures for a discrete random variable.

Suppose that Z is a discrete random variable with

$$P(Z = i) = p_i \text{ for } i = 1, \cdots m, \text{ and } \sum_{i=1}^{m} p_i = 1. \qquad (6.2.1)$$

Then, Shannon's entropy and the Gini index of the random variable Z are defined as

$$H_Z = -\sum_{i=1}^{m} p_i \log(p_i) \qquad (6.2.2)$$

and

$$G_Z = 1 - \sum_{i=1}^{m} p_i^2 \qquad (6.2.3)$$

respectively. In the following, we will summarize the construction procedure of the entropy-clustering tree.

For each $i = 1, \cdots k$, $j = 1, \cdots, h$, let $X_{i,j}$ denote the ith variable of the jth hospital, and let $H_{i,j}$ denote its entropy defined in (6.2.2).

A two-stage sampling procedure will be used for constructing a fixed-width confidence interval for $H_{i,j}$. Let n_0 denote the initial sample size, and let $N_{i,j}$, $i = 1, \cdots, k$, $j = 1, \cdots, h$ denote the sample size required for the ith variable of the jth hospital according to the equation (6.3.6) in Section 6.3. For each hospital $j = 1, \cdots, h$, let

$$N_j = N_{\max,j} = \max\{N_{1,j}, \cdots, N_{k,j}\} \qquad (6.2.4)$$

and let $\hat{H}_{i,j}$, $i = 1, \cdots, k$ denote the estimates obtained from this two-stage sampling procedure (that is, based on the two-stage sample of size N_j).

For a fixed variable i, based on these estimates of entropy we would like to divide the hospitals into 3 sub-groups. Suppose that there are already 3 sub-groups, and that each sub-group contains s_l subjects, $l = 1, 2, 3$. Then the empirical entropy of these 3 sub-groups is defined as

$$D_i = -\sum_{l=1}^{3} \frac{s_l}{\sum_{j=1}^{3} s_j} \log(s_l / \sum_{j=1}^{3} s_j). \qquad (6.2.5)$$

Now, for each variable i, according to the estimated entropy $\hat{H}_{i,j}$, $j = 1, \cdots, h$, where $\hat{H}_{i,j}$ is estimated in the spirit of (6.3.1), based on a sample of size N_j, we can divide the 1172 hospitals into 3 sub-groups according to the following rule:

Hospital j will be assigned to Group $1, 2$, or 3 according as

$$\widehat{H}_{i,j} = 0, \ 0 < \widehat{H}_{i,j} \leq C_i, \text{ or } \widehat{H}_{i,j} > C_i \text{ respectively,}$$

where C_i, $i = 1, \cdots, k$, is a threshold chosen such that the entropy D_i of the sub-groups (see equation (6.2.5)) is maximized.

That is, for each i, D_i is the maximum entropy when using the ith variable as the splitting variable. Therefore, at each node, we will select the variable with maximum D_i as the splitting variable. This kind of variable selection and splitting procedure will continue in each leaf-node for the rest of the variables until the differences of the entropy of variables among hospitals within each leaf-node are less than some given $\eta (> 0)$. This is, in some sense, a pre-specified precision criterion.

This procedure can be summarized in the following steps:

Step # 1 Estimation: To compute the estimate of the entropy $H_{i,j}$, say $\hat{H}_{i,j}$ for all $i = 1, \cdots, k$ and $j = 1, \cdots h$ sequentially, until the pre-specified precision criterion mentioned above is fulfilled.

Step # 2 Variable Selection: Based on the size of sub-groups obtained by using the corresponding entropies of each hospital, we compute D_i for each variable and then choose the one with the largest D_i as the splitting variable for this node.

Step # 3 Iteration: Repeat step # 2, for each leaf-node (sub-group) by using the rest of the variables that have not been used in its ancestor levels. If all the entropies D_i for all variables are less than a pre-specified real number $\eta > 0$, then no more sub-groups will be generated beneath this node.

Step # 4 Termination: If there are no more sub-groups generated for all the sub-nodes, the construction of the clustering tree stops.

Step # 5 Pruning: The leaf-nodes with number of hospitals less than 5 will be pruned.

The clustering tree based on the entropy variable selection scheme is given in Figure 6.2.1 and the one based on the Gini index is given in Figure 6.2.2. There are only slight differences between the two trees.

The order of variables selected for each node is different, as is the total number of leaf-nodes. According to the way that we construct the tree, Leaf-node 1 in Figure 6.2.1 represents a group of hospitals with no patients whose diseases belong to the ICD 9 codes 780-799 and 580-629. From the list of hospitals, we find that most of the hospitals belonging to Leaf-node 1 are maternity hospitals. Leaf-node 6 of Figure 6.2.1 represents hospitals with ability to serve patients whose diseases belong to the ICD 9 codes 001-139, 460-519, 580-629, and 780-799. From the list of hospitals, we find that these are major district hospitals which can handle a wide variety of diseases.

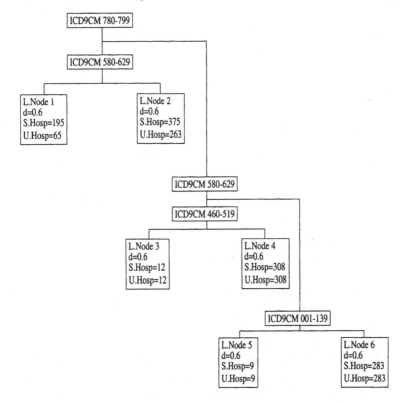

Figure 6.2.1: Entropy Tree.

There are 17 variables directly related to the proceeds claimed by hospitals, for example, diagnostic charges, ward, meal, X-ray, surgery, blood examination, and anesthesia. Since we are dealing with only in-patients and they may have stayed in hospitals for different lengths of time, we divided the itemized medical charges by the total. The following com-

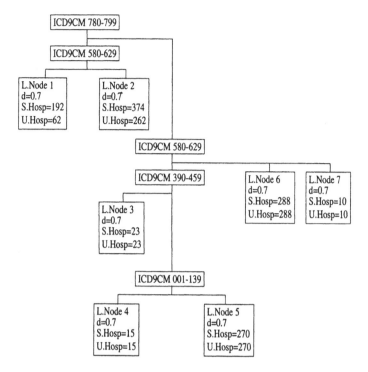

Figure 6.2.2: Gini Index Tree.

parisons among medical charges of all hospitals will be based on these proportions instead of the original charges.

For every leaf-node, we estimated the means of the proportional medical charges of each hospital in the leaf-node using a two-stage sampling procedure. We treat the estimated mean proportions as a representative vector associated with the hospitals, and project it into the plane spanned by the first two eigenvectors of the corresponding first two maximum eigenvalues in a PCA procedure. This way, we can view geometric relationships among the hospitals within each leaf-node. Figures 6.2.3 and 6.2.4 show the leaf-nodes from the entropy tree and the Gini index tree, respectively. The points marked by "X" indicate outliers found by using *Mahalanobis distance* (Weisberg (1985)).

Remark 6.2.1 The reason for choosing monothetic divisive methods is that most of the variables are categorical. It is hard for us to define any suitable measure of similarity in this case. Moreover, if we stratify the dataset according to the values of all categorical variables, then the dataset will be very sparse in the "data-cube". Usual dimension reduction techniques, such

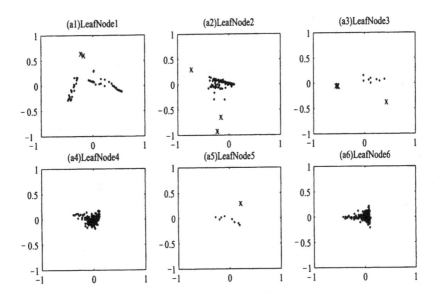

Figure 6.2.3: Leaf Nodes of Entropy Index Tree.

as the singular value decomposition (SVD), will be unlikely to succeed.

6.2.2 Stratification

From Figures 6.2.1 and 6.2.2, we can see that only the ICD 9 variables are selected in both cases. This suggests running the stratified analysis directly by using ICD 9 as the stratification variable.

According to the ICD 9 codes, we stratified patients from each hospital into 18 major disease groups (E800-E999 are not included). Then, for each disease group of each hospital, we applied the two-stage sequential method to estimate the mean vector of the proportional medical charges. We used this mean vector as the "representative." For each disease group, using these representative vectors for all hospitals, we conducted a PCA and projected those representative vectors into the space of the first two eigenvectors using the same procedure as for the leaf-nodes. Based on these pictures, and utilizing the *Mahalanobis distance*, we could identify the "outliers". These were the hospitals with dissimilar medical charge proportions under the same disease group. Figures 6.2.5 and 6.2.6 are based on the entropy for 18 major disease groups. The potential outliers are marked by "X" as before. To interpret these pictures more fully in search of further information, some dialogue with the medical administration experts will be necessary. We pursued the same procedure by using Gini indices, but the findings were si-

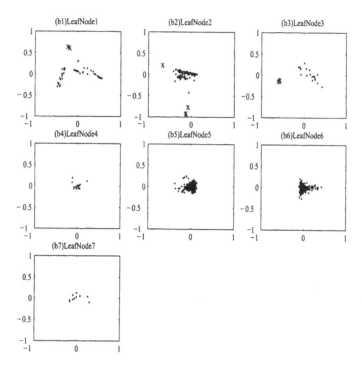

Figure 6.2.4: Leaf Nodes of Gini Index Tree.

milar. Therefore, we skip the pictures obtained by using Gini indices.

Remark 6.2.2 The dataset we have here corresponds to in-patients only. A patient is usually transferred to another hospital if the current hospital cannot provide appropriate medical services due to lack of suitable resources (for example, medical personnel, medical equipment, and ward space). Therefore, the itemized proceeds claimed by hospitals can be treated as indicators of the resources and services they provide.

Remark 6.2.3 When using two-stage sampling procedures to construct a fixed-width confidence interval for the entropy (or the Gini index) at both the variable selection step and the PCA step, we choose the coverage probability to be 95%, and the precision $\eta = 0.6$ and 0.7 for the entropy and Gini indices, respectively. In PCA, two-stage estimation is used to guarantee the stability of the estimated vector of the proportions, but the effect of precision has not been studied yet.

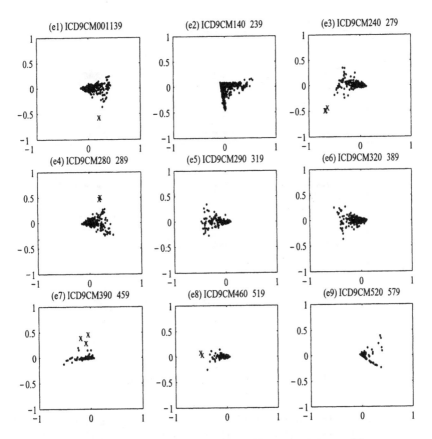

Figure 6.2.5: PCA When Stratified by ICD 9 (I).

6.2.3 Computational Complexity

In order to evaluate the performance of two-stage sampling procedures, we conducted similar analyses based on the whole dataset (that is, without two-stage sampling). The CPU times for all methods are recorded. From the clustering tree and the pictures obtained under PCA, we see that the results are very similar to those obtained by using two-stage sampling methods. From the entropy-clustering tree (Figures 6.2.1 and 6.2.7), we observe that there are only slight differences between trees obtained from full data and those based on the two-stage sampling method. By comparing Figures 6.2.5 and 6.2.6 with Figure 6.2.8, we note that the PCA results based on full data and those based on the two-stage sampling method are also very close to each other. But from Table 6.2.1, we see that the CPU times required for accomplishing the analyses are very different. The methods using two-stage

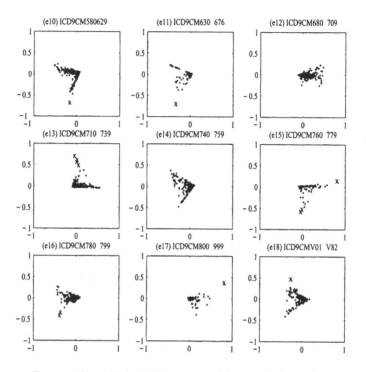

Figure 6.2.6: PCA When Stratified by ICD 9 (II).

sampling are substantially more efficient in terms of computational time. We also observe that the method using the Gini index required more CPU time in both analyses. The datasets were administered by using the Microsoft SQL server for PC. The CPU times reported here include the response time of the database software.

6.3   SEQUENTIAL FIXED-WIDTH INTERVAL ESTIMATION

Assume that Z is a discrete random variable with

$$P(Z = i) = p_i, i = 1, ..., m, \text{ and } \sum_{i=1}^{m} p_i = 1.$$

If we observe $Z_1, Z_2, \cdots$, i.i.d. with this distribution, define

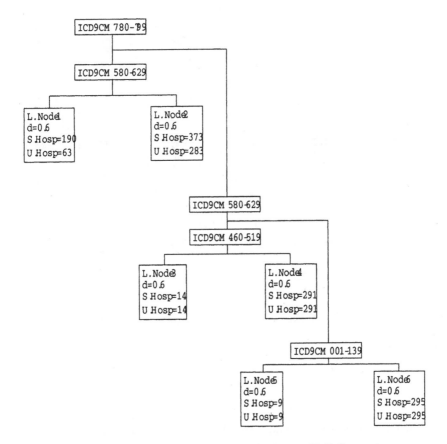

Figure 6.3.7: Entropy Tree Based on Full Data.

Table 6.3.1: CPU Times in Seconds

Method	Clustering Tree		PCA	
	Entropy	Gini Index	Entropy	Gini Index
Full Data	3300.0	7133.6	1859.5	4363.1
Two-Stage Sampling	1915.3	2081.0	1695.2	2083.7

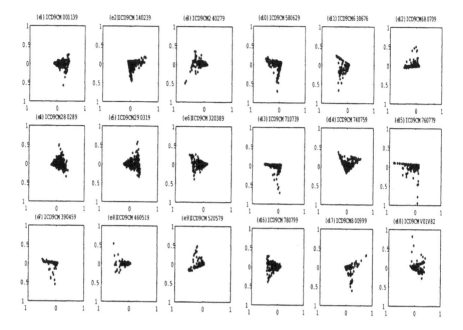

Figure 6.2.8: PCA Based on Entropy Index with Full Data When
Stratified by ICD 9.

$$Y_{in} = \text{ number of } Z_1, \cdots, Z_n \text{ in category } i.$$

With some given n, $(Y_{1n}, ..., Y_{mn})$ has a multinomial distribution with
parameters n and $p_1, p_2, ..., p_m$, and $\hat{p}_{in} = Y_{in}/n$ is the estimate of p_i, for
each i. Note that $Y_{in} = \sum_{k=1}^{n} \xi_{ik}$ where $\xi_{ik} = I_{\{Z_k \in category\ i\}}$. The vectors $(\xi_{1k}, ..., \xi_{mk}), k = 1, ..., n$, are i.i.d. with mean vector $(p_1, ..., p_m)$ and
covariance matrix $\Sigma = (\sigma_{ij})_{m \times m}$, where

$$\sigma_{ij} = \begin{cases} p_i(1 - p_i), & \text{if } i = j \\ -p_i p_j, & \text{if } i \neq j \end{cases}$$

The estimate of entropy,

$$\hat{H}_n = -\sum_{i=1}^{m} \hat{p}_{in} log(\hat{p}_{in}), \tag{6.3.1}$$

is a function of the random vector $(\hat{p}_{1n}, ..., \hat{p}_{mn})$ and it is continuously differentiable. Hence, by applying the multivariate delta method (Bickel and

Doksum (2001, Theorem 5.3.4)), we have

$$n^{1/2}(\hat{H}_n - H) \Rightarrow N(0, \sigma_H^2), \qquad (6.3.2)$$

where $\sigma_H^2 = W'\Sigma W$ and "$\Rightarrow$" means convergence in law, and

$$W' \equiv \partial H/\partial p = (-1 - log(p_1), \cdots, -1 - log(p_m)).$$

Hence, the asymptotic variance is equal to

$$\sigma_{H,n}^2 = \sum_{i=1}^{m}[1+log(p_i)]^2 p_i(1-p_i) - \sum\sum_{i \neq j}[1+log(p_i)][1+log(p_j)]p_i p_j. \quad (6.3.3)$$

Based on (6.3.1)–(6.3.3), for a given coverage probability $1 - \alpha$ ($0 < \alpha < 1$) and for a given precision criterion $d > 0$, we required that

$$P(|\hat{H}_n - H| \leq d) \geq 1 - \alpha. \qquad (6.3.4)$$

Then the smallest sample size ensuring that the inequality (6.3.4) is satisfied is approximately

$$N_d \equiv \text{ smallest integer } n \text{ such that } n \geq C(\alpha, d, \hat{\sigma}_{H,n}), \qquad (6.3.5)$$

where $C(\alpha, d, \hat{\sigma}_{H,n}) = (z_\alpha \hat{\sigma}_{H,n})^2/d^2$. N_d is a stopping rule for the sequential interval estimation procedure.

In Section 6.2, we applied a two-stage sampling procedure. Assuming that we have a first-stage sub-sample of size n_0, the second stage sample size can be estimated by $\langle C(\alpha, d, \hat{\sigma}_{H,n_0}) \rangle + 1$, where the notation $\langle u \rangle$ denotes the largest integer less than $u \in R$. Let

$$N_d \equiv \max \left\{ n_0, \ \langle C(\alpha, d, \hat{\sigma}_{H,n_0}) \rangle + 1 \right\}, \qquad (6.3.6)$$

and then the estimated entropy becomes $\hat{H}_{N_d}$.

The estimates $\hat{p}_{i,n} = Y_{i,n}/n$, $i = 1, \cdots, m$, are sample means of i.i.d. random variables and, therefore, can be shown to be strongly consistent estimates of p_i. This implies that $\hat{H}_n$ is also a strongly consistent estimate of H. Then, it can be shown by arguments similar to Woodroofe (1982, Example 1.8), that $\{(\hat{p}_{i,n} - p_i)\sqrt{n}\}$ is *uniformly continuous in probability* (u.c.i.p.). In addition, the entropy H is a bounded function of the p_i's, and so it follows from Woodroofe (1982, Lemma 1.4), that the estimate $\{\sqrt{n}(\hat{H}_n - H)\}$ is u.c.i.p. Hence, by applying Anscombe's theorem (Woodroofe (1982, Theorem 1.4)), we have

$$\lim_{d \to 0} P(|\hat{H}_N - H| \leq d) = 1 - \alpha. \qquad (6.3.7)$$

Now let's turn to the sequential estimate of the Gini index. By similar arguments, it can be shown that the estimate of Gini index, $\hat{G}_n = 1 - \sum_{i=1}^{m} \hat{p}_{i,n}^2$, is a strongly consistent estimate of G and is asymptotically normal, that is,

$$\sqrt{n}(\hat{G}_n - G) \Rightarrow N(0, \sigma_G^2), \qquad (6.3.8)$$

where $\sigma_G^2 = (\sum_{i=1}^{m} p_{i,n}^2)^2$ and "$\Rightarrow$" again means convergence in law.

Let n_0 be the initial sample size. It can be shown from (6.3.8) that the smallest fixed sample-size such that

$$P(|\hat{G}_n - G| \leq d) \geq 1 - \alpha,$$

is $(z_\alpha \sigma_G)^2/d^2$. Replacing σ_G by its estimate $\hat{\sigma}_G$, the sample-size combining the two stages is defined as

$$N = \max \left\{ n_0, \ \left\langle \frac{(z_\alpha \sigma_G)^2}{d^2} \right\rangle + 1 \right\}. \qquad (6.3.9)$$

It is easy to show that $\{\sqrt{n}(\hat{G}_n - G)\}$ is u.c.i.p., since G is continuously differentiable and bounded in $[0, 1]^m$. Hence, it can be shown that $\lim_{d \to 0} P(|\hat{G}_N - G| \leq d) = 1 - \alpha$.

6.4   CONCLUDING REMARKS

The use of sampling techniques in data mining procedures has been considered by many authors. In information technology, this is sometimes called "data reduction." The main idea of "data reduction" is to choose a subsample from the whole dataset that preserves the "structures" or "relationships" of the original dataset. The commonly used sampling methods are simple random sampling and stratified sampling. In this project, we applied a "stratified sampling" method within both the variable selection phase and the PCA procedure. Certainly, there are many possible ways of analysis that can be applied to the leaf-nodes of a clustering tree. The proportions of medical charges could be treated as continuous variables, and hence PCA would be a suitable tool to use. Without a fixed-width confidence interval estimation approach, the estimates of the "representative" points of each hospital will all have different variabilities associated with them, and the PCA based on them will be very unreliable. The fixed-width confidence interval estimation procedure not only improves the efficiency of a PCA procedure (in terms of smaller sample sizes), but also guarantees precision. Here we used an estimate of the mean vector of the medical charges as a representative of hospital charges. To achieve "robustness", we could use the estimated "centroid" of the medical charges instead of the mean vector. The sequential procedures we discussed here could still be implemented after some modification.

ACKNOWLEDGMENT

We would like to thank the editors for inviting us to contribute to this volume. We would also like to thank Dr. Chin-Hsin Chen for making the dataset available to us, and Mr. Sung-Chiang Lin, Dr. Chang's research assistant, for his efforts of data cleaning and programming. Finally, we would like to thank the referees for helpful comments that substantially improved the paper.

REFERENCES

[1] Bickel, P.J. and Doksum, K.A. (2001). *Mathematical Statistics, 2nd ed.* Prentice Hall: New Jersey.

[2] Breiman, L., Friedman, J., Olshen, R. and Stone, C. (1984). *Classification and Regression Trees.* Wadsworth: Belmont.

[3] Chakraborty, R. and Rao, C.R. (1991). Measurement of genetic variation for evolutionary studies. In *Handbook of Statistics,* **8,** Statistical Methods in Biological and Medical Sciences (C.R. Rao and R. Chakraborty, eds.), 271-316. Elsevier/North-Holland: New York.

[4] Chatterjee, S.K. and Sen, P.K. (2000). On stochastic ordering and a general class of poverty indexes. *Cal. Statist. Assoc. Bul.,* **50,** 137-155.

[5] Fayyad, U.M., Piatetsky-Shapiro, G., Smyth, P. and Uthurusamy, R. (1996). *Advances in Knowledge Discovery and Data Mining.* AAAI Press/ MIT Press: Cambridge.

[6] Han, J. and Kamber, M. (2000). *Data Mining: Concepts and Techniques.* Morgan Kaufmann: San Francisco.

[7] Hastie, T., Tibshirani, R. and Friedman, J. (2001). *Elements of Statistical Learning: Data Mining, Inference and Prediction.* Springer-Verlag: New York.

[8] Jongman, R.H.G., Ter Braak, C.J.F. and Van Tongeren, O.F.R. (1995). *Data Analysis in Community and Landscape Ecology.* Cambridge University Press: Cambridge.

[9] Mardia, K.V., Kent, J.T. and Bibby, J.M. (1979). *Multivariate Analysis.* Academia Press: New York.

[10] Mukherjee, S.P. (2000), Biodiversity–measurement and analysis. In *Handbook of Statistics*, **18**, Bioenvironmental and Public Health Statistics (P.K. Sen and C.R. Rao, eds.), 1061-1076. Elsevier/North-Holland: New York.

[11] Nayak, T.K. (1986). Sampling distributions in analysis of diversity. *Sankhya, Ser. B*, **48**, 1-9.

[12] Nayak, T.K. and Gastwirth, J.L. (1989). The use of diversity analysis to assess the relative influence of factors affecting the income distribution. *J. Bus. Econ. Statist.*, **7**, 453-460.

[13] Quinlan, J.R. (1993). *C4.5: Programs for Machine Learning.* Morgan Kaufmann: San Francisco.

[14] Rao, C.R. (1982). Diversity and dissimilarity coefficients: A unified approach. *Theor. Popu. Biol.*, **21**, 24-43.

[15] Rao, C.R. (1983). Diversity: Its measurement, decomposition, apportionment and analysis. *Sankhya, Ser. A*, **44**, 1-21.

[16] Rao, C.R. (1984). Gini-Simpson index of diversity: A characterization, generalization and applications. *Utilitas Math.*, **21**, 273-282.

[17] Sen, P.K. (1999). Utility-oriented Simpson-type indexes and inequality measures. *Cal. Statist. Assoc. Bul.*, **49**, 1-21.

[18] Weisberg, S. (1985). *Applied Linear Regression*, 2nd ed. Wiley: New York.

[19] Williams, W.T. and Lambert, J.M. (1960). Multivariate methods in plant ecology II: The use of an electronic digital computer for association-analysis. *J. Ecol.*, **48**, 689-710.

[20] Witten, I.H. and Frank, E. (2000). *Data Mining.* Morgan Kaufmann: San Francisco.

[21] Woodroofe, M. (1982). *Nonlinear Renewal Theory in Sequential Analy-*

sis. Society for Industrial and Applied Mathematics: Philadephia.

Addresses for communication:

YUAN-CHIN IVAN CHANG, Institute of Statistical Science, Academia Sinica, 128, Sec. 2, Academia Rd., Taipei, Taiwan 11529. E-mail: ycchang @sinica.edu.tw
ADAM T. MARTINSEK, Department of Statistics, University of Illinois, 725 South Wright Street, Urbana, IL 61801, U.S.A. E-mail: martinse@uiuc.edu

Chapter 7

Approximations and Bounds for Moving Sums of Discrete Random Variables

JIE CHEN
University of Massachusetts, Boston, U.S.A.

JOSEPH GLAZ
University of Connecticut, Storrs, U.S.A.

7.1 INTRODUCTION

Let $X_1, \cdots, X_N$ be independent and identically distributed (i.i.d) non-negative integer valued random variables. For integers $2 \leq m < N$, consider the moving sums of m consecutive observations. For $2 \leq m < N$, the *discrete scan statistic* is defined as the maximum of these moving sums:

$$S_{\mathbf{m}} = max\{X_i + \cdots + X_{i+m-1}; 1 \leq i \leq N - m + 1\}. \qquad (7.1.1)$$

In the special case of Bernoulli trials, the discrete scan statistic defined in (7.1.1) generalizes the notion of the longest run of 1's that has been investigated extensively in the statistical literature (Balakrishnan and Koutras (2001)). This discrete scan statistic has been used in testing the null hypothesis that the observations are identically distributed against an alternative hypothesis of clustering, that specifies an increased occurrence of events in a continuous subsequence of the observed data (Glaz and Naus (1991)). For a special class of alternatives it has been shown that the discrete scan statistic is a generalized likelihood ratio test statistic (Glaz and Naus (1991, Section 1.3)). Approximations and inequalities for the distribution and moments of S_m have been discussed in Chen and Glaz (1999) and Glaz et al. (2001).

In this article, we will discuss the use of discrete scan statistics in sequential experiments and review the methods that can be employed to implement the relevant inference procedures. Let $X_1, X_2, \cdots, X_j, \cdots$ be a sequence of i.i.d. non-negative integer valued random variables. Here, the random variable X_j represents the number of signals or events that have been detected or observed at the end of the jth discrete-time interval. Define

$$W_{k,m} = \inf \left\{ j \geq 1; \sum_{i=\max(1,j-m+1)}^{j} X_i \geq k \right\}, \qquad (7.1.2)$$

to be the waiting time for detecting or observing k or more signals in m consecutive discrete-time intervals. Then, $E(W_{k,m})$ is the mean recurrence time for observing at least k or more events in m consecutive discrete-time intervals. The quantities $W_{k,m}$ and $E(W_{k,m})$ play important roles in many applications. We now proceed to discuss briefly a few of these applications.

Suppose that a radar sweep is based on a dichotomous quantizer that records the digit 1 if the signal-plus-noise waveform exceeds a specified threshold, and records the digit 0 otherwise. In this case, a radar sweep is transformed into a sequence of Bernoulli trials. A pulse is generated at time j if $X_{j-m+1} + \cdots + X_j \geq k$, where k and m are chosen to achieve a specified level α of *false alarm probability:*

$$P(W_{k,m} > N) = P(S_m \leq k - 1) = \alpha \qquad (7.1.3)$$

under the assumption that no target is present within a specified number N of recorded observations. This target detection procedure is referred to as the *k-out-of-m moving window detector* (Dinneen and Reed (1956), Bogush (1972) and Nelson (1978)). Mirstik (1978) discussed various advantages of using a multistatic system in which pulses are received and processed by several dichotomous quantizers. In such a system the digital data transmitted to a detector consist of an infinite sequence of binomial random variables. A moving window detector will generate a pulse if the moving sum of binomial random variables exceeds a specified threshold. The length of a moving sum and a threshold are determined from a specified level of false alarm probability and the projected total number of observations.

In a time-sharing system, messages of constant length may be received by a traffic concentrator at discrete time intervals separated by t/m units, where t is the time required to transmit a message of constant length. At most k messages can be transmitted simultaneously from the concentrator. In this application, $W_{k,m}$ is the waiting time for a concentrator to stop transmitting messages. Chu (1970) and Fredrikson (1974) considered Poisson and binomial arrivals, respectively, at each discrete time mark. The distribution of $W_{k,m}$ and $E(W_{k,m})$ play important roles in the analysis and design of time-sharing systems.

In quality control schemes where one individual item is inspected at a time, we record $X_j = 1$ or 0 if the jth item is defective or not defective, respectively. Greenberg (1970) and Saperstein (1973) investigated statistical properties of *zone* tests. They defined the process to be *out of control* if a moving window of size m contains k or more observations outside a specified zone. $W_{k,m}$ is the waiting time between times when the process is declared to be out of control and $1/E(W_{k,m})$, the probability of a Type-I error for testing if the process is *in control*.

Saperstein (1976) investigated MIL-STD-105D reduced inspection plans that use moving sums of independent binomial or Poisson random variables. These plans are used in the analysis and design of quality control schemes for inspection of items arriving in "*lots*". In these inspection plans, accurate approximations for the distribution of $W_{k,m}$ and $E(W_{k,m})$ are of considerable importance.

In Section 7.2, we review approximations and inequalities for the distribution of $W_{k,m}$ and $E(W_{k,m})$ for binomial and Poisson models. Numerical results for selected examples are presented in Section 7.3. Concluding remarks and some open problems are given in Section 7.4.

7.2 THE DISTRIBUTION AND EXPECTATION OF $W_{k,m}$

It follows from the definitions of S_m and $W_{k,m}$ given in (7.1.1) and (7.1.2) respectively that

$$P(W_{k,m} > N) = P(S_{\mathbf{m}} \leq k - 1). \tag{7.2.1}$$

Therefore, $P(W_{k,m} \leq N) = P(S_\mathbf{m} \geq k)$. Since,

$$E(W_{k,m}) = \sum_{N=0}^{\infty} P(W_{k,m} > N), \qquad (7.2.2)$$

exact results, approximations, or inequalities for $E(W_{k,m})$ are obtained from exact results, approximations, or inequalities for $P(W_{k,m} > N)$ or $P(S_\mathbf{m} \leq k-1)$, respectively. We discuss shortly some methods one can use to evaluate $P(W_{k,m} > N)$ for binomial and Poisson models. Special attention is given to the important case involving the Bernoulli model. The results for $P(W_{k,m} > N)$ are presented equivalently in terms of the corresponding results for $P(S_\mathbf{m} \leq k-1)$.

7.2.1   Bernoulli Model

Exact results, approximations, and inequalities for $P(S_m \leq k-1)$ have been investigated extensively in the case of i.i.d. Bernoulli trials (Glaz et al. (2001, Chapter 13)). We present here the most efficient approach for evaluating $P(S_m \leq k-1)$. Fu (2001) employs a *finite Markov chain imbedding* method to derive an exact expression for $P(S_m \leq k-1)$ that can be easily computed for all values of N, m and k. This method has been developed in a series of articles by Fu (1986), Fu and Hu (1987), Fu and Koutras (1994), Koutras and Alexandrou (1996), Koutras (1996), and Boutsikas and Koutras (2000). Until now, $P(S_m \leq k-1)$ could be evaluated only for a limited set of values of N, m and k, since the dimension of the state space of the imbedded Markov chain becomes as high as $2^{m-1} + 1$. With this approach, for example, it is impossible to evaluate $P(S_m \leq k-1)$ when $m = 50$. For the problem at hand, Fu (2001) constructed an imbedded Markov chain with a state space that contained at most $2m$ elements. We present below a brief outline of his approach.

For $m \leq j \leq N$, let an *ending block* U_j be defined as

$$U_j = \left(\sum_{i=j-m+1}^{j} X_i, X_j - m + 1 \right).$$

Given k and j $(m \leq j \leq N)$, define a sequence of *index functions*:

$$I_j(k) = \begin{cases} 0 \text{ if } S_m(i) \leq k-1, \text{ for all } i, \; m \leq i \leq j \\[2mm] 1 \text{ if } S_m(i) \geq k, \text{ for some } i, \; m \leq i \leq j \end{cases}$$

where $S_m(i)$ is the scan statistic for a sequence of i Bernoulli trials. It follows that

$$P(S_m \leq k - 1) = P(I_N(k) = 0).$$

To evaluate $P(S_m \leq k - 1)$, the random variable $I_j(k)$, $m \leq j \leq N$, is imbedded as a component of a homogeneous Markov chain given by:

$$O_j = \begin{cases} (I_j(k), U_j) & \text{if } S_m(j) \leq k - 1 \text{ and } I_j(k) = 0 \\ \delta & \text{if } I_k(k) = 1 \end{cases}$$

with state space

$$\{(0, (0, 0)), (0, (1, 0)), (0, (1, 1)), ..., (0, (k-1, 0)), (0, (k-1, 1)), \delta\}.$$

This state space has $2k$ elements. We now present, without proof, the main result in Fu (2001): For $1 \leq k \leq m < N$,

$$P(S_m \leq k - 1) = \tau \mathbf{M}^{N-m} \mathbf{J}', \tag{7.2.3}$$

where τ is a $1 \times 2k$ vector of initial probabilities with the ith component $(2 \leq i = 2j \leq 2k - 1, 1 \leq j \leq k - 1)$ given by

$$\tau_i = \begin{cases} \binom{m-1}{j} p^j q^{m-j} & \text{if } i = 2j \\ \binom{m-1}{j-1} p^j q^{m-j} & \text{if } i = 2j + 1, \end{cases}$$

$\tau_1 = \binom{m-1}{0} q^m$ and $\tau_{2k} = 1 - \sum_{n=0}^{k-1} \binom{m}{n} p^n q^{m-n}$;
$\mathbf{J}'$ is the transpose of $\mathbf{J} = (1, 1, ..., 1, 0)_{1 \times 2k}$; and
$\mathbf{M} = [M_{(u,v),(x,y)}]$ is the transition probability matrix for the imbedded Markov chain of dimension $2k \times 2k$, where for $1 \leq u \leq k - 1$,

$$M_{(u,1),(x,y)} = \begin{cases} \frac{u-1}{m-1}p, & x = u, \quad y = 1 \\ \frac{m-u}{m-1}p, & x = u, \quad y = 0 \\ \frac{u-1}{m-1}q, & x = u - 1, \quad y = 1 \\ \frac{m-u}{m-1}q, & x = u - 1, \quad y = 0 \end{cases}$$

and for $0 \leq u \leq k - 1$,

$$
M_{(u,0),(x,y)} = \begin{cases}
\frac{u}{m-1}p, & x = u+1, \quad y = 1 \\[2mm]
\frac{m-u-1}{m-1}p, & x = u+1, \quad y = 0 \\[2mm]
\frac{u}{m-1}q, & x = u \qquad y = 1 \\[2mm]
\frac{m-u-1}{m-1}q, & x = u \qquad y = 0,
\end{cases}
$$

$M_{(k-1,0),\tau} = p$, $M_{\tau,\tau} = 1$, and $M_{(u,v),(x,y)} = 0$ elsewhere.

Fu (2001) presents numerical results for $P(S_m \leq k-1) = P(W_{k,m} > N)$ using S-Plus software. In Section 7.3, for selected values of the parameters p, k, m, and N we present numerical results for $P(W_{k,m} > N)$ and $E(W_{k,m})$.

7.2.2 Binomial and Poisson Models

Let $X_1, X_2, \cdots, X_j, \cdots$ be a sequence of i.i.d. binomial random variables with parameters n and p or i.i.d. Poisson random variables with mean λ. It follows from Glaz et al. (2001, Section 13.3) that, for $N \geq 2m$,

$$
\frac{q_{2m}}{\left[1 + \frac{q_{2m-1}-q_{2m}}{q_{2m}q_{2m-1}}\right]^{N-2m}} \leq P(W_{k,m} > N) \leq q_{2m}\left[1 - (q_{2m-1} - q_{2m})\right]^{N-2m},
$$

$$(7.2.4)$$

and, for $N \geq 3m$

$$
\frac{q_{3m}}{\left[1 + \frac{q_{2m-1}-q_{2m}}{q_{3m-1}}\right]^{N-3m}} \leq P(W_{k,m} > N) \leq q_{3m}\left[1 - (q_{2m-1} - q_{2m})\right]^{N-3m}.
$$

$$(7.2.5)$$

Here, for $1 \leq j \leq 3m$,

$$
q_j = P(W_{k,m} > j)
$$

can be efficiently evaluated using the algorithms in Karwe and Naus (1997). For $N \geq 3m$, the inequalities in (7.2.5) are tighter than the inequalities given in (7.2.4). It follows from Equation (7.2.2) that

$$
\sum_{j=0}^{3m-1} q_j + q_{3m}\left[1 + \frac{q_{3m-1}}{(q_{2m-1} - q_{2m})}\right] \leq E(W_{k,m}) \leq \sum_{j=0}^{3m-1} q_j + \frac{q_{3m}}{q_{2m-1} - q_{2m}}
$$

$$(7.2.6)$$

and

$$\sum_{j=0}^{3m-1} q_j + q_{3m} \left[1 + \frac{q_{3m-1}}{(q_{2m-1} - q_{2m})}\right] \le E(W_{k,m}) \le \sum_{j=0}^{3m-1} q_j + \frac{q_{3m}}{q_{2m-1} - q_{2m}}.$$

(7.2.7)

An immediate consequence of the bounds for $P(W_{k,m} > N)$ given in (7.2.4)-(7.2.5) and those for $E(W_{k,m})$ given in (7.2.6)-(7.2.7) is that an arithmetic average of these bounds can be used to approximate $P(W_{k,m} > N)$ and $E(W_{k,m})$, respectively.

A different approach to deriving approximations for $P(W_{k,m} > N)$ and $E(W_{k,m})$ is based on the method of product-type approximations discussed in Glaz and Naus (1991). We present below a refinement of this method. For $N = Lm + v$, $1 \le v \le m - 1$, we utilize the following representation: $P(W_{k,m} > N) =$

$$P\left(\bigcap_{i=1}^{L-1} B_i \cap B_L^*\right) = P(B_1) \prod_{i=2}^{L-1} P\left(B_i \Big| \bigcap_{j=1}^{i-1} B_j\right) P\left(B_L^* \Big| \bigcap_{j=1}^{L-1} B_j\right),$$

where, for $1 \le i \le L - 1$,

$$B_j = \bigcap_{j=1}^{m+1} \left(X_{(i-1)m+j} + \cdots + X_{im+j-1} \le k - 1\right),$$

and

$$B_L^* = \bigcap_{j=1}^{v+1} \left(X_{(L-1)m+j} + \cdots + X_{Lm+j-1} \le k - 1\right).$$

Using a Markov-like approximation (Glaz and Naus (1991), Chen and Glaz (1999)) for $P\left(\bigcap_{i=1}^{L-1} B_i \cap B_L^*\right)$, we get:

$$P(W_{k,m} > N)$$

$$\approx P(B_1) \prod_{i=2}^{L-1} P(B_i|B_{i-1}) P(B_L^*|B_{L-1})$$

$$= P(B_1 \cap B_2) \left[\prod_{i=3}^{L-1} \frac{P(B_{i-1} \cap B_i)}{P(B_{i-1})}\right] \frac{P(B_L^* \cap B_{L-1})}{P(B_{L-1})}$$

$$= q_{3m} \left(\frac{q_{3m}}{q_{2m}}\right)^{L-3} \left(\frac{q_{2m+v}}{q_{2m}}\right).$$

(7.2.8)

To derive a product-type approximation for $E(W_{k,m})$, we substitute in (7.2.2), for $N \geq 3m + 1$, the approximation for $P(W_{k,m} > N)$ given in (7.2.8) to get:

$$
\begin{aligned}
E(W_{k,m}) &\approx \sum_{j=0}^{3m} q_j + \sum_{v=1}^{m} \sum_{L=3}^{\infty} q_{3m} \left(\frac{q_{3m}}{q_{2m}} \right)^{L-3} \left(\frac{q_{2m+v}}{q_{2m}} \right) \\
&= \sum_{j=0}^{3m} q_j + \frac{q_{3m}}{q_{2m} - q_{3m}} \sum_{v=1}^{m} q_{2m+v} \\
&= \sum_{j=0}^{2m} q_j + \frac{q_{2m}}{q_{2m} - q_{3m}} \sum_{v=1}^{m} q_{2m+v}.
\end{aligned}
\tag{7.2.9}
$$

In Section 7.3, for selected values of n, p, k, m, and N, we evaluate the approximations and bounds for $P(W_{k,m} > N)$ and $E(W_{k,m})$.

7.3 NUMERICAL RESULTS

In this section we present numerical results for $P(W_{k,m} > N)$ and $E(W_{k,m})$ for selected values of the parameters for the binomial and Poisson models. We also discuss an example in quality control. In Tables 7.3.1–7.3.2 exact results are presented for the Bernoulli model. In Tables 7.3.3–7.3.4 and Tables 7.3.5–7.3.6, approximations and bounds are given for the binomial and Poisson models, based on (7.2.8)-(7.2.9) and (7.2.4)-(7.2.7) respectively. From the numerical results in Tables 7.3.3–7.3.6 it is evident that the approximations and bounds provide valuable information for the distribution of $W_{k,m}$. In Table 7.3.2, for the i.i.d. Bernoulli trials, $E(W_{k,m})$ could not be evaluated in two cases: $m = 10$, $k = 5$, $p = 0.05$ and $m = 50$, $k = 5$, $p = 0.01$. We used Fu (2001) to evaluate q_j, $1 \leq j \leq 3m$, and from inequalities (7.2.6) we get, for $p = 0.05$,

$$37040 \leq E(W_{5,10}) \leq 37132$$

and for $p = 0.01$

$$84317 \leq E(W_{5,50}) \leq 84429.$$

We now present a numerical example to illustrate the implementation of a quality control scheme based on moving sums of i.i.d. binomial random variables. Suppose that items arrive at an inspection facility in packed boxes, each containing $n = 25$ items. It is known that under normal circumstances the probability for a defective item is $p = 0.001$.

Assume that from practical considerations a quality control inspection scheme based on $m = 10$ is chosen and it has been decided that approximately $N = 200$ boxes will be inspected. We are interested in testing

$$H_0 : p = 0.001 \ vs \ H_a : p > 0.001.$$

Rejection of this null hypothesis will result in rejection of the shipped items. For the above testing problem one can employ the scan statistic S_m (Glaz and Naus (1991)). For a specified level α of probability of Type-I error, the above null hypothesis is rejected if $S_m \geq k$, where k is chosen so that

$$P(S_m \geq k) = P(W_{k,m} \leq N) = \alpha$$

or equivalently

$$P(W_{k,m} > N) = 1 - \alpha.$$

Now, for $\alpha = 0.01$ and $k = 4$, we get from (7.2.5):

$$0.9901 \leq P(W_{5,10} > 200) \leq 0.9901.$$

The accuracy of the bounds in (7.2.5) resulted in an accurate determination of the value of k.

7.4 CONCLUDING REMARKS

Numerical results presented in Tables 7.3.3–7.3.6 indicate that approximations and bounds for $P(W_{k,m} > N)$ and $E(W_{k,m})$ are quite accurate to be used in applications. From Tables 7.3.1–7.3.2 we can see that the exact results for i.i.d. Bernoulli trials based on Fu (2001) can be easily evaluated for most values of the parameters. There is some difficulty in evaluating $E(W_{k,m})$ when it is very large. In that case the inequalities given in (7.2.6) will provide accurate bounds.

In applications, the consecutive observations might not be independent. It would be of interest to extend these results to a sequence of dependent trials. One interesting model would be first-order homogeneous Markov dependent trials. For Bernoulli trials, the results in Fu (2001) can be extended to the case of Markov dependent trials. An algorithm has to be developed to evaluate efficiently $P(W_{k,m} > N)$ and $E(W_{k,m})$. For $0 - 1$ Markov dependent trials, Glaz (1983) derived simple bounds for $P(W_{k,m} > N)$ and $E(W_{k,m})$. The problem of deriving accurate bounds and approximations for Markov dependent binomial or Poisson random variables is much more complex.

Table 7.3.1: $P(W_{k,m} > N)$ for I.I.D. Bernoulli Trials

m	k	p	N	$P(W_{k,m} > N)$
10	3	.05	15	0.9745
			20	0.9617
		.10	15	0.8616
			20	0.8055
	5	.05	15	0.9998
			20	0.9997
		.10	15	0.9953
			20	0.9924
25	3	.05	37	0.7680
			50	0.6819
		.10	37	0.3374
			50	0.2217
	5	.05	37	0.9811
			50	0.9699
		.10	37	0.7977
			50	0.7158
50	3	.01	75	0.9695
			100	0.9545
		.05	75	0.3346
			100	0.2244
	5	.01	75	0.9995
			100	0.9992
		.05	75	0.7837
			100	0.7028
100	3	.01	150	0.8447
			200	0.7837
		.05	150	0.0306
			200	0.0098
	5	.01	150	0.9903
			200	0.9847
		.05	150	0.2187
			200	0.1277

Table 7.3.2: $E(W_{k,m})$ for I.I.D. Bernoulli Trials

m	k	p	$E(W_{k,m})$
10	3	0.05	399.13
		0.10	82.22
	5	0.05	
		0.10	1767.61
25	3	0.05	123.71
		0.10	37.31
	5	0.05	1209.48
		0.10	137.93
50	3	0.01	1718.25
		0.05	74.52
	5	0.01	
		0.05	263.93
100	3	0.01	747.47
		0.05	61.24
	5	0.01	9271.84
		0.05	119.23

Table 7.3.3: Approximations and Bounds for $P(W_{k,m} > N)$
for a Binomial Model

m	k	n	p	N	Lower Bound	$\hat{P}(W_{k,m})$	Upper Bound
5	2	5	0.01	10	0.9337	0.9337	0.9337
				20	0.8569	0.8583	0.8605
	3			10	0.9936	0.9936	0.9936
				20	0.9849	0.9849	0.9849
	2	10		10	0.7918	0.7918	0.7918
				20	0.5883	0.6000	0.6161
	3			10	0.9586	0.9586	0.9586
				20	0.9058	0.9065	0.9074
10	3	5		20	0.9561	0.9561	0.9561
				50	0.8717	0.8732	0.8755
	5			20	0.9993	0.9993	0.9993
				50	0.9977	0.9977	0.9977
	3	10		20	0.7884	0.7884	0.7884
				50	0.4776	0.5006	0.5328
	5			20	0.9861	0.9861	0.9861
				50	0.9557	0.9559	0.9562
5	2	5	0.05	10	0.3594	0.3594	0.3594
				20	0.0737	0.1119	0.1597
	3			10	0.6948	0.6948	0.6948
				20	0.4181	0.4418	0.4706
	2	10		10	0.0617	0.0617	0.0617
				20	0.0001	0.0029	0.0119
	3			10	0.2312	0.2312	0.2312
				20	0.0155	0.0416	0.0784
10	3	5		20	0.2216	0.2216	0.2216
				50	0.0015	0.0152	0.0597
	5			20	0.7102	0.7102	0.7102
				50	0.3225	0.3603	0.4087
	3	10		20	0.0086	0.0086	0.0086
				50	0.0000	0.0000	0.0006
	5			20	0.1249	0.1249	0.1249
				50	0.0000	0.0029	0.0251

Table 7.3.4: Approximations and Bounds for $E(W_{k,m})$
for a Binomial Model

m	k	n	p	Lower Bound	$\hat{E}(W_{k,m})$	Upper Bound
5	2	5	0.01	117.87	121.16	128.03
	3			1134.84	1138.39	1145.00
	2	10		35.64	38.13	44.01
	3			178.30	181.73	188.24
10	3	5		328.67	336.17	350.07
	5			18993.47	19001.29	19014.33
	3	10		65.44	70.85	82.45
	5			965.66	973.53	986.78
5	2	5	0.05	9.69	10.29	12.98
	3			22.60	24.56	29.30
	2	10		4.64	4.67	5.21
	3			7.58	7.84	9.46
10	3	5		14.35	14.88	18.20
	5			45.16	49.62	59.31
	3	10		6.57	6.58	6.79
	5			12.01	12.19	13.97

Table 7.3.5: Approximations and Bounds for $P(W_{k,m} > N)$
for a Poisson Model

m	k	λ	N	Lower Bound	$\hat{P}(W_{k,m})$	Upper Bound
5	2	0.10	10	0.7909	0.7909	0.7909
			20	0.6065	0.6333	0.6326
	3		10	0.9574	0.9574	0.9574
			20	0.9086	0.9144	0.9102
	2	0.25	10	0.3643	0.3643	0.3643
			20	0.0957	0.1456	0.1837
	3		10	0.6893	0.6893	0.6893
			20	0.4367	0.4777	0.4872
	2	0.50	10	0.0657	0.0657	0.0657
			20	0.0004	0.0060	0.0176
	3		10	0.2357	0.2357	0.2357
			20	0.0254	0.0611	0.0959
	2	1.00	10	0.0012	0.0012	0.0012
			20	0.0000	0.0000	0.0001
	3		10	0.0104	0.0104	0.0104
			20	0.0000	0.0002	0.0013
10	3	0.10	20	0.7870	0.7870	0.7870
			50	0.4854	0.5219	0.5392
	5		20	0.9853	0.9853	0.9853
			50	0.9546	0.9569	0.9551
	3	0.25	20	0.2263	0.2263	0.2263
			50	0.0026	0.0212	0.0676
	5		20	0.7018	0.7018	0.7018
			50	0.3227	0.3749	0.4091
	3	0.50	20	0.0097	0.0097	0.0097
			50	0.0000	0.0000	0.0009
	3		20	0.1287	0.1287	0.1287
			50	0.0000	0.0045	0.0298
	3	1.00	20	0.0000	0.0000	0.0000
			50	0.0000	0.0000	0.0000
	5		20	0.0004	0.0004	0.0004
			50	0.0000	0.0000	0.0000

Table 7.3.6: Approximations and Bounds for $E(W_{k,m})$
for a Poisson Model

m	k	λ	Lower Bound	$\hat{E}(W_{k,m})$	Upper Bound
5	2	0.10	36.20	44.77	44.81
	3		174.74	177.51	184.76
	2	0.25	9.93	11.02	13.72
	3		22.77	27.64	29.78
	2	0.50	4.71	4.76	5.53
	3		7.70	8.14	9.99
	2	1.00	2.51	2.51	2.56
	3		3.58	3.58	3.76
10	3	0.10	65.70	75.86	82.97
	5		922.45	930.23	943.63
	3	0.25	14.50	15.19	18.84
	5		44.47	51.39	58.85
	3	0.50	6.63	6.64	6.94
	5		12.10	12.33	14.40
	3	1.00	3.50	3.50	3.50
	5	5.00	5.52	5.52	5.56

ACKNOWLEDGMENT

The authors would like to thank the referees for their helpful suggestions.

REFERENCES

[1] Balakrishnan, N. and Koutras, M.V. (2001). *Runs and Scans with Applications*. Wiley: New York.

[2] Bogush, A.J. (1972). Correlated clutter and resultant properties of binary signals. *IEEE Trans. Aerosp. Elec. Sys.*, **9**, 20.

[3] Boutsikas, M.V. and Koutras, M.V. (2000). Reliability approximations for Markov chain imbeddable systems. *Method. Comp. Appl.*

Probab., **2**, 393-412.

[4] Chen, J. and Glaz, J. (1999). Approximations for the distribution and moments of discrete scan statistics. In *Scan Statistics and Applications* (J. Glaz and N. Balakrishnan, eds.), 27-66. Birkhäuser: Boston.

[5] Chu, W.W. (1970). Buffer behavior for Poisson arrivals and multiple synchronous constant outputs. *IEEE Trans. Comp.*, **19**, 530-534.

[6] Dinneen, G.P. and Reed, I.S. (1956). An analysis of signal detection and location by digital methods. *IRE Trans. Inform. Theory*, **2**, 29-39.

[7] Fredrikson, G.F.W. (1974). Buffer behavior for binomial input and constant service. *IEEE Trans. Commun.*, **22**, 1862-1866.

[8] Fu, J.C. (1986). Reliability of consecutive-k-out-of-n: F system with (k-1)-step Markov dependence. *IEEE Trans. Reliab.*, **35**, 603-602.

[9] Fu, J.C. (2001). Distribution of the scan statistic for a sequence of Bernoulli trials. *J. Appl. Probab.*, **38**, 908-916.

[10] Fu, J.C. and Hu, B. (1987). On reliability of a large consecutive-k-out-of-n: F system with (k-1)-step Markov dependence. *IEEE Trans. Reliab.*, **36**, 75-77.

[11] Fu, J.C. and Koutras, M.V. (1994). Distribution theory of runs: A Markov chain approach. *J. Amer. Statist. Assoc.*, **89**, 1050-1058.

[12] Glaz, J. (1983). Moving window detection for discrete data. *IEEE Trans. Inform. Theory*, **29**, 457-462.

[13] Glaz, J. and Naus, J.I. (1991). Tight bounds and approximations for scan statistic probabilities for discrete data. *Ann. Appl. Probab.*, **1**, 306-318.

[14] Glaz, J., Naus, J. and Wallenstein, S. (2001). *Scan Statistics.* Springer-Verlag: New York.

[15] Greenberg, I. (1970). On sums of random variables defined on a two-state Markov chain. *J. Appl. Probab.*, **13**, 604-607.

[16] Karwe, V. and Naus, J. (1997). New Recursive methods for scan statistic probabilities. *Comp. Statist. Data Anal.*, **23**, 389-404.

[17] Koutras, M.V. (1996). On a Markov chain approach for the study of reliability structures. *J. Appl. Probab.*, **33**, 357-367.

[18] Koutras, M.V. and Alexandrou, V.A. (1996). Runs, scans and urn model distributions: A unified Markov chain approach. *Ann. Inst. Statist. Math.*, **47**, 743-766.

[19] Mirstic, A.V. (1978). Multistatic radar binomial detection. *IEEE Trans. Aerosp. Elec. Sys.*, **14**, 103-108.

[20] Nelson, J.B. (1978). Minimal order models for false alarm calculations on sliding windows. *IEEE Trans. Aerosp. Elec. Sys.*, **15**, 352-363.

[21] Saperstein, B. (1973). On the occurrence of n successes within N Bernoulli trials. *Technometrics*, **15**, 809-818.

[22] Saperstein, B. (1976). The analysis of attribute moving averages: MIL-STD-105D reduced inspection plans. *Sixth Conf. Stoch. Proc. Appl.*, Tel Aviv.

Addresses for communication:

JIE CHEN, Department of Computing Services, University of Massachusetts, 100 Morrissey Boulevard, Boston, MA 02125, U.S.A. E-mail: jie.chen@umb.edu
JOSEPH GLAZ, Department of Statistics, University of Connecticut, UBox 4120, CLAS Building, 215 Glenbrook Road, Storrs, CT 06269-4120, U.S.A. E-mail: glaz@uconnvm.uconn.edu

Chapter 8

Estimation of the Slope in a Measurement-Error Model

SUJAY DATTA
Northern Michigan University, Marquette, Michigan

SAIBAL CHATTOPADHYAY
Indian Institute of Management, Calcutta, India

8.1 INTRODUCTION AND MOTIVATION

In this article, we discuss a real-life problem and its statistical solution via an interesting application of sequential analysis. First we describe the scenario. Early in the winter, one of the authors experienced problems with his home-heating system and thought that it might need replacement. At a local plumbing, heating and air-conditioning superstore, he was advised to look into the possibility that his home-insulation might not be functioning properly, particularly since the house was more than three decades old. Asked what might be a good indicator of the effectiveness of insulation, the store personnel suggested that one should check whether the indoor temperature was being affected significantly by the outdoor temperature. In other words, if the indoor temperature on a snowy and blustery day was significantly lower than that on a warmer day after the heating system stayed on for the same duration, starting from the same base temperature, it might mean that the house needed better insulation. The author was intrigued and thus began the data-collection phase of the present study. Over a period of nearly three months, both indoor and outdoor temperatures ($^{\circ}F$) were recorded once or twice a day with a handheld thermometer. Before recording the indoor temperature, the heating system would be turned off until the room-temperature dropped to $68^{\circ}F$ ($20^{\circ}C$) and then the heating system would be turned back on and kept running for nearly 3 hours. Meanwhile, the outdoor temperature at the *midpoint* of a three-hour interval would be measured using the same thermometer. Here is the resulting dataset:

Table 8.1.1: The Temperature Dataset

In	Out	In	Out	In	Out	In	Out	In	Out
76.0	37.9	76.0	40.5	75.0	34.0	72.0	39.0	74.2	29.2
79.5	40.5	73.0	42.0	78.0	39.5	76.0	36.4	75.3	31.8
77.0	35.0	71.5	29.0	69.0	26.0	73.5	37.0	76.5	33.1
73.5	34.0	77.8	37.0	74.0	35.0	75.6	32.3	74.5	30.5
71.0	26.5	75.6	38.5	77.7	37.0	74.9	39.0	72.8	28.8
75.5	36.5	71.0	26.4	76.5	35.8	69.5	24.5	72.6	28.5
71.8	22.0	76.0	32.0	73.5	28.0	74.2	31.5	75.0	30.1
69.5	22.0	73.0	28.0	71.0	28.0	74.0	30.0	75.6	30.8
74.5	37.0	79.6	44.5	75.0	39.1	78.5	38.5	76.2	32.0
75.0	34.0	72.3	36.2	72.2	24.0	72.2	30.5	75.2	30.5
71.5	28.0	74.0	33.5	72.2	26.5	70.5	23.5	74.3	29.1
78.0	40.5	71.0	25.0	72.8	29.5	71.5	26.5	72.5	27.8
77.2	34.5	70.3	25.7	71.8	29.0	73.0	28.6	72.8	28.5
69.7	25.0	69.5	23.2	73.1	26.5	75.0	28.0	73.6	29.0
75.8	30.4	75.5	30.0	76.6	32.0	73.4	29.2	73.4	28.7
72.8	28.0	73.0	28.0	74.8	29.5	75.3	30.8	76.1	32.2
75.9	32.1	75.9	32.0	76.0	33.5	77.0	34.4	76.5	32.8
76.8	33.5	77.0	34.0	74.5	31.7	77.6	35.1	78.0	37.1
74.8	34.1	74.1	33.6	73.8	33.0	73.8	32.8	73.4	32.2

Each day, temperatures were recorded either in the morning (outdoor around 7:30 A.M., indoor around 9:00 A.M.) or in the afternoon (outdoor around 3:00 P.M., indoor around 4:30 P.M.). Obviously, what we had in mind was a regression with the indoor temperature as the response variable (Y) and the outdoor temperature as the regressor (X). Under a regression setup, we might want to test a hypothesis that the slope is not zero. However, a confidence interval for the slope would be more useful. Such a confidence interval would serve not only as a tool for carrying out the hypothesis test, it would actually give us an idea as to how steeply the indoor temperature rises or falls with a unit change in the outdoor temperature. For example, if a 95% confidence interval for the slope turned out to be $(3.0°F, 4.5°F)$, it would be a more serious concern to a homeowner than a 95% interval such as $(0.3°F, 1.3°F)$. So, we decided to construct a confidence interval rather than performing a hypothesis test. In any case, both variables X and Y were *measured with errors*, since the thermometer used was not digital and was calibrated only to show whole-number temperature readings. That is, a temperature reading of, say, $79.5°F$ is only an eyeball approximation. So, a simple linear regression model with *errors in variables* (EIV) would be more appropriate rather than an ordinary regression model.

Regression models with errors in variables, also sometimes called *measurement-error models*, have been widely discussed in the literature. For a more detailed introduction, one may refer to Anderson (1976,1984), Fuller (1987), and Gleser (1981,1987). The basic setup is as follows: One observes pairs $\{(X_i, Y_i), i = 1, \ldots, n\}$ where

$$Y_i = \gamma + \beta\xi_i + \epsilon_i \; ; \; X_i = \xi_i + \delta_i \qquad (8.1.1)$$

with γ, β being constants, ϵ_i's i.i.d. distributed as $N(0, \sigma_\epsilon^2)$, δ_i's i.i.d. distributed as $N(0, \sigma_\delta^2)$, and ϵ_i's being independent of δ_i's. If the ξ_i's are fixed constants, this is called a *functional relationship* model, as opposed to a *structural relationship* where ξ_i's would also be assumed to have a probability distribution. Notice that under a functional relationship model, (X_i, Y_i) has a bivariate normal distribution with mean vector $(\xi_i, \gamma + \beta\xi_i)$ and dispersion matrix $\text{Diag}(\sigma_\delta^2, \sigma_\epsilon^2)$. On the other hand, if we assume in the structural relationship case that ξ_i's are i.i.d. $N(\xi, \sigma_\xi^2)$, then the *marginal* distribution of (X_i, Y_i) after ξ_i's are integrated out is bivariate normal with mean-vector $(\xi, \gamma + \beta\xi)$ and dispersion matrix $\begin{bmatrix} \sigma_\delta^2 + \sigma_\xi^2 & \beta\sigma_\xi^2 \\ \beta\sigma_\xi^2 & \sigma_\epsilon^2 + \beta^2\sigma_\xi^2 \end{bmatrix}$. Here, we adopt the latter formulation with ξ_i's following a Normal distribution. The reason is explained in the next section. Under this setup, assuming that $\gamma, \beta, \xi, \sigma_\xi^2, \sigma_\delta^2$ and σ_ϵ^2 are all unknown, we address the problem of estimating β, the slope-parameter. In order for β to be an identifiable parameter, we also need to assume that the ratio $\lambda = \sigma_\delta^2 / \sigma_\epsilon^2$ is known. How this assumption may be justified in the present case is explained in Section 8.2. Under this additional assumption, we consider *fixed-width* interval estimation of β. Intervals with fixed (or bounded) widths have been traditionally preferred since they eliminate the possibility of ending up with unacceptably wide intervals that are hardly of any use. One may refer to Finster (1983,1985) for further motivation.

8.2 JUSTIFICATION OF MODEL AND LITERATURE SURVEY

For the scenario described in Section 8.1, a *structural* EIV model seems more appropriate than a *functional* one, since there is no reason to believe that the outdoor temperature at 7:30 A.M. (or 3:00 P.M.) on a winter day is a *fixed* (unknown) constant plus an error term. It sounds more realistic if we postulate that the *true* (unknown) outdoor temperature at 7:30 A.M. is a random variable with some mean and variance and the *recorded* temperature at 7:30 A.M. on a particular day is the sum of a particular realization of this random variable and that of an independent error process. Now, since the existing literature about inference on a slope-parameter in a model of this kind relies heavily upon the assumption of normality for the two independent error processes, we decided to check this assumption. We did not find any blatant violation of this assumption (details are in the appendix). Also, since the same person was measuring both Y and X with the same instrument, it would be safe to assume that λ (that is, the variance-ratio mentioned in Section 8.1) is 1. Under this setup, we started looking for ways to construct a $100(1 - \alpha)\%$ confidence interval for the slope-parameter that would have a pre-specified bound for its width. This latter feature (that is, bounded width) is highly desirable for our purpose, even if we have to sacrifice an *exact* $100(1 - \alpha)\%$ confidence level and settle for an *approximately* $100(1 - \alpha)\%$ interval instead. Considering the main purpose of this study, it will be more useful to know that, for example, β lies in the range $(1.5, 2.5)$ rather than knowing that β is somewhere between 1 and 20 or simply that β exceeds 1.

Similar interval estimation problems have been discussed by a number of authors including Creasy (1956), Williams (1959), Gleser (1987), and Gleser and Hwang (1987). While the Creasy-Williams confidence set for β is exact, it has infinite length. This latter feature turns out to be inevitable for an *exact* confidence interval in a fixed-sample-size situation. Gleser and Hwang (1987) proved that no $100(1 - \alpha)\%$ confidence interval for β with finite (a.s.) length could maintain its nominal confidence level $(1 - \alpha)$ for all values of the unknown parameters. They showed that as $(\sigma_\xi^2/\sigma_\delta^2) \to 0$, the confidence coefficient of any finite-length confidence interval for β also tends to 0. Similar is the case when a finite-stage sampling scheme (for example, the two-stage sampling scheme of Mukhopadhyay (1980) or the three-stage scheme of Hall (1981)) is used. However, Gleser (1987) considered an *approximate* finite-length confidence interval centered at $\hat{\beta}_{MLE}$ by utilizing a consistent estimator of the variance of $\hat{\beta}_{MLE}$. He provided simulation-based lower bounds for the coverage-probability of the interval as a function of $\sigma_\xi^2/\sigma_\delta^2$, showing that the coverage-probability improved as $\sigma_\xi^2/\sigma_\delta^2$ and n increased.

Although Gleser (1987) has shown how to construct an *approximately* $100(1 - \alpha)\%$ confidence interval with finite width, unfortunately there is no known fixed-sample-size procedure for constructing a *fixed-width* or *bounded-width* confidence interval. More recently, Datta (1996) adopted a modified version of Gleser's (1987) approach and proposed a *sequential* one-at-a-time sampling scheme for this purpose. Using results from Aras and Woodroofe (1993), he derived asymptotic second-order expansions for the ASN of this procedure and for a lower bound of the associated coverage probability, as the width of the interval shrank to zero. However, a close examination of the article will reveal a serious drawback. The leading term in the second-order expansion was significantly less than the desired coverage probability $1 - \alpha$ and the simulated coverage probabilities almost always appeared to be lower than the target of 95%. So, in the context of our present problem, we set out to improve upon this procedure and found three different approaches to do away with the above-mentioned drawback. The first approach is based on an application of the Tchebysheff's inequality in much the same way as Mukhopadhyay and Datta (1996) or Datta and Mukhopadhyay (1997) did. The other two approaches involve a combination of the Tchebysheff's inequality and some kind of a resampling mechanism (bootstrap or jackknife) built into the boundary condition of a sequential stopping rule. These approaches, along with the heuristics of why they should work, are described in Section 8.3. Only for the first approach, we provide (Subsection 8.3.1) a brief sketch of how the asymptotic second-order properties can be derived. Similar derivations for the other two approaches have yet to be obtained. The emphasis of this article is on simulated performances of these three procedures for moderately large sample sizes and the results of applying them to a real-life problem. Tables summarizing the results of our extensive simulation studies are given in the appendix.

8.3 METHODOLOGIES AND ANALYSES

Under a *structural relationship* model, we are interested in constructing fixed-width confidence intervals for β. Given $d > 0$, Datta (1996) considered intervals of the form $I_n = (\hat{\beta}_{MLE}^{(n)} - d, \hat{\beta}_{MLE}^{(n)} + d)$ where $\hat{\beta}_{MLE}^{(n)}$ has the expression:

$$(2\lambda S_{XY}^{(n)})^{-1}\left[-(S_{XX}^{(n)} - \lambda S_{YY}^{(n)}) + \left\{(S_{XX}^{(n)} - \lambda S_{YY}^{(n)})^2 + 4\lambda (S_{XY}^{(n)})^2\right\}^{\frac{1}{2}}\right],$$

(8.3.1)

with

$$S_{XX}^{(n)} = \sum_{i=1}^{n}(X_i - \bar{X}_n)^2, \; S_{YY}^{(n)} = \sum_{i=1}^{n}(Y_i - \bar{Y}_n)^2, \text{ and}$$

$$S_{XY}^{(n)} = \sum_{i=1}^{n}(X_i - \bar{X}_n)(Y_i - \bar{Y}_n).$$

Here λ is the known variance-ratio defined in Section 8.1. For $\lambda = 1$, the $\hat{\beta}_{MLE}^{(n)}$ in (8.3.1) is also the *orthogonal least-squares* estimator of β. A consistent estimator of $V(\hat{\beta}_{MLE}^{(n)}) = \sigma_{\hat{\beta},n}^2$ is given by

$$\hat{\sigma}_{\hat{\beta},n}^2 = \frac{\{1 + \lambda(\hat{\beta}_{MLE}^{(n)})^2\}^2 \{S_{XX}^{(n)} S_{YY}^{(n)} - (S_{XY}^{(n)})^2\}}{(n-2)(\{S_{XX}^{(n)} - \lambda S_{YY}^{(n)}\}^2 + 4\lambda\{S_{XY}^{(n)}\}^2)}.$$

(8.3.2)

This, along with the central limit theorem and Slutsky's theorem, led Gleser (1987) to consider the following *approximate* $100(1 - \alpha)\%$ confidence interval for β:

$$\left(\hat{\beta}_{MLE}^{(n)} - t_{\alpha/2,n-2}\hat{\sigma}_{\hat{\beta},n} \;,\; \hat{\beta}_{MLE}^{(n)} + t_{\alpha/2,n-2}\hat{\sigma}_{\hat{\beta},n}\right).$$

(8.3.3)

Along the lines of Gleser (1987), Datta (1996) argued that in order for an interval of the form $(\hat{\beta}_{MLE}^{(n)} - d, \hat{\beta}_{MLE}^{(n)} + d)$ to have a confidence coefficient approximately $1 - \alpha$, one must have

$$d \geq t_{\alpha/2,n-2}\{V(\hat{\beta}_{MLE}^{(n)})\}^{1/2} = t_{\alpha/2,n-2}\sigma_{\hat{\beta},n},$$

(8.3.4)

that is, the sample-size has to be at least the smallest integer n_0 (say) such that the inequality in (8.3.4) holds with $n = n_0$. But this n_0

is of no practical importance, since it involves the unknown variance of $\hat{\beta}_{MLE}^{(n)}$. So, Datta (1996) proposed the following purely sequential procedure: One starts with $\{(X_i, Y_i) : i = 1, \ldots, m\}$ where

$$m \equiv m(d) = \max\{2 , \langle z_{\alpha/2}/d\rangle + 1\}, \qquad (8.3.5)$$

with $\langle x \rangle$ standing for the largest integer $< x$. Then one continues by observing one additional vector (X_i, Y_i) at a time until there are N_0 of them where $N_0 \equiv N_0(d)$ is given by:

$$N_0(d) = \inf\left\{ n \geq m : \frac{z_{\alpha/2}^2\{1 + \lambda(\hat{\beta}_{MLE}^{(n)})^2\}^2\{S_{XX}^{(n)}S_{YY}^{(n)} - (S_{XY}^{(n)})^2\}}{d^2(\{S_{XX}^{(n)} - \lambda S_{YY}^{(n)}\}^2 + 4\lambda\{S_{XY}^{(n)}\}^2)} \right\}.$$
$$(8.3.6)$$

At the stopped stage, using N_0 pairs of observations, one estimates β by the fixed-width confidence interval I_{N_0} mentioned at the beginning of this section, namely,

$$I_{N_0} = [\hat{\beta}_{MLE}^{(N_0)} - d , \hat{\beta}_{MLE}^{(N_0)} + d].$$

Because of the consistency of $\hat{\sigma}_{\hat{\beta},n}^2$ for the true variance of $\hat{\beta}_{MLE}^{(n)}$, it is easy to see from (8.3.6) that N_0 is finite a.s. and it diverges to ∞ a.s. as the semi-width d shrinks to 0. Also, taking limits as $d \to 0 (\Longleftrightarrow n_0 \to \infty)$ in the string of inequalities

$$\frac{(N_0 - 2)z_{\alpha/2}^2\hat{\sigma}_{\hat{\beta},N_0}^2}{d^2 n_0} \leq \frac{N_0}{n_0} \leq \frac{(N_0 - 3)z_{\alpha/2}^2\hat{\sigma}_{\hat{\beta},N_0-1}^2}{d^2 n_0} + \frac{m}{n_0},$$

which is a direct consequence of the way N_0 was defined, it can be easily shown that $N_0/n_0 \to 1$ a.s.. These are asymptotic *first-order* properties of N_0. Using tools of Aras and Woodroofe (1993), Datta (1996) also derived the following asymptotic *second-order* expansions.

Theorem 8.3.1 *Under the present setup, as $d \to 0$:*

(i) $E(N_0 - n_0) = \nu_0 + o(1)$;

(ii) $P(\beta \in I_{N_0}) \geq 1 - z_{\alpha/2}^{-2} - n_0^{-1}K_0 + o(d^2)$;

for suitable constants K_0 and ν_0.

As pointed out earlier, the leading term $1 - z_{\alpha/2}^{-2}$ is considerably lower than the target $1 - \alpha$. So, we decided to use the fact that the coverage

probability $P[\beta \in (\hat{\beta}_{MLE}^{(n)} - d , \hat{\beta}_{MLE}^{(n)} + d)] = 1 - P[| \hat{\beta}_{MLE}^{(n)} - \beta | \geq d]$
satisfies the Tchebysheff's inequality:

$$1 - P[| \hat{\beta}_{MLE}^{(n)} - \beta | \geq d] \geq 1 - \frac{E(\hat{\beta}_{MLE}^{(n)} - \beta)^2}{d^2}. \qquad (8.3.7)$$

As a result, in order for the coverage probability to be at least $1 - \alpha$, it is sufficient to have $\frac{E(\hat{\beta}_{MLE}^{(n)} - \beta)^2}{d^2} \leq \alpha$. Now, $E(\hat{\beta}_{MLE}^{(n)} - \beta)^2 = \text{MSE}(\hat{\beta}_{MLE}^{(n)}) = V(\hat{\beta}_{MLE}^{(n)}) + \text{bias}^2$, and if we decide to ignore the bias term (as it approaches 0 at a faster rate), then all we need to ensure is that $\frac{V(\hat{\beta}_{MLE}^{(n)})}{d^2} \leq \alpha$. So, if the *true* variance of $\hat{\beta}_{MLE}^{(n)}$ were known, the smallest $n(= n_1$, say) that satisfies this inequality would, in some sense, serve as the *optimal* sample size for this fixed-width interval estimation problem. But the *true* variance is not known. A way around this stumbling block would again be a *sequential* sampling scheme similar to (8.3.6):
One starts with $\{(X_i, Y_i) : i = 1, \ldots, m'\}$ where

$$m' \equiv m'(d) = \max\{2 , \langle 1/(\alpha^{1/2}d)\rangle + 1\}, \qquad (8.3.8)$$

with $\langle x \rangle$ once again meaning the largest integer $< x$. Then one continues by observing one vector (X_i, Y_i) at a time until there are N_1 of them where $N_1 \equiv N_1(d)$ is defined as

$$N_1(d) = \inf\left\{n \geq m : n \geq \frac{\{1 + \lambda(\hat{\beta}_{MLE}^{(n)})^2\}^2\{S_{XX}^{(n)}S_{YY}^{(n)} - (S_{XY}^{(n)})^2\}}{(\alpha d^2)(\{S_{XX}^{(n)} - \lambda S_{YY}^{(n)}\}^2 + 4\lambda\{S_{XY}^{(n)}\}^2)}\right\}. \qquad (8.3.9)$$

At the stopped stage, using N_1 pairs of observations, one estimates β by the fixed-width interval I_{N_1}. This new stopping rule enjoys similar asymptotic *first-order* properties as its predecessor. But using techniques from Aras and Woodroofe (1993), one can also prove that

Theorem 8.3.2 *Under the present setup, as $d \to 0$:*

(i) $E(N_1 - n_1) = \nu_1 + \text{o}(1);$

(ii) $P(\beta \in I_{N_1}) \geq 1 - \alpha - n_1^{-1}K_1 + \text{o}(d^2);$

for suitable constants ν_1 and K_1.

A brief outline of the proof is provided in Subsection 8.3.1. The constant K_1 involves various population parameters and, as will be clear from our simulation studies, it will be sometimes positive and sometimes negative, depending on the set of values of those parameters.

An alternative approach to ignoring the "bias^2" term in the MSE of $\hat{\beta}_{MLE}^{(n)}$ is to find an estimator of it and use it together with the variance-estimator (8.3.2) in computing an estimate of the MSE. This estimated MSE will then be used in forming a sequential stopping rule. We had a hunch that such a sequential procedure might perform at least as well as the one described above. Since we do not know a closed-form expression for the bias term, we consider a *bootstrap* estimator. From a sample of size n, we draw a large number of random subsamples of size $k (= 2, \ldots, n-1)$ with replacement and for each k, compute the average of all $\hat{\beta}_{MLE}^{(k)}$'s obtained from the corresponding subsamples (let us call it $\hat{\beta}_{AVG}^{(k,n)}$). The grand average $\hat{\beta}_{AVG}^{(n)}$ of all $\hat{\beta}_{AVG}^{(k,n)}$'s $(k = 2, \ldots, n-1)$ will be used in estimating the bias. The estimate of bias^2 will be $\widehat{\text{bias}}^2{}_{n,boot} = (\hat{\beta}_{AVG}^{(n)} - \hat{\beta}_{MLE}^{(n)})^2$. The new sequential stopping rule will now be as follows:

We start with $\{(X_i, Y_i) : i = 1, \ldots, m'\}$ where m' is as in (8.3.8). Then, we continue by observing one pair (X_i, Y_i) at a time until there are N_2 of them where

$$N_2 \equiv N_2(d) = \inf\left\{ n \geq m : \alpha \geq \frac{\hat{V}(\hat{\beta}_{MLE}^{(n)}) + \widehat{\text{bias}}^2{}_{n,boot}}{d^2} \right\}. \quad (8.3.10)$$

At the stopped stage, using N_2 pairs of observations, we estimate β by the fixed-width interval I_{N_2}. When we ran simulations to compare the performance of this procedure with that of the previous one for moderate sample sizes, we were pleased to find that it was indeed producing intervals with coverage probabilities comparable to (and sometimes higher than) those obtained from its competitor with significantly smaller sample sizes on an average. But on a moderately powerful computer (see the appendix for specific details), the implementation of the bootstrap-based procedure was consuming a great deal of computation-time. Although a more efficiently written code and a more powerful computer would probably reduce the run-time and render this promising procedure really useful in practice, we decided to look for an alternative procedure which shows a comparable performance and is easier to implement.

The quest for an alternative procedure made us think of the *jack-knife* estimate of the bias term. From a sample of size n, we *remove* one observation at a time and compute the MLE of β from the corresponding subsample of size $n-1$ (say, $\hat{\beta}_{MLE}^{(n-1),-i}$). The grand average of all $\hat{\beta}_{MLE}^{(n-1),-i}$'s (say, $\hat{\beta}_{AVG}^{(n-1)}$) will now be used in estimating the "bias²" term. The estimate bias²$_{n,jack} = (\hat{\beta}_{MLE}^{(n)} - \hat{\beta}_{AVG}^{(n-1)})^2$ will replace bias²$_{n,boot}$ in the boundary crossing condition of the sequential stopping rule (8.3.10) and give rise to a new stopping rule N_3. When we put this new procedure to test for moderate sample sizes, we were delighted to find that it took considerably less computing time than its bootstrap-based counterpart (ranging between several minutes to a couple of hours for 1000 iterations). Also, this procedure used even smaller sample sizes on an average, and yet produced intervals with coverage probabilities comparable to (often higher than) those produced by its competitors.

Although the theoretical derivations of the asymptotic properties of the two procedures based on bootstrap and jackknife techniques have yet to be accomplished, it is clear that they work. See the tables summarizing simulation results in the Appendix for a comparison among them.

8.3.1   A Sketch of the Proof of Theorem 8.3.2

The idea is to verify the six conditions of Aras and Woodroofe (1993) so that the results therein can be applied to this situation. We first define i.i.d. random vectors $U_1, U_2, \ldots$ such that

$$U_i' = (X_i - \mu_X, Y_i - \mu_Y, (X_i - \mu_X)^2 - \sigma_X^2, (Y_i - \mu_Y)^2 - \sigma_Y^2,$$

$$(X_i - \mu_X)(Y_i - \mu_Y) - \sigma_{XY}).$$

$$(8.3.11)$$

If G denotes the common distribution of the U_i's, then G has mean-vector $(0,0,0,0,0)$ and dispersion matrix $M = \begin{bmatrix} M_1 & O \\ O & M_2 \end{bmatrix}$ where

$$M_1 = \begin{bmatrix} \sigma_X^2 & \sigma_{XY} \\ \sigma_{XY} & \sigma_Y^2 \end{bmatrix} \text{ and}$$

$$M_2 = \begin{bmatrix} 2\sigma_X^4 & 2\sigma_{XY}^2 & 2\sigma_X^2\sigma_{XY} \\ 2\sigma_{XY}^2 & 2\sigma_Y^4 & 2\sigma_Y^2\sigma_{XY} \\ 2\sigma_X^2\sigma_{XY} & 2\sigma_Y^2\sigma_{XY} & \sigma_X^2\sigma_Y^2 + \sigma_{XY}^2 \end{bmatrix},$$

with $\sigma_X^2 = \sigma_\delta^2 + \sigma_\xi^2$, $\sigma_Y^2 = \sigma_\epsilon^2 + \beta^2\sigma_\xi^2$, $\sigma_{XY} = \beta\sigma_\xi^2$ and $\mathbf{O}$ denoting a null matrix of appropriate size. Next, on a suitable subset D of $\mathcal{R}^5$ (the choice of which will be made precise shortly), define a function $g : D \to \mathcal{R}$ in the following way:

$$(x_1, \ldots, x_5)$$

$$\longrightarrow \quad \left(\frac{(1+\lambda\beta^2)^2(\sigma_X^2\sigma_Y^2 - \sigma_{XY}^2)}{(\sigma_X^2 - \lambda\sigma_Y^2)^2 + 4\lambda\sigma_{XY}^2}\right)\left(\frac{g^{(1)}(x_1, \ldots, x_5)}{g^{(2)}(x_1, \ldots, x_5)}\right)$$

$$= \quad C_g \, g^{(1)}(x_1, \ldots, x_5)\{g^{(2)}(x_1, \ldots, x_5)\}^{-1}, \text{ say,}$$

$$(8.3.12)$$

where $g^{(1)}(x_1, \ldots, x_5) =$

$$\frac{\{(x_3 - x_1^2 + \sigma_X^2) - \lambda(x_4 - x_2^2 + \sigma_Y^2)\}^2 + 4\lambda(x_5 - x_1 x_2 + \sigma_{XY})^2}{(x_3 - x_1^2 + \sigma_X^2)(x_4 - x_2^2 + \sigma_Y^2) - (x_5 - x_1 x_2 + \sigma_{XY})^2}$$

$$(8.3.13)$$

and $g^{(2)}(x_1, \ldots, x_5) =$

$$\{2\lambda(x_5 - x_1 x_2 + \sigma_{XY})^2\}^{-2}\left[\{4\lambda(x_5 - x_1 x_2 + \sigma_{XY})^2 + \{(x_3 - x_1^2 + \sigma_X^2) - \lambda(x_4 - x_2^2 + \sigma_Y^2)\}^2 - \{(x_3 - x_1^2 + \sigma_X^2) - \lambda(x_4 - x_2^2 + \sigma_Y^2)\}\{(x_3 - x_1^2 + \sigma_X^2)(x_4 - x_2^2 + \sigma_Y^2) - (x_5 - x_1 x_2 + \sigma_{XY})^2\}^{1/2}\{g^{(1)}(x_1, \ldots, x_5)\}^{1/2}\right]^2,$$

$$(8.3.14)$$

and the domain D of the function g is that subset of $\mathcal{R}^5$ where both $g^{(1)}$ and the inverse of $g^{(2)}$ are defined. It is easy to observe that $g(0, 0, 0, 0, 0) = 1$ and g is twice continuously differentiable on some neighborhood of the origin. Finally, we define a sequence of real-valued functions $\{g_n\}_{n\geq 1}$ from D to $\mathcal{R}$ as:

$$(x_1, \ldots, x_5) \longrightarrow \begin{cases} C_g \, g^{(1)}(x_1, \ldots, x_5)\{g^{(2)}(x_1, \ldots, x_5)\}^{-1} & \text{for } n \leq \frac{1}{C_g} \\ C_g\left[\max\left\{\frac{1}{n}, \frac{g^{(2)}(x_1, \ldots, x_5)}{g^{(1)}(x_1, \ldots, x_5)}\right\}\right]^{-1} & \text{for } n > \frac{1}{C_g}. \end{cases}$$

$$(8.3.15)$$

Then $g_n \equiv g$ for all $n \geq 1$ on a sufficiently small neighborhood of the origin. In view of (8.3.9) and (8.3.11)-(8.3.15), one immediately recognizes that the stopping variable N_1 in (8.3.9) is similar in form to t_a in equation (2) of Aras and Woodroofe (1993) for suitably defined a and Z_n ($n \geq 1$) that satisfy equation (17) of their paper, with the

g_n's being as in (8.3.15). Also, since (X, Y) has a bivariate normal distribution, equation (12) of Aras and Woodroofe (1993) holds here for any $q > 0$. Hence, due to their Proposition 4, conditions (C4)-(C6) of Aras and Woodroofe (1993) are automatically taken care of with $\alpha = q/2$ (that is, any $\alpha > 0$) and $\xi = \frac{1}{2}\langle W , D^2(g) \mid_0 W \rangle$ where $W = (W_1, \ldots, W_5)' \sim N_5(0, M)$ and $D^2(g) \mid_0$ is the 5x5 matrix of second-order derivatives of the function g. Notice also that(C1) of Aras and Woodroofe (1993) is obviously true here for any positive p as the distribution G has mean 0 and the joint distribution of (X,Y) has finite joint moments of all positive orders. In order to verify condition (C2) of Aras and Woodroofe (1993) in this situation, observe that $Z_n = ng_n(\bar{U}_n)$ and hence $Z_n \leq n^2$ for all but a few small values of n. Further details are omitted. See, for example, the way in which condition (C2) has been verified in Datta (1996) or Mukhopadhyay and Datta (1996). In these two papers, as well as in Datta (1995), one will find how condition (C3) of Aras and Woodroofe (1993) can be verified using a result of Katz (1963) on the tail-probability of a distribution.

Having verified the conditions, we are now in a position to apply Theorem 4 of Aras and Woodroofe (1993) with a real-valued function h defined as:

$$(x_1, \ldots, x_5) \longrightarrow -h^{(1)}(x_1, \ldots, x_5)/h^{(2)}(x_1, \ldots, x_5) + h^{(1)}(x_1, \ldots, x_5)/\lambda^{1/2} - \beta^2 \qquad (8.3.16)$$

where we write $h^{(1)}(x_1, \ldots, x_5) = \{(x_3 - x_1^2 + \sigma_X^2) - \lambda(x_4 - x_2^2 + \sigma_Y^2)\}$, $h^{(2)}(x_1, \ldots, x_5) = 2\lambda(x_5 - x_1 x_2 + \sigma_{XY})$. Thus we obtain the second-order expansion promised in our Theorem 8.3.2 and this completes the proof of our theorem.

8.4   CONCLUDING REMARKS

We have three procedures to apply on our temperature data and get confidence intervals for the slope. However, we admit that in this case, data were *not* gathered using a sequential design. This is because we gathered data before deciding how we would analyze it, and the methodologies described in Section 8.3 did not exist at the time of data collection. But now that the methodologies are documented, one may plan ahead and gather data following an appropriate sampling design in future applications. Nevertheless, what we have done here is not unprecedented in the literature. We believe that it is a legitimate approach in a situation like this where the very nature of a problem necessitates the development of a new methodology involving a sequential design — a fact that might be unknown at the data collection stage.

In any case, since we had only a limited amount of data, we sought a confidence interval with semi-width 1.0 so that with $\lambda = 1$, the *optimal* fixed sample-size for the first sequential procedure (8.3.6) would not be "too large". If we had the opportunity to collect more data, we certainly would have liked to construct a narrower interval. Two of the three methodologies *stopped* (only the bootstrap-based one did not) and we obtained two 95% confidence intervals for β. The stopping rule (8.3.9) stopped with the 89th observation whereas the jackknife-based procedure stopped with the 86th observation. The resulting intervals were (-0.703, 1.297) and (-0.686, 1.314), respectively. Looking at these intervals, it appears that the outdoor temperatures do *not* affect the indoor temperatures seriously enough for us to conclude that the house immediately needs additional insulation.

Finally, just for the sake of curiosity, we wanted to examine if using a different permutation of our original dataset, we could terminate the bootstrap-based procedure. This is a legitimate idea to explore, since the observation-pairs in the original dataset were i.i.d. (in particular, exchangeable). Luckily, we did not have to grope in the darkness to find an appropriate permutation — just using the original dataset in the reverse order did it! The bootstrap-based procedure stopped at the 92nd observation and the resulting interval was (-0.741, 1.259) which seems to be in general agreement with the other two intervals.

8.5 APPENDIX

Here, we first explain how we verified the normality assumption for the outdoor and indoor temperatures. For each set of temperatures separately, we created a histogram and a normal Q-Q plot. None of these showed any blatant violation of the normality assumption. Then, in order to reinforce the conclusion from the diagrams, we carried out a number of different normality tests (namely, the Anderson-Darling test, the Shapiro-Wilk test and the Kolmogorov-Smirnov test). In each case, the p-value turned out to be greater than 0.1.

Next, we provide summaries of simulation studies for the stopping rules N_1, N_2 and N_3 respectively. FORTRAN-77 codes were written using Visual Fortran 5.0 on an IBM Thinkpad 1300i computer and IMSL subroutines were used to generate standard normal random deviates. Mentioned on top of each table are the set of parameter values that were used. It should be noted that $\tau'^2 = \sigma_\xi^2/\sigma_\delta^2$. Each table displays a number of different values of the semi-width d and for each value of d, the sequential sampling scheme with stopping rule N_i ($i = 1, 2, 3$) was replicated 1000 times. The average sample size ($\bar{N_i}$) from those 1000 iterations is reported (along with the standard deviation), and so is the estimated coverage probability $\bar{p}$ (along with its standard deviation). Also reported are the starting sample sizes (m') and, only for procedure (8.3.9), the *optimal* fixed sample-size n_1 for each value of d (to be compared with the corresponding $\bar{N}$ value in order to adjudge the efficiency of a sequential procedure). For the bootstrap-based procedure, due to much longer run-times, extensive simulation results could not be reported but there is enough to draw general conclusions.

A quick comparison among simulated performances of the three sequential procedures for moderate sample sizes reveals that they generally produce intervals with coverage probabilities at least $1 - \alpha$. We saw an exception in a few cases where slightly lower coverage probabilities were obtained. There is no evidence suggesting that one of the procedures is a clear winner in terms of coverage probabilities, but both the bootstrap-based and jackknife-based procedures appear to use somewhat *smaller* sample sizes on an average than the sampling scheme (8.3.9). All three procedures, however, are using considerably smaller sample sizes than n_1. The jackknife-based procedure seems to be performing the best in terms of maintaining a low sample size, but again, these are just observations based on 1000 simulations for each set of parameter-values and the superiority (or otherwise) of any procedure to its competitors will ultimately have to be established analytically.

ACKNOWLEDGMENT

We thank two referees for their helpful remarks.

REFERENCES

[1] Anderson, T.W. (1976). Estimation of linear functional relationships: approximate distributions and connection with simultaneous equations in econometrics (with discussions). *J. Roy. Statist. Soc.*, *Ser. B*, **38**, 1-36.

Table 8.5.1: The Sequential Procedure (8.3.9)
$$\sigma_\epsilon^2 = 3.0, \ \xi = 5.0, \ (\gamma, \beta) = (5.0, \ -2.0)$$

λ	τ'^2	n_1	d	m'	$\bar{N}_1$	$s(\bar{N}_1)$	$\bar{p}$	$s(\bar{p})$
1.0	0.5	100	1.6733	3	63.5550	2.0908	0.946	0.0071
1.0	0.5	200	1.1832	4	155.5280	3.6564	0.923	0.0084
1.0	0.5	400	0.8367	6	345.1230	5.3882	0.933	0.0079
1.0	0.5	800	0.5916	8	782.5090	6.6799	0.991	0.0029
1.0	1.0	100	1.0954	5	77.4140	1.7502	0.944	0.0073
1.0	1.0	200	0.7746	6	185.2410	2.5967	0.958	0.0063
1.0	1.0	400	0.5477	9	386.8400	3.2356	0.995	0.0022
1.0	1.0	800	0.3873	12	794.6170	4.2920	1.000	0.0000
1.0	2.0	100	0.7416	7	89.8970	1.3210	0.978	0.0046
1.0	2.0	200	0.5244	9	197.0070	1.7444	0.998	0.0014
1.0	2.0	400	0.3708	13	393.0100	2.3092	1.000	0.0000
1.0	2.0	800	0.2622	18	797.3650	3.1341	1.000	0.0000
1.0	4.0	100	0.5123	9	95.1260	1.0006	0.996	0.0020
1.0	4.0	200	0.3623	13	198.6980	1.3578	0.999	0.0010
1.0	4.0	400	0.2562	18	395.6220	1.8026	1.000	0.0000
1.0	4.0	800	0.1811	25	797.9190	2.5989	1.000	0.0000
4.0	0.5	100	1.3784	4	65.0930	2.0201	0.938	0.0076
4.0	0.5	200	0.9747	5	162.0470	3.4556	0.903	0.0094
4.0	0.5	400	0.6892	7	369.1470	4.6414	0.959	0.0063
4.0	0.5	800	0.4873	10	790.1390	5.6500	1.000	0.0078
4.0	1.0	100	0.9487	5	79.1200	1.7044	0.935	0.0078
4.0	1.0	200	0.6708	7	189.2090	2.3989	0.969	0.0055
4.0	1.0	400	0.4743	10	390.4270	3.0114	0.998	0.0014
4.0	1.0	800	0.3354	14	794.0260	4.2487	1.000	0.0000
4.0	2.0	100	0.6614	7	91.1010	1.2970	0.977	0.0047
4.0	2.0	200	0.4677	10	197.8700	1.6392	0.998	0.0014
4.0	2.0	400	0.3307	14	394.3200	2.2269	1.000	0.0000
4.0	2.0	800	0.2338	20	796.8150	3.1958	1.000	0.0000
4.0	4.0	100	0.4643	10	95.3370	1.0060	0.998	0.0014
4.0	4.0	200	0.3283	14	199.0670	1.3033	0.999	0.0009
4.0	4.0	400	0.2322	20	395.6660	1.8323	1.000	0.0000
4.0	4.0	800	0.1642	28	797.7740	2.5514	1.000	0.0000

Table 8.5.2: The Bootstrap-Based Sequential Procedure (8.3.10)
$\sigma_\epsilon^2 = 3.0$, $\xi = 5.0$, $(\gamma, \beta) = (5.0, -2.0)$

λ	τ'^2	d	m'	$\bar{N}_2$	$s(\bar{N}_2)$	$\bar{p}$	$s(\bar{p})$
1.0	0.5	1.6733	3	61.0330	2.1376	0.888	0.00009
1.0	0.5	1.1832	4	152.3970	3.5806	0.932	0.00003
1.0	0.5	0.8367	6	344.0990	5.0067	0.960	0.00002
1.0	0.5	0.5916	8	779.0280	8.0973	0.998	0.00001

Table 8.5.3: The jackknife-Based Sequential Procedure
$\sigma_\epsilon^2 = 3.0$, $\xi = 5.0$, $(\gamma, \beta) = (5.0, -2.0)$

λ	τ'^2	d	m'	$\bar{N}_3$	$s(\bar{N}_3)$	$\bar{p}$	$s(\bar{p})$
1.0	0.5	1.6733	3	50.749	2.0086	0.875	0.0001
1.0	0.5	1.1832	4	137.577	3.5886	0.876	0.0001
1.0	0.5	0.8367	6	360.792	5.2035	0.950	0.00004
1.0	0.5	0.5916	8	782.826	6.6519	0.992	0.000008
1.0	1.0	1.0954	5	74.776	1.7342	0.935	0.00006
1.0	1.0	0.7746	6	180.090	2.4955	0.966	0.00003
1.0	1.0	0.5477	9	388.875	3.1665	0.996	0.000004
1.0	2.0	0.7416	7	89.907	1.2617	0.984	0.000016
1.0	2.0	0.5244	9	190.778	1.7754	0.995	0.000005
1.0	2.0	0.3708	13	393.963	2.3013	1.000	0.000000
1.0	4.0	0.5123	9	93.840	1.0293	0.994	0.000006
1.0	4.0	0.3623	13	194.605	1.3810	0.999	0.000001
1.0	4.0	0.2562	18	395.813	1.8124	1.000	0.000000
4.0	0.5	1.3784	4	63.750	2.0113	0.942	0.000055
4.0	0.5	0.9747	5	158.007	3.2762	0.907	0.000084
4.0	0.5	0.6892	7	369.147	4.6414	0.959	0.000039
4.0	1.0	0.9487	5	79.334	1.6573	0.938	0.000058
4.0	1.0	0.6708	7	184.122	2.3230	0.968	0.000031
4.0	1.0	0.4743	10	390.842	2.9967	0.999	0.000001
4.0	2.0	0.6614	7	89.518	1.2694	0.978	0.000021
4.0	2.0	0.4677	10	191.525	1.7235	0.994	0.000006
4.0	2.0	0.3307	14	394.345	2.2212	1.000	0.000000

[2] Anderson, T.W. (1984). Estimating linear statistical relationships. *Ann. Statist.*, **12**, 1-45.

[3] Aras, G. and Woodroofe, M. (1993). Asymptotic expansions for the moments of a randomly stopped average. *Ann. Statist.*, **21**, 503-519.

[4] Creasy, M.A. (1956). Confidence limits for the gradient in the linear functional relationship. *J. Roy. Statist. Soc., Ser. B*, **18**, 65-69.

[5] Datta, S. (1995). On multistage parametric inference-procedures: The "fine-tuning" aspect and the distribution-free scenario. *Ph.D. Dissertation*, Dept. of Statist., Univ. of Conn., Storrs.

[6] Datta, S. (1996). Sequential fixed-precision estimation in stochastic linear regression and errors-in-variables models. *Tech. Rep., Dept. of Statist., Univ. of Mich.*, Ann Arbor, Michigan.

[7] Datta, S. and Mukhopadhyay, N. (1997). On sequential fixed-size confidence regions for the mean vector. *J. Multivar. Anal.*, **60**, 233-251.

[8] Finster, M. (1983). A frequentistic approach to sequential estimation in the general linear model. *J. Amer. Statist. Assoc.*, **78**, 403-407.

[9] Finster, M. (1985). Estimation in the general linear model when the accuracy is specified before data collection. *Ann. Statist.*, **13**, 663-675.

[10] Fuller, W.A. (1987). *Measurement Error Models*. Wiley: New York.

[11] Gleser, L.J. (1981). Estimation in a multivariate "errors in variables" regression model: Large sample results. *Ann. Statist.*, **9**, 24-44.

[12] Gleser, L.J. (1987). Confidence intervals for the slope in a linear errors-in-variables regression model. In *Advances in Multivariate Statistical Analysis* (K. Gupta, ed.), 85-109. D. Reidel: Dordrecht.

[13] Gleser, L.J. and Hwang, J.T. (1987). The nonexistence of $100(1 - \alpha)\%$ confidence sets of finite expected diameter in errors-in-variables

and related models. *Ann. Statist.*, **15**, 1351-1362.

[14] Hall, P. (1981). Asymptotic theory of triple sampling for estimation of a mean. *Ann. Statist.*, **9**, 1229-1238.

[15] Katz, M.L. (1963). The probability in the tail of distribution. *Ann. Math. Statist.*, **34**, 312-318.

[16] Mukhopadhyay, N. (1980). A consistent and asymptotically efficient two-stage procedure to construct fixed-width confidence intervals for the mean. *Metrika*, **27**, 281-284.

[17] Mukhopadhyay, N. and Datta, S. (1996). On sequential fixed-width confidence intervals for the mean and second-order expansions of the associated coverage probabilities. *Ann. Inst. Statist. Math.*, **48**, 497-507.

[18] Williams, E.J. (1959). *Regression Analysis*. Wiley: New York.

Addresses for communication:

SUJAY DATTA, Department of Mathematics, Statistics & Computer Science, Northern Michigan University, 1401 Presque Isle Avenue, Marquette, MI 49855, U.S.A. E-mail: sdatta@euclid.nmu.edu
SAIBAL CHATTOPADHYAY, Indian Institute of Management Calcutta, Joka, Diamond Harbour Road, Kolkata 700104, West Bengal, India. E-mail: chattopa@iimcal.ac.in

Chapter 9

Kernel Density Estimation of Wool Fiber Diameter

BASIL M. DE SILVA
RMIT University, Melbourne, Australia

NITIS MUKHOPADHYAY
University of Connecticut, Storrs, U.S.A.

9.1   INTRODUCTION

During auctions, farmers bring bales of wool fiber to the Australian marketplace and hope to get a good price which largely depends on the fiber diameter (X). Buyers normally pay a higher price for wool with smaller fiber diameter. Prior to auction every lot made up of wool receives a tag displaying the mean and the coefficient of variation (standard deviation/mean) of the fiber diameter X. Usually, buyers get a sense of the quality or "fineness" of wool sample by looking at the mean and the coefficient of variation of the fiber diameter.

Wool is often sold in one "lot" that may consist of many bales. Before the auction may begin, farmers hand over their lots to the *Australian Wool Testing Authority* (AWTA) for inspection and calibration. AWTA takes samples from the cores of the randomly selected bales from each lot and measures the wool fiber diameter. The fiber diameters are determined with the help of machines using laserscan. The measurement data generated by laserscan machines are then summarized and displayed alongside a lot of wool. The displayed information (Appendix C) normally provides a histogram of fiber diameter together with the average and coefficient of variation of fiber diameter. Information on some covariates such as the "percentage of vegetable matter" and "greasyness" are also displayed before auction but the price for a bale seems to be driven mostly by the distribution of fiber diameter X.

For an overview, we mention that spinning fineness of wool proposed by Butler and Dolling (1995) is a more involved measure of wool's quality than its mean fiber diameter X alone. Using the model given in Butler and Dolling (1995), *spinning fineness* (F) is defined as:

$$F = \sqrt{\frac{125}{161}\mu_X^2 \left(1 + 5V_X^2\right)}, \qquad (9.1.1)$$

where μ_X is the mean diameter and V_X is the coefficient of variation. However, since this measure has not yet been adopted by AWTA and F is currently not used in the determination of wool's quality, we continue to work exclusively with the variable X.

In 2001, a fourth year student, Angela Pezic, from the RMIT University in Melbourne did a project on *"Counting Precision for Spinning Fineness Under Non-Normal Distribution"*. She managed to gather and analyze a number of large datasets with the cooperation of the AWTA. We consider only one such dataset to illustrate our analysis.

The data under scrutiny consists of sample size 4000 $(= n)$ and the percentage $(= p)$ of wool fiber of certain diameters in a lot. These percentages are given for fiber of certain diameter of $5, 6, \ldots, 80$ microns. Many lots do not have any wool with fiber diameter less than 7 microns or greater than 48 microns. We converted the percentages into frequencies (np) and focused on wool fiber diameter ranging between 7 and 48 microns. Then, we looked upon this large dataset to constitute our relevant population. With the help of S-PLUS, a random permutation of the data was first chosen and this was then treated as a source of the incoming x values as needed. From this data, we obtained x values by simple random sampling and without replacement. Since the size of the data is large, we take liberty in ignoring the finite-population-correction term. The goal is to approximate the associated population distribution via density estimation with a sense of preassigned accuracy. These are explained shortly.

Let us think of $X_1, X_2, \cdots, X_n, \ldots$, a sequence of independent random variables or observations on wool fiber diameter from a large population of such measurements having some unknown continuous density function $f(x)$. We aim at simultaneously estimating $f(x)$ for $x = 7, \ldots, 48$.

9.2 THE FORMULATION

The well-known kernel estimator of the density function $f(x)$ has the form

$$\widehat{f}_{n,h_n}(x) = \frac{1}{nh_n} \sum_{i=1}^{n} K\left(\frac{x - X_i}{h_n}\right) \quad \text{for } -\infty < x < \infty. \quad (9.2.1)$$

Let us make the standard assumptions regarding the kernel density $K(\cdot)$ as in Isogai (1987a), namely that (i) $K(\cdot)$ is a bounded probability density function on the real line, (ii) $\lim_{|u| \to \infty} |u| K(u) = 0$, (iii) $\int_{-\infty}^{\infty} uK(u)du = 0$, and (iv) $\int_{-\infty}^{\infty} u^2 K(u)du < \infty$. Let us denote

$$B = \int_{-\infty}^{\infty} K^2(u)du. \quad (9.2.2)$$

Here $K(\cdot)$ is known as the kernel and h_n is called the band-width. The accuracy of the estimator $\widehat{f}_{n,h_n}(x)$ for a large sample size n depends upon the choices of the kernel $K(\cdot)$ and the band-width h_n as well as the unknown nature of the density $f(x)$. We suppose that the band-width $h_n = n^{-r}$ for $\frac{1}{5} < r < 1$ which is a standard practice.

Our goal is to estimate the unknown density function $f(x)$ simultaneously at $q(\geq 1)$ distinct points $x_1, \ldots, x_q$ with the preassigned *simultaneous confidence coefficient* $1 - \alpha, 0 < \alpha < 1$. Obviously, $f(x_j)$ will be estimated by the estimator $\widehat{f}_{n,h_n}(x_j), j = 1, \ldots, q$. Let us first define three q-dimensional column vectors $\boldsymbol{x}, \boldsymbol{f}(\boldsymbol{x})$ and $\widehat{\boldsymbol{f}}_{n,h_n}(\boldsymbol{x})$ where

$$\boldsymbol{x}' = (x_1, x_2, \ldots, x_q), \boldsymbol{f}'(\boldsymbol{x}) = (f(x_1), f(x_2), \ldots, f(x_q)), \text{ and}$$
$$\widehat{\boldsymbol{f}}'_{n,h_n}(\boldsymbol{x}) = \left(\widehat{f}_{n,h_n}(x_1), \widehat{f}_{n,h_n}(x_2), \ldots, \widehat{f}_{n,h_n}(x_q)\right). \quad (9.2.3)$$

From equations (4.73) and (4.75) of Hall (1992), we can immediately claim that

$$\mathbf{E}\left[\widehat{f}_{n,h_n}(x_j)\right] = f(x_j) + O(h_n^2) \text{ and}$$
$$\mathbf{V}\left[\widehat{f}_{n,h_n}(x_j)\right] = B(nh_n)^{-1}f(x_j) + o\left((nh_n)^{-1}\right), \text{ for } j = 1, \ldots, q, \quad (9.2.4)$$

whatever be the kernel function $K(\cdot)$ satisfying the properties (i)-(iv). The first part compares favorably with Isogai's (1987a) equation (3.1), but Isogai's (1987a) variance expression was different from ours. One should note, however, that Isogai (1987a) had used a more complicated *recursive density estimator* rather than the simple kernel estimator defined in (9.2.1). For the sake of simplicity, we continue working with the more traditional estimator from (9.2.1).

Suppose that we employ the standard normal kernel $K(u) = \phi(u)$ where $\phi(u) = \{\sqrt{2\pi}\}^{-1}\exp(-u^2/2)$ with $-\infty < u < \infty$. This special choice of the kernel function satisfies the standing assumptions (i)-(iv) with $B = \{2\sqrt{\pi}\}^{-1}$. Next, we claim that

$$
\begin{aligned}
\mathrm{Cov}\left(\widehat{f}_{n,h_n}(x_j), \widehat{f}_{n,h_n}(x_l)\right) &= o\left((nh_n)^{-1}\right) + O(n^{-1}) \\
&= o\left((nh_n)^{-1}\right) \text{ for } j \neq l = 1, \ldots, q,
\end{aligned}
\tag{9.2.5}
$$

which is a new result as far as we can tell. For completeness, we sketch proofs of (9.2.4) and (9.2.5) in Appendix A. Now, using the multivariate central limit theorem we can write

$$
\frac{(nh_n)^{1/2}\left[\widehat{f}_{n,h_n}(x) - f(x)\right]}{\sqrt{B}} \xrightarrow{\mathcal{L}} N_q(0, \Sigma),
\tag{9.2.6}
$$

where Σ is a $q \times q$ diagonal matrix, namely $\mathrm{diag}\,(f(x_1), \ldots, f(x_q))$, which is assumed positive definite. The positive definiteness of the Σ matrix is equivalent to assuming that $f(x_1), \ldots, f(x_q)$ are all positive. A sequential version was proved for a sequence of recursive density estimators by Isogai (1984).

We have mentioned before that AWTA typically displays histograms of observed wool diameters. Obviously, such pictures do not provide any associated sense of accuracy or the level of confidence one may be able to assign while estimating $f(x)$ for some fixed x. So, for fixed x, one may opt for a confidence interval for $f(x)$ having a predetermined confidence coefficient. But, it is well-known that a customary confidence interval can be wider than what one can be fruitfully used for making inferences. So, from the very beginning, one would rather ask for a confidence interval having (i) a predetermined confidence coefficient *and* (ii) a predetermined width. That way, in practice, one would expect to come up with a short confidence interval with a high probability of coverage.

In a practical application, one may like to estimate $f(x_j)$ precisely for each $j = 1, \ldots, q$. Let $d(> 0)$ be a preassigned number and we propose the following simultaneous fixed-size spherical confidence region for $f(x)$:

$$\mathcal{R}_n = \left\{ f(x) : \sum_{j=1}^q \left[\widehat{f}_{n,h_n}(x_j) - f(x_j) \right]^2 \le d^2 \right\}. \qquad (9.2.7)$$

We also require that the simultaneous confidence coefficient associated with the confidence region $\mathcal{R}_n$ is approximately at least $1 - \alpha$ where $0 < \alpha < 1$ is also preassigned, that is we wish to conclude:

$$\mathcal{P}_f \left\{ \sum_{j=1}^q \left[\widehat{f}_{n,h_n}(x_j) - f(x_j) \right]^2 \le d^2 \right\} \gtrsim 1 - \alpha, \qquad (9.2.8)$$

for a large sample size, n.

Now, we note that for any fixed $d(> 0)$, the statement

$$\left(\widehat{f}_{n,h_n}(x) - f(x) \right)' \Sigma^{-1} \left(\widehat{f}_{n,h_n}(x) - f(x) \right) \le d^2 \qquad (9.2.9)$$

is obviously equivalent to the following statement:

$$\sum_{j=1}^q \left\{ \left[\widehat{f}_{n,h_n}(x_j) - f(x_j) \right]^2 / f(x_j) \right\} \le d^2. \qquad (9.2.10)$$

It is also clear, however, that (9.2.10) implies

$$\sum_{j=1}^q \left\{ \left[\widehat{f}_{n,h_n}(x_j) - f(x_j) \right]^2 \right\} \le d^2 \max_{1 \le j \le q} f(x_j), \qquad (9.2.11)$$

and hence we combine (9.2.9)-(9.2.11) to claim that

$$\mathcal{P}_f \left\{ \left(\widehat{f}_{n,h_n}(x) - f(x) \right)' \Sigma^{-1} \left(\widehat{f}_{n,h_n}(x) - f(x) \right) \le d^2 \right\}$$
$$\le \mathcal{P}_f \left\{ \sum_{j=1}^q \left[\widehat{f}_{n,h_n}(x_j) - f(x_j) \right]^2 \le d^2 \max_{1 \le j \le q} f(x_j) \right\}. \qquad (9.2.12)$$

Now, let Y be a random variable that has the chi–square distribution with q degrees freedom and suppose that $\chi^2_{q,\alpha}$ denotes its upper $100\alpha\%$ point. Then, denoting $f_{\max} = \max_{1 \le j \le q} f(x_j)$, we can rewrite (9.2.12) as follows:

$$\mathcal{P}_f \left\{ \sum_{j=1}^q \left[\widehat{f}_{n,h_n}(x_j) - f(x_j) \right]^2 \leq d^2 \right\}$$

$$\geq \mathcal{P}_f \left\{ \left(\widehat{\boldsymbol{f}}_{n,h_n}(\boldsymbol{x}) - \boldsymbol{f}(\boldsymbol{x}) \right)' \Sigma^{-1} \left(\widehat{\boldsymbol{f}}_{n,h_n}(\boldsymbol{x}) - \boldsymbol{f}(\boldsymbol{x}) \right) \leq [d^2/f_{\max}] \right\}$$

$$\approx \mathcal{P} \left\{ Y \leq \frac{n h_n}{B} [d^2/f_{\max}] \right\}, \text{ for large } n, \text{ in view of the limiting}$$

distribution from (9.2.6) and with $B = \{2\sqrt{\pi}\}^{-1}$

$$\geq 1 - \alpha.$$

$$(9.2.13)$$

Now, let $n = n_0$ be the smallest positive integer such that

$$n h_n B^{-1} \left[d^2 / \max_{1 \leq j \leq q} f(x_j) \right] \geq \chi^2_{q,\alpha}$$

$$\Leftrightarrow n_0 \equiv n_0(d) \approx \left\{ \chi^2_{q,\alpha} B \left(\max_{1 \leq j \leq q} f(x_j) \right) d^{-2} \right\}^{\frac{1}{1-r}}, \qquad (9.2.14)$$

since $h_n = n^{-r}$. It should be obvious that

$$\lim_{n_0 \to \infty} \mathcal{P}_f \left\{ \sum_{j=1}^q \left[\widehat{f}_{n_0,h_{n_0}}(x_j) - f(x_j) \right]^2 \leq d^2 \right\} \geq 1 - \alpha, \qquad (9.2.15)$$

but we do realize that the magnitude of n_0 remains unknown.

9.3 LITERATURE REVIEW AND SPECIFIC AIMS

Silverman's (1986) monograph gave an up-to-date development in the non-sequential area of nonparametric density estimation. For a general overview of bootstrapping methodology, one may refer to Efron and Tibshirani (1993) and Hall (1992).

We refer back to (9.2.14) and recall that an expression for the "optimal" fixed sample-size $n_0(d)$ is known but its magnitude remains unknown. This is why a multi-stage sampling methodology is called for in the first place. We show how an appropriate two-stage methodology can be implemented. In doing this, we also examine closely the role of a smooth bootstrapped version of the two-stage estimation methodology.

The problem of sequential density estimation was first considered by Yamato (1971). Wegman and Davies (1975) presented some sequential procedures which satisfied some type of error control. Carroll (1976) considered the problem of estimating an unknown density function $f(x)$ at one fixed point x_0, which may or may not be known. Carroll's paper proposed two new classes of sequential stopping rules to construct a fixed-width confidence interval for $f(x_0)$. Martinsek (1993) constructed confidence bands with prescribed width and confidence for an unknown density function. Xu and Martinsek (1995) investigated a fully sequential procedure for an unknown density function on a finite interval. Martinsek and Xu (1996) developed a fixed-width confidence band for the density of data observed under right censoring. Kundu and Martinsek (1997) had worked under the L_1 distance. A systematic overview of the sequential density estimation problems was given in Mukhopadhyay (1997).

de Silva et al. (2000) extended the ideas from Carroll (1976) and Isogai (1987a) to develop a fixed-width confidence interval procedure and critically examined its consistency property for estimating $f(x_0)$. Subsequently, de Silva and Mukhopadhyay (2001) compared fixed-width confidence intervals for $f(x_0)$ obtained by means of purely sequential, accelerated sequential, two-stage and three-stage procedures.

In a series of papers, Isogai (1980/81,1984,1987a,b,1988) had developed important aspects of sequential nonparametric density estimation problems. In these papers, Isogai handled a recursive form of the kernel density estimator which is more complicated than the traditional form described in (9.2.1).

9.3.1 Specific Aims

Our aim remains modest. We focus on demonstrating practical usefulness of a two-stage kernel density estimator of the form described in (9.2.1). The present article first explores some of the sticky aspects of choosing the design parameters that are essential in practical implementation of a two-stage procedure for constructing the confidence region $\mathcal{R}_n$ from (9.2.7). Recall that we require the confidence coefficient $\mathcal{P}_f \{ f(x) \in \mathcal{R}_n \}$ to be approximately at least $1 - \alpha$. The problem of "optimal" band-width (h_n) selection is also addressed and we do so via bootstrapping. Subsequently, we exploit the idea of smooth bootstrapping to significantly reduce oversampling that is inherent within nearly all two-stage methodologies.

Initially, we examine performances of the proposed methodology with the help of computer simulations. Then, we turn around and use these methodologies to simultaneously estimate the wool fiber density $f(x_j)$ where $x_j = 6 + j, j = 1, 2, \ldots, 42(= q)$.

9.4 A TWO-STAGE PROCEDURE

Recall that we had fixed $K(u) = \phi(u)$. Then, in view of the expression of $n_0(d)$ from (9.2.14), we propose the following stopping rule for a two-stage procedure along the line of de Silva et al. (2000). Let us denote

$$N \equiv N(d)$$

$$= \max \left\{ m, \left\langle \left\{ \chi^2_{q,\alpha} B \left(\max_{1 \leq j \leq q} \widehat{f}_{m,h_m}(x_j) + m^{-1} \right) d^{-2} \right\}^{\frac{1}{1-r}} \right\rangle + 1 \right\}$$

$$(9.4.1)$$

where $< u >$ stands for the largest integer $< u$ and $B = \int_{u=-\infty}^{\infty} K^2(u) du = \{2\sqrt{\pi}\}^{-1}$.

In the initial stage, we start with the observations $X_1, \ldots, X_m$ of size m and estimate the required final sample size $n_0(d)$ by N. If $N = m$, then we need no more observations in the second stage. But, if $N > m$, then we record additional observations $X_{m+1}, \ldots, X_N$ of size $N - m$ in the second stage. Finally, we propose the following simultaneous fixed-size spherical confidence region for $f(x)$:

$$\mathcal{R}_N \equiv \mathcal{R}_N(d) = \left\{ f(x) : \sum_{j=1}^{q} \left[\widehat{f}_{N,h_N}(x_j) - f(x_j) \right]^2 \leq d^2 \right\}. \quad (9.4.2)$$

In an application of the two-stage procedure (9.4.1), the experimenter needs to select suitable values for the design constants r and m for fixed preassigned values of d, α and $x_1, \ldots, x_q$. It is very important to use the best available choices for these two design constants. The following two subsections address these issues.

9.4.1 Band-Width Selection

As a general practice, having observed a fixed sample of size n, the band-width $h \equiv h_n$ is chosen in such a way that the *mean integrated squared error* (MISE), defined by

$$\text{MISE}(h) = \mathbf{E} \int_{-\infty}^{\infty} \left[\widehat{f}_{n,h}(x) - f(x) \right]^2 dx, \qquad (9.4.3)$$

is minimized. If $f(x)$ was normal with variance σ^2, then Taylor (1989) proved that the choice $h = 1.06\sigma n^{-1/5}$ indeed minimized the MISE given in (9.4.3). One could perhaps use another approach to determine an appropriate band-width. While it may be of some value to compare effects of different ways to choose band-width on density estimation, that is not the primary focus here. The MISE criterion is conceptually easy to understand and it has been used extensively within the density estimation literature. So, we continue with the MISE related criteria to select band-width.

In the present situation, however, we go along bootstrap band-width selection technique suggested by Taylor (1989) and Faraway and Jhun (1990) and minimize the bootstrap estimator of MISE, namely,

$$\widehat{\text{MISE}}^*(h) = \mathbf{E}_* \int_{-\infty}^{\infty} \left[\widehat{f}_{n,h}^*(x) - \widehat{f}_{n,h}(x) \right]^2 dx. \qquad (9.4.4)$$

Let $X_1^*, \cdots, X_n^*$ be a bootstrap sample from the original data $X_1, \cdots, X_n$ and we write $\widehat{f}_{n,h}^*(x)$ to denote $(nh)^{-1}\sum_{i=1}^{n} K\left[(x - X_i^*)/h\right]$. An explicit expression for (9.4.4) is given in Shao and Tu (1995).

Taylor (1989) gave a simplified expression of $\widehat{\text{MISE}}^*(h)$ when a standard normal kernel was used. In the case when $K(u) = \phi(u)$, Taylor (1989) obtained:

$$\begin{aligned}
\widehat{\text{MISE}}^*(h) = & \frac{1}{2nh\sqrt{\pi}} + \frac{1}{2n^2 h}\sum_{i=1}^{n}\sum_{j=1}^{n}\left[\phi\left(\frac{X_i - X_j}{2h}\right) \right. \\
& \left. - \frac{4}{\sqrt{3}}\phi\left(\frac{X_i - X_j}{\sqrt{3}h}\right) + \sqrt{2}\phi\left(\frac{X_i - X_j}{\sqrt{2}h}\right) \right].
\end{aligned} \qquad (9.4.5)$$

Using the Taylor series expansion, the expression of $\widehat{\text{MISE}}^*(h)$ given in (9.4.5) can be expanded as

$$\widehat{\text{MISE}}^*(h) = \frac{1}{2nh\sqrt{\pi}} + \frac{1}{2n^2\sqrt{2\pi}}\sum_{p=0}^{\infty}\frac{(-1)^p a_p}{h^{2p+1}}g_p(X_1,\ldots,X_n) \qquad (9.4.6)$$

where

$$g_p(X_1,\ldots,X_n) = \sum_{i=1}^{n}\sum_{j=1}^{n}(X_i - X_j)^{2p} \text{ and}$$
$$a_p = 2^{-3p} - (4/\sqrt{3})6^{-p} + 2^{(1-4p)/2}$$

Clearly, $g_p(X_1, \ldots, X_n)$ and a_p are positive entities and these are both decreasing as p increases. Hence, for any given set of data $X_1, \ldots, X_n$, it can be proved that $\widehat{\text{MISE}}^*(h)$ decreases as h increases. That is, if

$$\min_h \widehat{\text{MISE}}^*(h) = \widehat{\text{MISE}}^*(h^*) \qquad (9.4.7)$$

then we view h^* as the "optimal" choice for h. Note that in a sequential framework, $h \equiv h_n = n^{-r}$ for $5^{-1} < r < 1$ so that $h^* \equiv \max_r h$ corresponds to the choice $r \equiv r^*$, the minimum possible value of r. Thus r^*, the "optimal" value of r is given by

$$r^* = \tfrac{1}{5} + \varepsilon = 0.2 + \varepsilon \qquad (9.4.8)$$

where ε is a small positive number. In the practical application of the procedure, we set $\varepsilon = 0.0001$.

9.4.2   Initial Sample Size Selection

Mukhopadhyay (1980) gave a specific choice for the initial sample size m. One may refer to Mukhopadhyay and Solanky (1994, Section 2.2.2) or Ghosh et al. (1997, Section 6.2.2) for more details about this procedure. In this procedure the initial sample size m is determined using the following expression

$$m \equiv m(d) = \max \left\{ 2, \left\langle \left(\chi_{q,\alpha}^2 B d^{-2} \right)^{1/\{(1-r)(1+\gamma)\}} \right\rangle + 1 \right\}, \qquad (9.4.9)$$

where $d(> 0)$ is held fixed. Here, the design constant γ is a positive number.

Here $m(d)$ is only the pilot sample size whereas the final sample size $N(d)$ is meant to estimate $n_0(d)$ given by (9.2.14). We note that both $m(d), n_0(d)$ converge to infinity as $d \to 0$. But since we have used fixed $\gamma(> 0)$ in (9.4.9), it is clear that the ratio $m(d)/n_0(d)$ converges to zero as $d \to 0$. In other words, when d is small, ultimately one is going to have a large final sample size so that in such a case, one may very well start with a large pilot size $m(d)$. But then it should be reassuring to know that $m(d)$ would be large too and yet it would be negligible compared with $n_0(d)$.

In the context of wool fiber density f, we frequently find that estimating it at a point x within appropriate $\pm d_0$ value serves the purpose. We remain mindful of such d_0 and then let $d = d_0\sqrt{q}$. For the fixed value of $d_0 = 0.005$ where $d = d_0\sqrt{q}$, we select $r = r^* = 0.2001$ and $q = 42$, that is, $d = 0.005\sqrt{42} = 0.032$. Figure 9.4.1 shows how m decreases as a function of γ. From this figure, we note that the choice of m remains practically unchanged if γ exceeds 2.

We remind ourselves that at this point, AWTA is unlikely to favor sampling techniques involving many steps. Hence, we focus on implementing a two-stage procedure that takes a practitioner away from sampling in one step. The choice of γ or equivalently that of the initial sample size m is very crucial. We realize, however, that m should not be too large or too small.

Using Figure 9.4.1, we find that a choice of $\gamma \in [1, 2]$ may be associated with some reasonable magnitude of m that can be used in implementing the two-stage procedure (9.4.1). In our current exercise, fixing $\gamma = 1.5$ together with $r = 0.2001, q = 42, \alpha = 0.05, \chi^2_{42,0.05} = 58.12, B = \{2\sqrt{\pi}\}^{-1}$ and $d_0 = 0.005$, we came up with $m = 126$. Note that $m = 126$ may not be the optimal initial sample size but it will give a reasonable starting point. For different preassigned values of α, q, r, γ and $d = d_0\sqrt{q}$, one can determine the magnitude of suitable m using the same approach. The S-PLUS function **m.fun** given in Appendix B.1, will compute m for given α, q, r, γ and d_0 values.

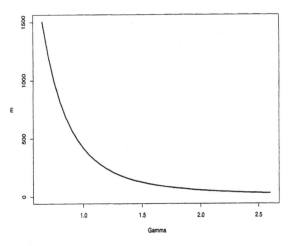

Figure 9.4.1: Stage One Sample Size Versus γ.

9.4.3   Properties of the Methodology

Combining the standard techniques from Mukhopadhyay (1980) and those of Isogai (1987a,b), we can verify the following large-sample properties of the proposed two-stage nonparametric density estimation methodology (9.4.1):

As the sample size $n_0 \equiv n_0(d) \to \infty$, we can *conclude*,

$$
\begin{aligned}
&\text{(i)} \quad \frac{N(d)}{n_0(d)} \xrightarrow{\mathcal{P}} 1, \mathbf{E}_f\left[\frac{N(d)}{n_0(d)}\right] \to 1; \text{ and} \\
&\text{(ii)} \quad \liminf \mathcal{P}_f\left\{(f(x_1), \ldots, f(x_q)) \in \mathcal{R}_{N(d)}\right\} \geq 1 - \alpha,
\end{aligned}
\qquad (9.4.10)
$$

where the fixed-size spherical confidence region $\mathcal{R}_{N(d)}$ was defined in (9.4.2). To prove part (ii), we need (9.2.6) plus the material from equations (9.2.9) - (9.2.15), and the verification that the sequence of density estimators $\left\{\widehat{f}_{n,h_n}(x); n \geq 1\right\}$ satisfies Anscombe's (1952) tightness condition. See Isogai (1987a,b).

The property (i) suggests that the final sample size N is expected to be in a close proximity of the "optimal" fixed sample size n_0 whereas the achieved confidence coefficient is expected to exceed $1 - \alpha$. But, these conclusions hold when both n_0 and N are extremely large. In practice, for moderate sample sizes, however, $\mathbf{E}_f[N]$ is seen to far exceed n_0 in computer simulations. This is referred to as "oversampling". In the next section, we apply bootstrapping techniques to reduce "oversampling".

9.5   SMOOTH BOOTSTRAPPING

Two-stage stopping variable given in (9.4.1) is based on the normal approximation from (9.2.6) and we obtain (9.4.1) by approximating

$$
Y = \frac{mh_m}{B} \sum\nolimits_{j=1}^{q} \left(\widehat{f}_{m,h_m}(x_j) - f(x_j)\right)^2 / f(x_j)
\qquad (9.5.1)
$$

by chi–square distribution with q degrees of freedom. Here, we replace the chi–square critical value, $\chi^2_{q,\alpha}$ in (9.4.1) by the corresponding bootstrap critical value.

In this method, one first draws a bootstrap sample $X'_1, X'_2, \ldots, X'_m$ with replacement from the set of initial observations of size m where m is chosen as in (9.4.9). Then, we apply the smoothing technique proposed by Grund and Hall (1995) and Silverman and Young (1987). A smooth bootstrap sample $X_1^*, X_2^*, \ldots, X_m^*$ is given by

$$X_i^* = X_i' + \delta_i h_m^* \quad \text{for } i = 1, 2, \ldots, m. \tag{9.5.2}$$

Here δ_i's are independent random observations from the kernel density $K(\cdot)$. Define

$$U_i^* = m h_m \sum_{j=1}^{q} \left(\widehat{f^*}_{m, h_m^*}(x_j)_i - \widehat{f}_{m, h_m^*}(x_j) \right)^2 \tag{9.5.3}$$

where $\widehat{f^*}_{m, h_m^*}(x_j)_i$ is the ith smooth bootstrap estimate of $\widehat{f}_{m, h_m^*}(x_j)$ for $i = 1, 2, \ldots, n_b$ and n_b is the number of the bootstrap replications. If

$$\xi_\alpha = [n_b(1 - \alpha)]^{\text{th}} \text{ largest value of } \left\{ U_{(1)}^*, U_{(2)}^*, \ldots, U_{(n_b)}^* \right\}. \tag{9.5.4}$$

where $U_{(i)}^*$'s are the ordered values of U_i^* for $i = 1, 2, \ldots, n_b$. Then, ξ_α is the upper $50\alpha\%$ non-standardized bootstrap critical value. Thus, from (9.2.14) and (9.4.1), we have

$$\xi_\alpha \text{ is a bootstrap estimator of } \chi_{q, \alpha}^2 B \max_{1 \le j \le q} \widehat{f}_{m, h_m}(x_j).$$

Hence, a bootstrap estimator of the final sample size, is given by

$$N^* \equiv N^*(d) = \max \left\{ m, \left\langle \{ \xi_\alpha / d^2 \}^{\frac{1}{1-r}} \right\rangle + 1 \right\}. \tag{9.5.5}$$

If $N^* = m$, then we need no more observations in the second stage. But, if $N^* > m$, then we record additional observations $X_{m+1}, \ldots, X_{N^*}$ of size $N^* - m$ in the second stage. Finally, we propose the following simultaneous fixed-size spherical confidence region:

$$\mathcal{R}_{N^*} \equiv \mathcal{R}_{N^*}(d) = \left\{ f(x) : \sum_{j=1}^{q} \left[\widehat{f}_{N^*, h_{N^*}}(x_j) - f(x_j) \right]^2 \le d^2 \right\} \tag{9.5.6}$$

as we did in (9.4.2).

9.6 A SIMULATION STUDY

Frequently, data collected from engineering or biological investigations tend to be positively skewed. Experimenters often use Weibull, gamma, or lognormal distributions to model such data. In some cases, customary goodness-of-fit tests fail to reject these distributions, indicating that each distribution under consideration may be used as a model for the data. Such a situation was discussed in Lim (1991). Wool fiber diameter data described in Section 9.1 also fall into this category. So, our aim is to estimate the density of wool fiber diameter using the proposed methodology.

Before we apply the nonparametric kernel density estimation procedure to wool data, we test it using simulated data from a Weibull distribution (shape parameter $= 1.3$, scale parameter $= 4.0$), a gamma distribution (shape parameter $= 1.668$, rate $= 1/2.223$), and a lognormal distribution (normal mean $= 1.0712$, normal variance $= 0.47109$). These distributions have the same mean ($= 3.6943$) and variance ($= 8.215$). Figures 9.6.1–9.6.3 show that the probability densities associated with these three distributions are very close to each other.

As explained in Section 9.4.2, by setting $\gamma = 1.5, r = 0.2001, \alpha = 0.05, q = 42$ and then picking $d_0 = 0.005$ (that is, $d = 0.032$) we have the initial sample size $m = 126$. Thus, we generated initial observations of size 126 from the chosen distributions in the first stage using S-PLUS. Recall that N^* (or N) stands for the final sample size of the two-stage procedure with (or without) bootstrapping. Let $n_b = 1000$ be the number of bootstrap replications. For this part of our investigation with simulated data, we fixed $x_j = 0.2j$ with $j = 1, ..., 42(= q)$. Table 9.6.1 gives the observed values of N and N^* for the generated data. It is clear that the bootstrap procedure from Section 9.5 significantly reduced oversampling when we compared it with the non-bootstrapped version from Section 9.4.

The S-PLUS function **Two.fun** given in Appendix B.2 determines N using a two-stage procedure without bootstrapping whereas the S-PLUS function **TwoBoot.fun** given in Appendix B.3 determines N^* for the bootstrapped version of our two-stage procedure. In these simulations, we found that 1000 bootstraps were enough for the desired results to converge. The S-PLUS function **Density.fun** given in Appendix B.4 computes the kernel density function for given data set and x_j values. Note that here $\gamma = 1.5$, $r = 0.2001$, $\alpha = 0.05$, $q = 42$ and $d_0 = 0.005$, thus using the S-PLUS function **m.fun** we again have $m = 126$.

Inclusion of these S-PLUS functions is expected to make practical implementation of the proposed sampling designs a little smoother. These user-friendly functions may also be modified quite easily, if necessary.

Table 9.6.1: Final Sample Sizes

Distribution	n_0	N	N^*
Weibull	21028	25787	16031
Gamma	22630	23297	14527
Lognormal	31205	35149	15510

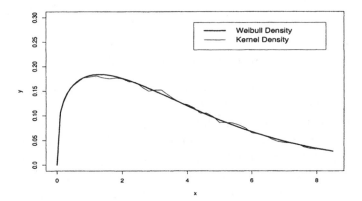

Figure 9.6.1: Weibull Function.

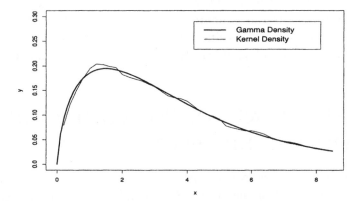

Figure 9.6.2: Gamma Function.

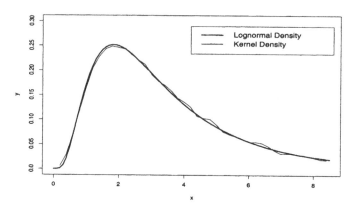

Figure 9.6.3: Lognormal Function.

In the case of each theoretical distribution, since N^* came out sizably smaller than N (see Table 9.6.1), we decided to implement only the bootstrapped two-stage methodology from Section 9.5. Figures 9.6.1–9.6.3 present the theoretical probability density functions and their estimated kernel density functions obtained by using the simulated data. These figures clearly show that the estimated and theoretical densities stay "close" to each other as required by the preset level of accuracy. Table 9.6.1 also displays the optimal sample size n_0 from (9.2.14) for the three distributions. While comparing these values with the corresponding two-stage sample size (N), we obviously saw that the two-stage procedure over-sampled by considerable amount. Note that both n_0 and N utilized $\max_{1 \leq j \leq q} f(x_j)$, and therefore these could be considerably higher than what we might have required for the analysis. On the other hand, the bootstrap estimate of the sample size (N^*) utilizes only the critical point and not specifically $\max_{1 \leq j \leq q} f(x_j)$. Thus, N^* values in Table 9.6.1 are found significantly lower than the corresponding n_0 values.

9.7   APPLICATION TO WOOL FIBER DIAMETER DATA

Our main thrust of this exercise has been to estimate the probability densities of wool fiber diameter at $7, 8, \ldots, 48$ microns. As in the case of our simulation study in Section 9.6, we take initial observations of size

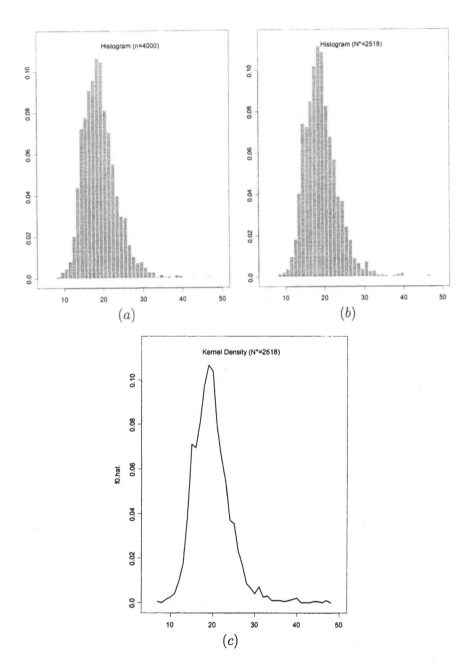

Figure 9.7.1: Wool Fiber Diameter Data.

$m = 126$, and we fix $q = 42$, $d_0 = 0.005$, and $d = 0.032$. Next, we determined N, the sample size without bootstrapping, using the S-PLUS function **Two.fun** and obtained the required observations. Similarly, we determined N^*, the sample size with bootstrapping, using the S-PLUS function **TwoBoot.fun** and obtained the required observations. The results are given in Table 9.7.1. Clearly, the application of bootstrapping significantly reduced oversampling in the two-stage procedure.

Table 9.7.1: Final Sample Size for Wool Data

Two-Stage without Bootstraps	$N = 13627$
Two-Stage with 1000 Bootstraps	$N^* = 2518$

Since N^* turned out much smaller than N, we report on the bootstrapped methodology alone. In the second stage, $2392 (= 2518 - 126)$ new observations were recorded in a single batch and the density function was estimated using the combined data of size $N^* = 2518$.

A histogram of fiber diameter values (dataset of size $n = 4000$) recorded by AWTA from a lot consisting of a number of bales that had arrived from the same agricultural farm is given in Figure 9.7.1.a. We emphasize that AWTA actually recorded these 4000 measurements. A histogram of the data obtained from the bootstrap two-stage procedure is given in Figure 9.7.1.b and a smoothed version of the estimated kernel density is shown in Figure 9.7.1.c. The three figures appear incredibly similar!

9.8 CONCLUDING REMARKS

In the case of the particular fiber data under investigation, our finding suggests that AWTA could estimate the density of fiber diameter having prespecified accuracies with a sample of size $N^* = 2518$ instead of making $n = 4000$ measurements. That amounts to 37.05% savings in the sample size alone which immediately translates into significant savings in expended real time needed for data collection and other costly resources. We have checked several sets of data from other lots of wool fiber and we found that the bootstrapped two-stage methodology from Section 9.5 saves between 30% and 40% of AWTA's measurements. This gain in efficiency, seen in the context of data collection and density estimation, is certainly substantial. Whether AWTA will be inclined to experiment with an appropriately truncated (at 4000) version of the bootstrapped two-stage methodology from Section 9.5 remains to be seen.

APPENDICES

Appendix A. Biases, Variances, and Covariances of Density Estimator

Appendix A1. Expansion of Bias from (9.2.4)

For brevity, we simply write x instead of x_j. We assume that the kernel $K(u)$ is not necessarily given by $\phi(u)$, but $K(u)$ satisfies the basic conditions (i)-(iv). We can express $\mathrm{E}\left[\widehat{f}_{n,h_n}(x)\right]$ as

$$
\begin{aligned}
\mathrm{E}&\left[\frac{1}{nh_n}\sum\nolimits_{i=1}^{n}K\left(\frac{x-X_i}{h_n}\right)\right] \\
&= \frac{1}{h_n}\int_{y=-\infty}^{\infty}K\left(\frac{x-y}{h_n}\right)f(y)dy \\
&= \frac{1}{h_n}\int_{u=\infty}^{-\infty}K\left(u\right)f(x-h_n u)(-h_n)du \text{ with } x-y=h_n u \\
&= \int_{u=-\infty}^{\infty}K\left(u\right)f(x-h_n u)du \\
&= \int_{u=-\infty}^{\infty}K\left(u\right)\left[f(x)-\tfrac{h_n u}{1!}f^{(1)}(x)+\tfrac{(-h_n u)^2}{2!}f^{(2)}(x)-\ldots\right]du \\
&= f(x)+\tfrac{1}{2}h_n^2 f^{(2)}(x)\int_{u=-\infty}^{\infty}u^2 K\left(u\right)du-\ldots,
\end{aligned}
\tag{A.1}
$$

since $\int_{u=-\infty}^{\infty}K\left(u\right)du = 1$ and $\int_{u=-\infty}^{\infty}uK\left(u\right)du = 0$. In other words, from (A.1), we can write

$$
\mathrm{E}\left[\widehat{f}_{n,h_n}(x)\right] = f(x)+O(h_n^2) = f(x)+O(n^{-2r}), \tag{A.2}
$$

for a single fixed point x.

Appendix A2. Expansion of Variance from (9.2.4)

For brevity, we simply write x instead of x_j. We assume that the kernel $K(u)$ is not necessarily given by $\phi(u)$, but $K(u)$ satisfies the basic conditions (i)-(iv). Next, we can express $\mathrm{V}\left[\widehat{f}_{n,h_n}(x)\right]$ as

$$\mathbf{V}\left[\frac{1}{nh_n}\sum_{i=1}^{n}K\left(\frac{x-X_i}{h_n}\right)\right]$$

$$=\frac{1}{nh_n^2}\left\{\mathbf{E}\left[K^2\left(\frac{x-X_1}{h_n}\right)\right]-\mathbf{E}^2\left[K\left(\frac{x-X_1}{h_n}\right)\right]\right\}. \tag{A.3}$$

Observe from (A.1) that we have already shown $\mathbf{E}\left[K\big((x-X_1)/h_n\big)\right]=O(h_n)$. Now, we obtain

$$\mathbf{E}\left[K^2\left(\frac{x-X_i}{h_n}\right)\right]=\int_{y=-\infty}^{\infty}K^2\left(\frac{x-y}{h_n}\right)f(y)dy$$

$$=\int_{u=\infty}^{-\infty}K^2\left(u\right)f(x-h_nu)(-h_n)du \text{ with } x-y=h_nu$$

$$=h_n\int_{u=-\infty}^{\infty}K^2\left(u\right)f(x-h_nu)du$$

$$=h_n\int_{u=-\infty}^{\infty}K^2\left(u\right)\left[f(x)-\tfrac{h_nu}{1!}f^{(1)}(x)+\tfrac{(-h_nu)^2}{2!}f^{(2)}(x)-\ldots\right]du$$

$$=Bf(x)h_n+O(h_n^2). \tag{A.4}$$

At this point, we combine (A.3)-(A.4) to write

$$\mathbf{V}\left[\widehat{f}_{n,h_n}(x)\right]=\tfrac{1}{nh_n^2}[Bf(x)h_n+O(h_n^2)]=\tfrac{1}{nh_n}Bf(x)+O(\tfrac{1}{n})$$

$$=\tfrac{1}{nh_n}Bf(x)+o\left(\tfrac{1}{nh_n}\right), \tag{A.5}$$

for a single fixed point x.

Appendix A3. Expansion of Covariance from (9.2.5)

We assume that the kernel $K(u)$ is given by $\phi(u)$ and address the covariance term between $\widehat{f}_{n,h_n}(x_j)$ and $\widehat{f}_{n,h_n}(x_l)$ for any fixed pair j, l

such that $1 \leq j \neq l \leq q$. We write

$$
\begin{aligned}
&\mathrm{Cov}\left(\widehat{f}_{n,h_n}(x_j), \widehat{f}_{n,h_n}(x_l)\right) \\
&= \frac{1}{(nh_n)^2}\mathrm{Cov}\left(\sum_{i=1}^{n} K\left(\frac{x_j - X_i}{h_n}\right), \sum_{s=1}^{n} K\left(\frac{x_l - X_s}{h_n}\right)\right) \\
&= \frac{1}{nh_n^2}\mathrm{Cov}\left(K\left(\frac{x_j - X_1}{h_n}\right), K\left(\frac{x_l - X_1}{h_n}\right)\right) \\
&= \frac{1}{nh_n^2}\left[\mathbf{E}\left\{K\left(\frac{x_j - X_1}{h_n}\right) K\left(\frac{x_l - X_1}{h_n}\right)\right\} + O(h_n^2)\right] \\
&\quad \text{since } \mathbf{E}\left[K\big((x_p - X_1)/h_n\big)\right] = O(h_n), p = j, l.
\end{aligned}
\tag{A.6}
$$

Now, we denote $x^* = \frac{1}{2}(x_j + x_l)$ and proceed to evaluate the expression $\mathbf{E}\left\{K\left((x_j - X_1)/h_n\right) K\left((x_l - X_1)/h_n\right)\right\}$ as follows:

$$
\begin{aligned}
&\mathbf{E}\left\{K\big((x_j - X_1)/h_n\big)K\big((x_l - X_1)/h_n\big)\right\} \\
&= \frac{1}{2\pi}\int_{u=-\infty}^{\infty} \exp\left\{-\frac{1}{2h_n^2}\left((x_j - u)^2 + (x_l - u)^2\right)\right\} f(u)du \\
&= \frac{1}{2\pi}\exp\left\{-\frac{1}{4h_n^2}(x_j - x_l)^2\right\} \int_{u=-\infty}^{\infty} \exp\left\{-\frac{1}{h_n^2}(u - x^*)^2\right\} f(u)du.
\end{aligned}
\tag{A.7}
$$

Next, to arrive at its rate of convergence, we obtain

$$
\begin{aligned}
&\int_{u=-\infty}^{\infty} \exp\left\{-\frac{1}{h_n^2}(u - x^*)^2\right\} f(u)du \\
&= \int_{v=-\infty}^{\infty} \exp(-v^2)f(x^* + h_n v)(h_n)dv \text{ with } u - x^* = h_n v \\
&= h_n \int_{v=-\infty}^{\infty} \exp(-v^2)\left[f(x^*) + \sum_{s=1}^{\infty} \frac{(h_n v)^s}{s!}f^{(s)}(x^*)\right] dv = O(h_n).
\end{aligned}
\tag{A.8}
$$

We combine (A.7) - (A.8) and obtain

$$
\begin{aligned}
&\mathbf{E}\left\{K\big((x_j - X_1)/h_n\big)K\big((x_l - X_1)/h_n\big)\right\} \\
&= O(h_n)O\left(\exp\left\{-\frac{1}{4}(x_j - x_l)^2/h_n^2\right\}\right) \\
&= o(h_n),
\end{aligned}
\tag{A.9}
$$

which implies (using (A.6)) that

$$
\begin{aligned}
\mathrm{Cov}\left(\widehat{f}_{n,h_n}(x_j), \widehat{f}_{n,h_n}(x_l)\right) &= o\left((nh_n)^{-1}\right) + O(n^{-1}) \\
&= o\left((nh_n)^{-1}\right) \text{ for } j \neq l = 1, \ldots, q,
\end{aligned}
\tag{A.10}
$$

Appendix B. The Computer Programs
Appendix B1. S-PLUS Function for m

```
m.fun <- function(r=0.2001,q=42,gam=1.5,alpha=0.05,d0=0.005)
{
# File Name: m.fun
# This function computes the initial sample size using
# modified two-stage procedure with N(0,1) kernel.
#
#   INPUT VALUES
# r     = parameter r in kernel density estimation (input)
# q     = number of confidence intervals (input)
# gam   = gamma in modified two-stage procedure (input)
# alpha = confidence coefficient (input)
# d0    = half width of the confidence interval (input)
#
#   PARAMETERS
# d   = d0 * sqrt(q)  and  d2 = d^2
# h   = band width
#
#   OUTPUT
# m = Stage I  sample size
# -------------------------------------------------
# Obtain B value for N(0,1) Kernel
b0 <- 2 * sqrt(pi)
B <- 1/b0
#-------------------------------------------------
# Compute d2 = q*d^2
d2 <- q * d0 * d0
#-------------------------------------------------
# Compute Qa=Chi^2 (q,alpha) and factor QBd
Qa <- qchisq(1 - alpha, q)
QBd <- (Qa * B)/d2
# -------------------------------------------------
# Compute the initial sample size, m
rg <- (1 - r) * (1 + gam)
rg0 <- 1/rg
m1 <- QBd^rg0
m <- max(2, ceiling(m1))
list(m = m)
}
```

Appendix B2. S-PLUS Function for N

```
Two.fun <- function(X=Wool0.dat,x=c(7:48),alpha=.05,d0=.005)
{
```

```
# Function Name: Two.fun
# This function estimates the final sample size for q
# fixed-width confidence intervals using Two-stage
# ------------------------------------------------------
#    INPUT VALUES
# X      = given data set (input)
# x      = points where density to be estimated (input)
# alpha = confidence coefficient (input)
# d0      = half width of the confidence interval (input)
#
#    PARAMETERS
# m = Stage I sample size (# observations in X)
# q = number of x values (# confidence intervals)
# d = d0* sqrt(q)    and    d2 = d^2
# r = parameter r in kernel density estimation
# h = band width
#
#    OUTPUT
# N = Final sample size
# ------------------------------------------------------
# Fixed parameters
r <- 0.2001
#
# Obtain m=Stage one sample size & q=# of x-values
m <- length(X)
q <- length(x)
#
# Compute d2 =  q*d0^2
d2 <- q * d0 * d0
# ------------------------------------------------------
# Compute B for N(0,1) Kernel
b0 <- 2. * (1. + r) * sqrt(pi)
B <- 1./b0
# ------------------------------------------------------
# Compute h and mh
h <- 1./(m^r)
mh <- m * h
#------------------------------------------------------
# Compute Qa=chi^2 (q,alpha) and factor QBd
Qa <- qchisq(1 - alpha, q)
QBd <- (Qa * B)/d2
# ------------------------------------------------------
# Compute f0.hat at x0[i] using std normal kernel
f0.hat <- numeric()
for(i in 1:q) {
    w <- (x[i] - X)/h
    K <- dnorm(w)
    f0.hat[i] <- sum(K)/mh
```

```
}
f0.max <- max(f0.hat)
n1 <- QBd * (f0.max + 1/m)
r1 <- 1./(1. - r)
n2 <- n1^r1
N <- max(m, ceiling(n2))
list(d0 = d0, d = round(sqrt(d2), 3), alpha = alpha,
    m = m, q = q, N = N)
}
```

Appendix B3. S-PLUS Function for N^*

```
TwoBoot.fun<-function(X=Wool0.dat,x=c(7:48),alpha=.05,d0=.005)
{
# Function Name: TwoBoot.fun
# This function estimates the final sample size for q
# fixed-width confidence intervals using Two-stage
# procedure with Bootstrapping.
# --------------------------------------------------
#   INPUT VALUES
# X     = given data set (input)
# x     = points where density to be estimated (input)
# alpha = confidence coefficient (input)
# d0    = half width of the confidence interval (input)
#
#   PARAMETERS
# nB = number of Bootstrap replication
# m  = Stage I sample size (# observations in X)
# q  = number of x values (# confidence intervals)
# d  = d0* sqrt(q)    and    d2 = d^2
# r  = parameter r in kernel density estimation
# h  = band width
#
#   OUTPUT
# N = Final sample size
# --------------------------------------------------
# Fixed Parameters
r <- 0.2001
nB <- 1000
#
# Obtain m=Stage one sample size & q=# of x values
m <- length(x)
q <- length(x)
#
# Compute d2 =   q*d0^2
d2 <- q * d0 * d0
# --------------------------------------------------
```

```
# Compute B for N(0,1) Kernel
b0 <- 2. * (1. + r) * sqrt(pi)
B <- 1./b0
# ------------------------------------------------------
# Compute h and mh
h <- 1./(m^r)
mh <- m * h
# ------------------------------------------------------
# N = Final sample sizes for replacing chi-square
#     critical value by bootstrap critical value
# ------------------------------------------------------
# Compute f.hat(x) where x[i] for N(0,1) kernel
# and f.max = max(f.hat) for all q points
f.hat <- numeric()
for(i in 1.:q) {
    w <- (x[i] - X)/h
    K <- dnorm(w)
    f.hat[i] <- sum(K)/mh
}
# Compute maximum of f.hat
f.max <- max(f.hat)
#-------------------------------------------------------
# Set the seed for simulation
set.seed(234)
# ------------------------------------------------------
# Compute f.star[x], Smooth Bootstrap estimator for f(x)
# and replicate for nB times.
Y.boot <- numeric()
for(iB in 1.:nB) {
    delta <- rnorm(m)
    x.star <- sample(X, replace = T) + h * delta[ii]
    #
    # Bootstrap Estimate for all q points
    f.star <- numeric()
    for(i in 1.:q) {
        w <- (x[i] - x.star)/h
        K <- dnorm(w)
        f.star[i] <- sum(K)/mh
    }
    Y.boot[iB] <- mh * sum((f.star - f.hat)^2)
}
# Compute xi, bootstrap critical point
xi <- sort(Y.boot)[nB * (1 - alpha)]
#
# Estimate N using two-stage procedure with Smooth Bootstrap Critical Po
n1 <- xi/d2
r1 <- 1/(1 - r)
n2 <- n1^r1
```

```
N <- max(m, ceiling(n2))
list(d0 = d0, d = round(sqrt(d2), 2), alpha = alpha,
    m = m, q = q, nB = nB, N = N)
}
```

Appendix B4. S-PLUS Function to Compute $f(x)$ for Data Object x

```
Density.fun <- function(X=Wool.dat, x=c(7:48))
{
# File Name: Density.fun
# This function estimates pdf using N(0,1) kernel
# -----------------------------------------------
#   INPUT VALUES
# X     = given data set (input)
# x     = points where density to be estimated (input)
#
#   PARAMETERS
# r = parameter r in kernel density estimation
# q = number of x value;   h = band width
# n = sample size (number of observations in X)
#
#   OUTPUT
# f.hat = Estimated kernel density values at x
# -------------------------
# Fixed parameters
r <- 0.2001
q <- length(x)
n <- length(X)
# -------------------------------------------------
# Estimate f0.hat[i] at x=x[i] (i=1,2,...,q), using N(0,1)
# -------------------------------------------------
# Estimate Density
n <- length(X)
f.hat <- numeric()
h <- 1./(n^r)
nh <- n * h
for(i in 1:q) {
    w <- (x[i] - X)/h
    K <- dnorm(w)
    f.hat[i] <- round(sum(K)/nh, 3.)
}
PDF <- cbind(x, f.hat)
list(n = n, PDF = PDF)
}
```

Appendix C. A Typical AWTA Ltd Wool Certificate

AWTA Certificate

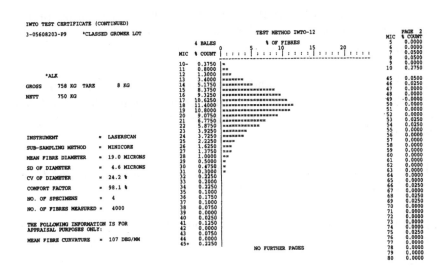

(Published with the kind permission of AWTA Ltd.)

ACKNOWLEDGMENT

Two referees raised several important issues and also suggested including more explanations in a number of places in an earlier version. Their critical assessments have led to a considerably clearer presentation for which we remain grateful and we take this opportunity to thank them. Further we would like thank the Australian Wool Testing Authority (AWTA) for making their commercial wool data available for RMIT University projects.

REFERENCES

[1] Anscombe, F.J. (1952), Large sample theory of sequential estimation. *Proc. Camb. Phil. Soc.*, **48**, 600-607.

[2] Butler, K.L. and Dolling, M. (1995). Spinning fineness for wool. *J. Textile Inst.*, **86**, 164-166.

[3] Carroll, R.J. (1976). On sequential density estimation. *Z. Wahrs. verw. Gebiete*, **36**, 137-151.

[4] de Silva, B.M. and Mukhopadhyay, N.(2001). Sequential procedures for nonparametric kernel density estimation. *Bul. Int. Statist. Inst.*, 53rd Session. Seoul, Republic of Korea, 8-9.

[5] de Silva, B.M., Roy, A. and Mukhopadhyay, N. (2000). Two-stage nonparametric density estimation with bootstrapping. *Int. Conf. on Recent Development in Statist. and Probab. and Their Appl.* New Delhi, India.

[6] Efron B. and Tibshirani, R.J. (1993). *An Introduction to the Bootstrap.* Chapman & Hall: New York.

[7] Faraway, J.J. and Jhun, M. (1990). Bootstrap choice of band-width for density estimation (with discussion). *J. Amer. Statist. Assoc.*, **85**, 1119-1122.

[8] Ghosh, M., Mukhopadhyay, N. and Sen, P.K. (1997). *Sequential Estimation.* Wiley: New York.

[9] Grund, B. and Hall, P. (1995). On the minimisation of L^p error in the mode estimation. *Ann. Statist.*, **23**, 2264-2284.

[10] Hall, P. (1992). *The bootstrap and Edgeworth Expansion.* Springer -Verlag: New York.

[11] Isogai, E. (1980/81). Stopping rules for sequential density estimation. *Bul. Math. Statist.*, **19**, 53-67.

[12] Isogai, E. (1984). Joint asymptotic normality of nonparametric recursive density estimators at a finite number of distinct points. *J. Japan Statist. Soc.*, **14**, 125-135.

[13] Isogai, E. (1987a). The convergence rate of fixed-width sequential confidence intervals for a probability density function. *Sequential*

Anal., **6**, 55-69.

[14] Isogai, E. (1987b). On the asymptotic normality for nonparametric sequential density estimation. *Bul. Inform. Cybernet.*, **22**, 215-224.

[15] Isogai, E. (1988). A note on sequential density estimation. *Sequential Anal.*, **7**, 11-21.

[16] Kundu, S. and Martinsek, A.T. (1997). Bounding the L_1 distance in nonparametric density estimation. *Ann. Inst. Statist. Math.*, **49**, 57-78.

[17] Lim, S.C. (1991). Simulation of the Melbourne Underground Railway Loop Operation. *M.S. Thesis*, Dept. of Math., RMIT, Melbourne.

[18] Martinsek, A.T. (1993). Fixed width confidence bands for density functions. *Sequential Anal.*, **12**, 169-177.

[19] Martinsek, A.T. and Xu, Y. (1996). Fixed width confidence bands for densities under censoring. *Statist. Probab. Lett.*, **30**, 257-264.

[20] Mukhopadhyay, N. (1980). A consistent and asymptotically efficient two-stage procedure to construct fixed-width confidence intervals for the mean. *Metrika*, **27**, 281-284.

[21] Mukhopadhyay, N. (1997). An overview of sequential nonparametric density estimation. *Nonlin. Anal., Theory, Meth., Appl.*, **30**, 4395-4402.

[22] Mukhopadhyay, N. and Solanky, T.K.S. (1994). *Multistage Selection and Ranking Procedures*. Marcel Dekker: New York.

[23] Shao, J. and Tu, D. (1995). *The Jackknife and Bootstrap*. Springer -Verlag: New York.

[24] Silverman, B.W. (1986). *Density Estimation for Statistics and Data Analysis*. Chapman & Hall: London.

[25] Silverman, B.W. and Young, G.A. (1987). The bootstrap: To smooth or not to smooth? *Biometrika*, **74**, 469-479.

[26] Taylor, C.C. (1989). Bootstrap choice of the smoothing parameter in kernel density estimation. *Biometrika*, **76**, 705-712.

[27] Wegman, E.J. and Davies, H.I. (1975). Sequential nonparametric density estimation. *IEEE Trans. Inform. Theory*, **21**, 619-628.

[28] Xu, Y. and Martinsek, A.T. (1995). Sequential confidence bands for densities. *Ann. Statist.*, **23**, 2218-2240.

[29] Yamato, H. (1971). Sequential estimate of a continuous probability density function and mode. *Bul. Math. Statist.*, **14**, 1-12.

Addresses for communication:

BASIL M. DE SILVA, Department of Mathematics and Statistics, RMIT University City Campus, GPO Box 2476V, Melbourne, Victoria 3001, Australia. E-mail: desilva@rmit.edu.au
NITIS MUKHOPADHYAY, Department of Statistics, University of Connecticut, UBox 4120, CLAS Building, 215 Glenbrook Road, Storrs, CT 06269-4120, U.S.A. E-mail: mukhop@uconnvm.uconn.edu

Chapter 10

Financial Applications of Sequential Nonparametric Curve Estimation

SAM EFROMOVICH

University of New Mexico, Albuquerque, U.S.A.

10.1   INTRODUCTION

The core topic in the finance literature is how to analyze risky assets, correctly price them, form an optimal portfolio of risky assets, choose a right hedging, and quantify the trade-back between risk and expected return. One of the main technical tools to solve these problems is the *capital asset pricing model* (CAPM) that is used by both professional portfolio managers and individual investors. Given certain simplifying assumptions outlined below, the CAPM states that the relationship between the expected return on a risky asset (such as a stock, a venture or a portfolio of securities) and the expected return on the entire security market is described by a single parameter β (beta). Empirically, β is defined as the slope of the ordinary least squares linear regression where the excess return on the market over the risk–free rate is the predictor and the excess return on the asset over the risk–free rate is the response. As an example, according to the theory, when $\beta = 1$, the excess return on the asset tends to mirror the excess return on the market; when $\beta = 0$, there is no correlation between these excess returns.

If the model is correct and security markets are efficient, stock returns should on an average confirm the linear relationship. Early statistical tests, made in the seventies, supported the theory, but later a growing number of empirical studies suggested that β's of common stocks do not adequately explain stock returns (see Fama and French (1996) and Hawawini and Keim (1999)). Assumptions of the CAPM are also under heavy scrutiny. See, for example, Fama and French (1996), Hawawini and Keim (1999), Luenberger (1998) and the references therein. The CAPM assumes that investors have complete and, therefore, identical information about the market, they buy and sell securities according to the same mean–variance optimization strategy, and thus the security market is in an equilibrium and security prices fully reflect all available information. In other words, the security market is efficient; see Hawawini and Keim (1999), Lo (2000), and Luenberger (1998). Regardless of these drawbacks and the criticism, according to the recent survey (Lo (2000)), the CAPM remains at the center of attention in the financial, economical, business and statistical literature. See also the splendid discussion in Fama and French (1996).

This article is devoted to a statistical analysis of historical data and developing a robust nonparametric sequential analog of the empirical CAPM that can be used when the underlying assumptions fail. In other words, the nonparametric part of the model means that the relationship between the excess returns of an asset and the market can be of any shape (not necessarily linear). This approach allows us to analyze inefficient security markets and markets where investors use different utility functions for creating optimal portfolios; see the discussions in Long (1990) and Luenberger (1993,1998) that justify the possibility of a nonlinear relationship. Also, since the empirical study conducted so far has been based on the analysis of scattergrams, it seems quite natural to try to employ modern statistical methods of nonparametric curve estimation. See the discussion in Efromovich (1999).

The reasoning behind employing a sequential approach is also intuitively clear. Recall that in an empirical study, a fixed sample-size (traditionally $n = 60$ or 5 years) is used to test the CAPM and to calculate the β's. There is no prudent justification of this or any other fixed sample-size. The sample-size may be too large for some periods of time and too small for others. Note that, for any security market, the larger the period is, the more outdated is the information that is being used. Clearly, the current situation in the US market is different from what it was in the early 1990's. Also, to avoid the effect of such factors as firm size, dividend yield, geopolitical environment and labor disputes, the period of the statistical analysis should be as short as possible. This is precisely where sequential analysis appears to be promising, because its primary goal is to achieve a given accuracy by using the smallest possible sample-sizes. We shall also see that even a simple sequential tool such as the ladder of scattergrams overlaid by estimates for an increasing sequence of sample-sizes (or levels of errors) becomes a valuable financial tool.

The thrust of this article is a completely data–driven sequential statistical procedure (software) that allows the investor to analyze the relationship between the excess rate of returns on an asset and the excess rate of returns on the market using the shortest period of historical data.

The article is organized as follows. To make the article interesting and readable for both statisticians unfamiliar with the CAPM and investors unfamiliar with nonparametric curve estimation, Sections 10.2 and 10.3 are devoted to brief introductions to these topics. Applications are discussed in Section 10.4. Specific details of a statistical procedure involving nonparametric estimation that may be of interest to a reader, are given in Section 10.5. Conclusions are presented in Section 10.6.

10.2 CAPITAL ASSET PRICE MODEL

One of the main financial problems that dominates the discipline of investment science is to determine the arbitrage–free (or equilibrium) price of an asset.

The well-known approach that deduces the correct price of a risky asset is the capital asset pricing model (CAPM). See the discussions in the books by Campbell et al. (1997) and Luenberger (1998). This approach follows from the familiar Markowitz mean–variance portfolio theory. The main assumption is that everyone is a mean–variance optimizer, everyone agrees on the means, variances and covariances of the asset-returns, and everyone assumes the same risk–free rate of borrowing and lending that is available to all with no transaction-fees. Under this assumption, the CAPM asserts that

$$r_A = r_f + \beta(r_M - r_f), \tag{10.2.1}$$

where r_A and r_M are the expected rate of returns from an asset and from the market respectively, r_f is the risk–free rate of return and β is the corresponding *beta* of the asset. The value $r_A - r_f$ is called the *expected excess rate of return* of the asset, and it is the amount by which the rate of return is expected to exceed the risk–free rate. Likewise, $r_M - r_f$ is the expected excess rate of return of the market portfolio.

We may conclude that the CAPM asserts that the expected excess rate of return of an asset is proportional to the expected excess rate of return of the market portfolio, with β being the proportionality factor.

The corresponding random "one-period" rates R_A and R_M, considered in the finance literature, are related as

$$R_A = r_f + \beta(R_M - r_f) + \epsilon_A, \qquad (10.2.2)$$

where $r_A = E\{R_A\}$, $r_M = E\{R_M\}$ and ϵ_A is assumed to be zero–mean, finite-variance and independent of R_M.

Using the assumed independence of R_M and ϵ_A, we get

$$\mathrm{Var}(R_A) = \beta^2 \mathrm{Var}(R_M) + \mathrm{Var}(\epsilon_A). \qquad (10.2.3)$$

If we think of the variance of a security rate of return as the risk of that security, then the equation (10.2.3) states that the risk of a security is the sum of two terms. The first term, $\beta^2 \mathrm{Var}(R_M)$, is referred to as the *systematic risk*. It is a combination of the security's β and the inherent risk in the market. The second term, $\mathrm{Var}(\epsilon_A)$, is called the *specific risk* that is due to the specific asset being considered.

These facts explain why β is considered one of the main financial characteristics of a risky asset. Values of β are available for all widely held stocks and mutual funds. An interested reader can find them, for instance, at *www.money.cnn.com*. Traditionally, these values are calculated using the classical least squares regression method based on the record of past stock prices (usually 5 years of monthly data). There is a belief in the financial literature that β-values for a particular stock drift around somewhat over time, but unless there are drastic changes in a company's situation, β tends to be relatively stable. See Hull (1999) and Luenberger (1998).

Let us now explain why the relation for the rates is called a pricing model. Suppose that an asset is purchased at price P_0 and later sold at price P_1. The (one-interval) rate of return is then $R = (P_1 - P_0)/P_0$. Here P_1 is random so the CAPM implies

$$P_0 = \frac{E\{P_1\}}{1 + r_f + \beta(r_M - r_f)}. \qquad (10.2.4)$$

This gives us the "right" price of the asset according to the CAPM. The reader familiar with the discounting formula might notice that (10.2.4) generalizes it for the case of a stochastic market. The rate $r_f + \beta(r_M - r_f)$ can be considered as a risk–adjusted interest rate.

Formula (10.2.4) is one of many applications of CAPM. Let us mention one more application that will shed light on the accuracy needed in the estimation of beta. This application is about hedging on the basis of index futures (or index options). If the hedger likes stocks being held in the portfolio but feels some uncertainty about the performance of the market as a whole, then under mild assumptions shorting $\beta V/F$ index futures contracts removes the risk arising from market moves. Here, V is the value of the portfolio, F is the value of assets underlying a futures contract, and it is assumed that the maturity of the futures contract is close to the maturity of the hedge. Similarly, the portfolio can be insured using index options.

One can find more about CAPM and its applications in Campbell et al. (1997), Fama and French (1996), Hawawini and Keim (1999), Hull (1999), Lo (2000), and Luenberger (1998).

10.3 NONPARAMETRIC REGRESSION ANALYSIS

Suppose that we observe n pairs $\{(R_{M1}, R_{A1}), \ldots, (R_{Mn}, R_{An})\}$ where R_{Ml} is the excess rate of return from the market (the market rate minus a Treasury Bill rate) during the lth period and R_{Al} is the excess rate of return from an asset. Then, the nonparametric regression model we consider here is:

$$R_{Al} = f(R_{Ml}) + \sigma(R_{Ml})\epsilon_l, \quad l = 1, 2, \ldots, n, \qquad (10.3.1)$$

where $f(x)$ is called the (nonparametric) regression function and $\sigma(x)$ is called the volatility (scale) function. The regression errors $\epsilon_1, \epsilon_2, \ldots, \epsilon_n$ are random variables (errors) with zero mean and unit variance. They may be dependent and have different distributions. Since the predictors R_{Ml} are random and the volatility function is not necessarily constant, the model (10.3.1) is called a *random design heteroscedastic regression* model. See the discussion in Efromovich (1999, Chapter 4).

The main aim of nonparametric regression is to highlight an important structure in the data without any assumption about the shape of an underlying regression function. In other words, the nonparametric approach allows the data to speak for itself. Note that the situation is quite different if the statistician uses a parametric approach. For instance, by using a linear regression, the statistician assumes that $f(x) = \alpha + \beta x$ regardless of the data on hand.

It is a good tradition in the statistical literature to begin data analysis with the visualization of a scattergram of the data. Unfortunately, this important step is "traditionally" skipped in the financial literature that typically presents tables of technical quantities such as p-values, t-statistics or some estimated parameters of a regression model. Scattergram is a plot that exhibits pairs of observations as points in the xy-plane. As an example, the top diagram in Figure 10.3.1 exhibits a scattergram for the excess rate of monthly returns of the Microsoft stock (the sticker symbol is $msft$) during a 60-month period that ended on December 1, 2001, versus the excess rate of monthly returns of the market whose proxy is Standard and Poor's 500 index (the sticker symbol is $\hat{s}pc$). The 13-weeks Treasury Bill (the sticker symbol is $\hat{\imath}rx$) serves as the proxy for the risk–free asset. All further numerical examples will follow the same approach. If CAPM is correct, then a linear relationship between these two rates with zero y-intercept should be observed. On the other hand, if the model is incorrect or its assumptions are invalid (for instance, when there is no market equilibrium), then a more complicated relationship may be visible. See discussions on this issue in Campbell et al. (1997), Fama and French (1996), and Hawawini and Keim (1999).

Do we see any pronounced "relationship" between R_A and R_M in this scattergram? The dataset is challenging, and so the answer "no" may not be surprising. Let us examine whether a classical parametric regression analysis helps in our understanding of this dataset. A widely used parametric linear regression model assumes that $f(x) = \alpha + \beta x$, and then the problem reduces to estimating the y-intercept, α, and the slope, β, by minimizing the sum of squared errors $\sum_{l=1}^{n}(R_{Al} - \alpha - \beta R_{Ml})^2$. The result leads to the familiar least–squares regression.

The least-squares regression line is shown next in Figure 10.3.1 (it is superimposed on the scattergram). Does it help us realize that there is a relationship between R_M and R_A? The answer is probably "yes". The fitted line indeed highlights the fact that larger returns from the market typically imply larger returns from the stocks. Parameters of the fitted regression model are as follows: the slope β is 1.83 which is close to the "official" 1.81, and the y-intercept is 1.40. Notice that the

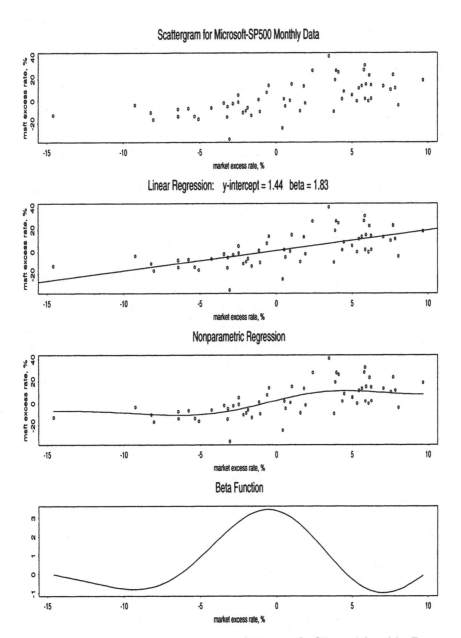

Figure 10.3.1: Statistical Analysis of Microsoft–SP500 Monthly Data.

presence of a non-zero y–intercept contradicts CAPM.

We shall continue the discussion of a linear regression in the next section. For the time being, let us consider nonparametric estimation. In this article a *series approach* is used whose idea is to approximate f by a partial sum:

$$f_J(x) = \sum_{j=0}^{J} \theta_j \varphi_j(x), \qquad \theta_j = \int \varphi_j(x) f(x) dx, \qquad (10.3.2)$$

where θ_j's are Fourier coefficients, $\varphi_0, \varphi_1, \ldots$ are known elements of a basis and J is a cut-off. Because the rate of market's returns is bounded, it is convenient to use a basis supported on a finite interval, say $[0,1]$. Then, the simplest basis consists of the polynomials: $\varphi_0(x) = 1$, $\varphi_1(x) = (x - 1/2)/(1/12)^{1/2}$, $\varphi_2(x) = [x^2 - 1/3 - (x - 1/2)]/(4/45 - 1/12)^{1/2}, \ldots$. In this case, notice that the cutoff $J = 1$ implies a linear regression $f_1(x)$, but if a particular dataset requires a larger cut-off then linear regression will not be useful. Another convenient and simple basis is a *cosine basis* with elements $\varphi_0(x) = 1$, $\varphi_j(x) = \sqrt{2} \cos(\pi j x)$, $j = 1, 2, \ldots$. The series approach is closely related to the classical linear approach and it is the best approach for analyzing smooth regression functions. A discussion of other types of nonparametric estimators can be found in Efromovich (1999, Chapter 8).

According to (10.3.1)–(10.3.2), there is a simple procedure for estimating Fourier coefficients:

$$\tilde{\theta}_j = n^{-1} \sum_{l=1}^{n} R_{Al} \varphi_j(R_{Ml})/p(R_{Ml}) \qquad (10.3.3)$$

where $p(x)$ is the probability density function of R_M. It is easy to see that the estimate is unbiased. Indeed

$$E\{\tilde{\theta}_j\} = E\{R_A \varphi_j(R_M)/p(R_M)\} = \int f(x) \varphi_j(x) dx = \theta_j. \qquad (10.3.4)$$

The cut-off J is estimated using the procedure of empirical mean integrated squared error minimization:

$$\hat{J} = \operatorname{argmin}_J \Big(2dn^{-1}(J+1) - \sum_{j=0}^{J} \tilde{\theta}_j^2 \Big), \qquad (10.3.5)$$

where $d = \int (\sigma^2(x)/p(x)) dx$ is the coefficient of difficulty. Note that both $p(x)$ and d are easily estimated (Efromovich (1999, Chapter 4)), and thus the data–driven nonparametric estimator is constructed. Note that the companion software package is available. See instructions in Efromovich (1999, Appendix B) on how to download and use this software.

The nonparametric estimate is then shown in Figure 10.3.1 and, to highlight its shape, the bottom diagram shows its derivative that can be referred to as the *beta function* and it generalizes the notion of β. The nonparametric estimate reveals the interesting sigmoidal (s-shaped) relationship between R_M and R_A where larger absolute values of the market returns imply smaller (compared to those predicted by CAPM) absolute values of the asset returns. This reflects the fact that investors refuse to pay beta-fold premiums for the Microsoft stock when the market does exceptionally well, and the shareholders do not dump the stock beta-fold when the market plunges. Note that this conclusion does not necessarily contradict CAPM because a security market jumping more than 5% per month cannot be in an equilibrium. In other words, when the US Central Bank describes the security market as "exuberant", the investor should be skeptical about the validity of the CAPM's assumptions.

Now we are in a position to explain the sequential nonparametric approach. Let us denote the nonparametric estimate based on k pairs of observations, where the first pair is the most recent historical observation, by $\hat{f}_k$. Then a sequential estimate couples this estimate with a so-called *stopping time* τ. It is basically a statistic taking on integer values and such that if $\tau = n$, then the stopping time is based only on the first n observations. The stopping time τ satisfies

$$E\{\int (\hat{f}_\tau(x) - f(x))^2 dx\} \le \delta. \tag{10.3.6}$$

Here δ represents the level of accuracy of the estimation process. We explain in the next section how to choose it. The aim is to find a sequential estimate with the minimal expected stopping time. This approach allows an investor's decision to be based on the most recent data without compromising the accuracy.

The method of finding the optimal stopping time is based on the empirical risk approximation procedure suggested in Chaudhuri et al. (1997) and Efromovich (1989,1994,1995). The Parseval identity implies that

$$\int (\hat{f}_k(x) - f(x))^2 dx = \sum_{j=0}^{\hat{J}} (\tilde{\theta}_j - \theta_j)^2 + \sum_{j>\hat{J}} \theta_j^2. \tag{10.3.7}$$

The first sum is approximated by $dk^{-1}(\hat{J}+1)$ and the second sum, by $\sum_{j=\hat{J}+1}^{S}(\tilde{\theta}_j^2 - dk^{-1})$ where $S = S_\delta$ is a sufficiently large deterministic number. See Efromovich (1994,1999) in this context. Then,

$$\tau = \operatorname{argmin}_k \left(dk^{-1}(\hat{J}+1) + \sum_{j=\hat{J}+1}^{S} (\tilde{\theta}_j^2 - dk^{-1}) \right) \le \delta. \qquad (10.3.8)$$

Specific details of this sequential estimator will be presented in Section 10.5.

It is important to note that a series nonparametric regression estimator is robust against dependency in the errors. See the discussion in Efromovich (1999). For the case of error distributions with very heavy tails special modifications may be useful. See the discussions in Chaudhuri et al. (1997) and Efromovich (1999).

Finally, let us note the following interesting fact. Typically, sequential estimation is used when there is a price attached to each observation. For the empirical analysis considered in the article, all the observations are free and available to the statistician, but there is a "price" to be paid for the use of outdated information. We may think about the data as nonstationary with a distribution that is slowly changing over time. This is what creates a penalty for using larger sample-sizes, and thus makes a sequential approach so appealing and so natural.

10.4 Applications

The suggested sequential nonparametric analog of the CAPM is a new tool for the analysis of "fair" prices of risky assets. This empirical model is robust because it can be used for the analysis of both efficient and inefficient security markets while the CAPM can be used only for the analysis of markets in equilibrium. Moreover, the sequential part of the model allows an investor to make decisions based on most recent historical observations.

Let us begin with the analysis of some more examples similar to the one exhibited in Figure 10.3.1. This will allow the reader to get used to the nonparametric analysis of standard historical datasets. Figure 10.4.1 shows us results of the analysis for 3 widely held stocks. Each of the diagrams is a condensed version of Figure 10.3.1. The top diagram shows the scattergram for Nokia corporation; it is superimposed by the linear regression estimate (the dotted line) and also the nonparametric

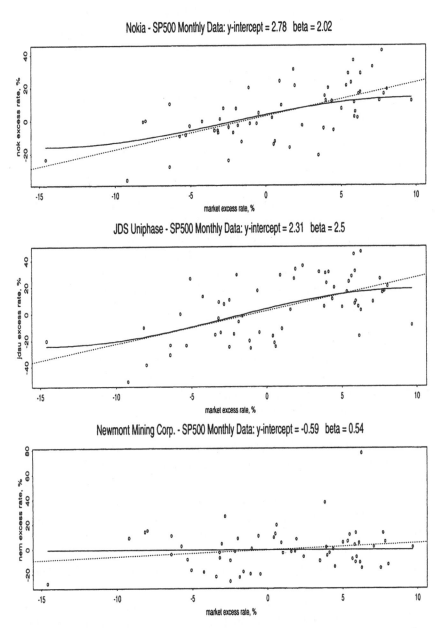

Figure 10.4.1: Linear and Nonparametric Regressions Based on the Traditional 5 Years of Historical Monthly Data. The Dotted and Solid Lines Show the Linear and Nonparametric Regression, Respectively.

estimate (the solid line). The value of beta is close to the "official" number 2.2, and the nonzero y-intercept indicates a large deviation from the CAPM (with respect to the underlying risk-free rate of about 0.15%).

As we see, for the smallest market's returns, the shape of the nonparametric curve resembles the classical linear one. However, based on the visualization of this graphic, we should not rush our judgment. The issue is that the range of rates of stock returns is big in comparison with the risk-free rate which is the financial instrument to calibrate the returns. The top diagram in Figure 10.4.2 shows us the "zoomed in" version of the curves. Let us recall that the CAPM considers larger rates of returns on risky assets as a reward for the risk taken. In other words, everything should be compared with monthly risk-free rates. Keeping in mind the above-mentioned value of the underlying risk-free rate, we conclude that the numerical difference between the linear and the nonparametric curves is dramatic. This clearly reflects the fact that the security markets are extremely volatile and it is difficult to expect that the assumption underlying the CAPM holds in such an environment.

Returning to the top diagram in Figure 10.4.1, note that the curves lead to different conclusions for the extreme values of the market returns when the market exhibits the "exuberant" behavior. In particular, a hedger using the nonparametric curve will need a smaller exposure to the market than a hedger using the empirical β.

The middle diagram in Figure 10.4.1 exhibits an even more appealing example. Here JDS Uniphase optical and networking corporation (the sticker symbol is *jdsu*) is analyzed. Again the nonparametric curve has a smaller y-intercept and shapes of the curves look similar for the smallest market returns (but again, compare this impression with the analysis of the middle diagram in Figure 10.4.2). For larger market rates the shapes of the estimates are different, and the nonparametric regression makes less dramatic conclusion about the shareholders' risk and reward associated with the market movements than the CAPM does.

The bottom diagrams in Figures 10.4.1-10.4.2 show us the analysis of Newmont Mining Corporation (the sticker symbol is *nem*). The official beta is 0.55 and this means that the stock price is positively correlated with the market and it can be hedged using the market. Conclusion of the nonparametric regression is absolutely different: there is no relationship between excess returns of the stock and of the index.

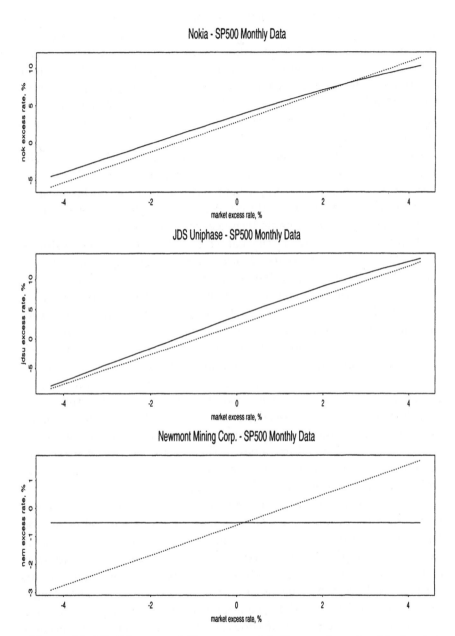

Figure 10.4.2: Linear and Nonparametric Regressions for Moderate Rates of the Market Return. The Dotted and Solid Lines Show the Linear and Nonparametric Regression, Respectively.

Actually, the reader familiar with the analysis of the scattergrams may notice that the positive correlation is created by a single outlier: look at that approximately 80% monthly return. The fact that the company is engaged in the production and acquisition of gold properties sheds an additional light on the scattergram. Note that the nonparametric regression indicates the average monthly return of approximately -0.5%. This together with the scattergram sheds light on the reward–risk pattern of the stock.

Now let us take another step toward the sequential approach. This step will be a familiar method of analyzing the ladder of historical datasets with increasing sample-sizes. We shall see that this method, motivated by sequential analysis, sheds new light on the empirical CAPM and can be a valuable technical tool for investors.

Figure 10.4.3 shows us the ladder for Microsoft–SP500 historical data. The ladder is in years (2 through 8) of monthly returns implying that $n = 24, 36, \ldots, 96$. These diagrams are more informative than the single scattergram shown in Figure 10.3.1 which is based on the traditionally considered 5 years of historical data. The title of each diagram indicates n and the corresponding empirical mean integrated squared error of the nonparametric estimate as well as the parameters of the linear regression. The top diagram indicates that over the period of the most recent two years shapes of the linear and nonparametric regressions are very close. Note that this was the period of the largest monthly market's return over the last 8 years. A quick glance on the other diagrams shows that this return can be considered as an outlier. Let us also add that all extremes in the stock's returns occurred during the most recent year. This explains the monotone decrease in the empirical error as the sample-size increases.

Let us look at the corresponding ladder of calculated errors because they indicate possible values of delta used by the sequential estimator. For the first two years ($n = 24$) the error is huge due to the small sample-size and the extremely large returns discussed earlier. Addition of the third year ($n = 36$) reduces the error threefold. This occurs because we have the same range of the market rates, no new "huge" returns are added and the nonparametric curve better fits the data. Adding one more year ($n = 48$) does not decrease the error a lot because the range of the market's returns is increased (look at the approximately -15% monthly return). There is a large decrease in the empirical error after adding the 6th and then the 7th year, after that the error is practically the same.

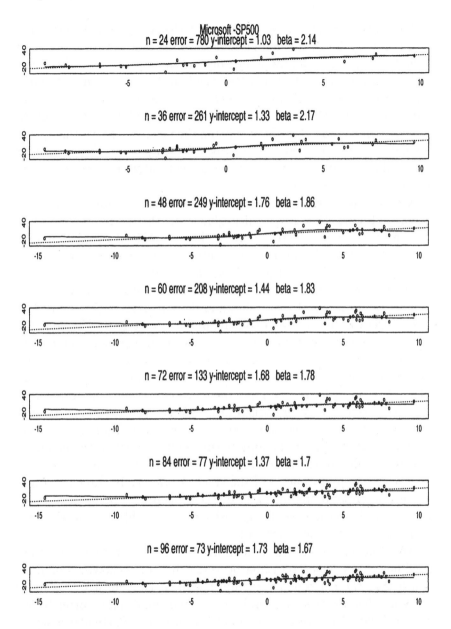

Figure 10.4.3: The Ladder of Scattergrams Overlaid by the Estimates for Microsoft–SP500 Historical Data. Each Diagram Is Based on n Most Recent Months. Values of n and the Corresponding Empirical Errors are Shown in the Titles. The Solid and Dotted Lines Are the Nonparametric and Linear Regressions, Respectively.

Another interesting observation, that sheds light on the empirical beta, is the ladder of calculated betas. We see that both 4 and 5 years of the historical data imply practically the same beta. Also, the longer the historical data used, the smaller the beta becomes. On first glance this empirical observation contradicts to the general opinion that the golden era of the Microsoft's shareholders is over, but let us recall the discussion in Section 10.2 about β being the measure of investment risk. The ladder indicates that the risk taken by shareholders has gradually increased over the past years and the last several years were exceptionally risky.

Now let us look at the ladder for Newmont Mining corporation shown in Figure 10.4.4. A quick glance on the ladder reveals when and why this gold–industry company has the official positive β equal to 0.55. Note that this positive correlation with the market contradicts the historical opinion that gold (and as a result everything related to its production) serves as a natural hedge for the stocks market.

The two most recent years, shown in the top diagram (Figure 10.4.4), support the validity of the historical opinion about gold. The nonparametric curve indicates that there is no relationship between the returns while the more "aggressive" linear regression indicates a small negative correlation; this is what everyone might expect from investment in the gold industry. Adding one more year makes the outcome even more interesting (see the second diagram). We see the pronounced shape of the nonparametric regression and the neutral position of the linear regression. But look at the almost threefold increase in the empirical error that is due to the huge jumps in the stock's monthly returns. After 4 years the performance of the stock is practically stable, and now we can understand the meaning of the official $\beta = 0.55$. It is solely created by the outlier, that impressive 80% monthly return when the stock's price was almost doubled.

After the analysis of this ladder of the diagrams, it is easy to understand how the sequential estimator works. It simply chooses a minimal n such that the corresponding empirical error is at most the assigned δ. Figure 10.4.5 shows us several sequential estimates for Nortel–SP500 historical data (the sticker symbol of Nortel is nt). Particular assigned values of δ (delta) and the corresponding values n of the stopping time τ are shown in the titles.

As we see, the Nortel's stock is a very special one where the time passes by but the conclusions of the nonparametric and linear regressions do not change. The official β of the company is 2.2, and we also

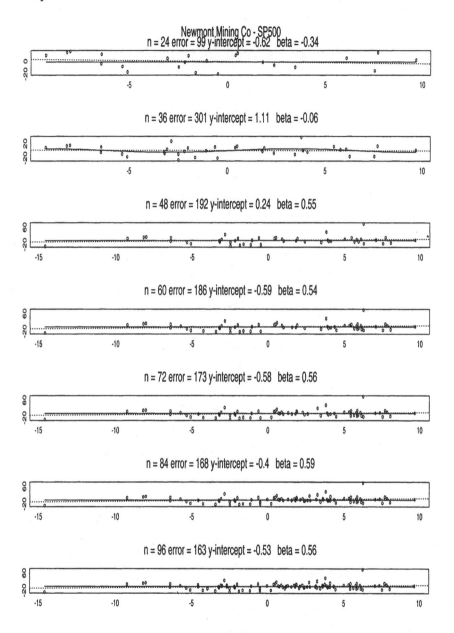

Figure 10.4.4: The Ladder of Scattergrams Overlaid by the Estimates for Newmont Mining Corp.– SP500 Historical Data. Each Diagram Is Based on n Most Recent Months. Values of n and the Corresponding Empirical Errors Are Shown in the Titles. The Solid and Dotted Lines Are the Nonparametric and Linear Regressions, Respectively.

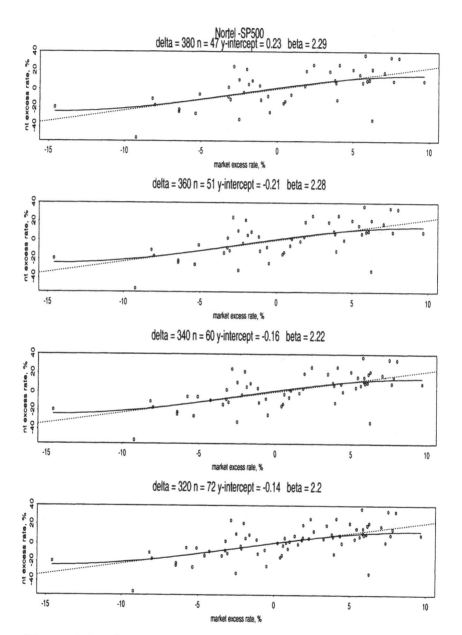

Figure 10.4.5: Sequential Nonparametric Estimates for the Historical Data of Nortel Corporation. The Solid and Dotted Lines Are the Nonparametric and Linear Estimates, Respectively.

see that the calculated betas are stable over these years. It is also clear that the sample size used by the empirical CAPM can be reduced without any dramatic consequences. Another interesting remark is the practically zero y-intercepts over all these years, this indicates that the CAPM fairly well describes the "right" price of the stock. The only exclusion from the CAPM's rule are the tails where the nonparametric model outperforms the CAPM.

The figures presented allow the investor to choose an appropriate value of δ. Recall that it should reflect the familiar compromise between the investor's belief in the relevance of the historical data and the accuracy of the beta function estimation. In the Nortel example $\delta = 380$ is a fair choice. For the case of using beta for hedging (the case was discussed in Section 10.1), the accuracy of estimation is defined by the utility function used. See, for example, Luenberger (1998).

We may conclude that the sequential nonparametric regression sheds a new light on pricing risky assets. The software developed by the author may be an useful additional technical tool for investors and hedgers who wish to know more about assets than just the values of β's disclosed by companies. The software is added to the statistical package (Efromovich (1999)) and it is available via the Internet.

10.5 TECHNICAL DETAILS

In this section, we explain several important steps used in the construction of the data–driven estimator discussed earlier. The main point is the estimation of Fourier coefficients. Equation (10.3.3) explains the underlying idea of the estimation process, but the estimator (10.3.3) is not data-driven. To make it data-driven, we began with a pilot estimator

$$\check{\theta}_j = \sum_{l=1}^{n} R_{A(l)} (2s)^{-1} \int_{R_{M(l-s)}}^{R_{M(l+s)}} \varphi_j(x) dx. \qquad (10.5.1)$$

Here, $(R_{M(l)}, R_{A(l)})$ are pairs of observations arranged in ascending order according to the market returns. The estimator (10.5.1) can be viewed as a kind of naive numerical integration. To realize this, one simply needs to look at (10.3.1) and then note that the naive integration implies robustness of the estimate with respect to the joint distribution of the market returns. Alternatively, the same estimator can be viewed as an adaptive version of (10.3.3) because spacings of market returns are inversely proportional to their density.

The parameter s used by the estimator is defined as $[1 + \log(\log(n + 20))]/2$, rounded upward. Here "log" stands for natural logarithm. This choice for s is based on intensive Monte Carlo simulations.

Then the estimator (10.5.1) is shrunk using Efromovich–Pinsker algorithm:

$$\hat{\theta}_j = [(\breve{\theta}_j^2 - dn^{-1})_+/\breve{\theta}_j^2]\breve{\theta}_j. \qquad (10.5.2)$$

Here, $(\cdot)_+$ denotes the positive part.

The procedure of estimating the coefficient of difficulty, d, is borrowed from Efromovich (1995, Section 4.2). The s is $4 + \log(20 + 1/\delta)$, rounded downward. Again, Monte Carlo simulations have been used to choose this parameter.

Finally, in all the examples, the cosine basis is used. No significant improvement has been observed when the enriched cosine–polynomial basis is used instead. On the other hand, with outliers deleted and the range of studied market returns shrunk to more "reasonable" (not "exuberant") rates, preliminary results indicate the superiority of the enriched basis. These results will be presented elsewhere.

10.6 CONCLUDING REMARKS

Knowing the relationship between a risky asset's rate of returns and a security market's rate of returns allows an investor to solve a wide class of financial problems ranging from calculating the optimal cost of a venture and forming an optimal portfolio of risky assets to optimal hedging of risky assets. If everybody uses the mean–variance approach to invest and has the same information about security markets, then the capital asset pricing model (CAPM) states that there is a linear relationship between the rates that is solely defined by the asset's β. If the investor does not accept the assumption underlying the CAPM or the historical data rejects the CAPM's conclusions, the investor can use the empirical sequential nonparametric model developed in this article. The nonparametric nature of the model allows for any shape of the relationship between the rates In other words, the historical dataset speaks for itself. The sequential nature of the model allows one to employ only the most recent historical observations and thus it limits the effect of relatively slow-changing factors such as firm size and dividend yield. Moreover, the suggested sequential ladder of estimates helps the investor to gain a deeper understanding of the relationship between movements of the asset and the market instead of relying solely on β.

The estimator developed possesses very attractive statistical properties of being robust and completely data-driven. The investor needs no advanced statistical education to use the technical tool developed, and the examples presented illustrate its performance.

REFERENCES

[1] Campbell, J.Y., Lo, A.W. and MacKinlag, A.C. (1997). *The Econometrics of Financial Markets*. Princeton Univ. Press: Princeton.

[2] Chaudhuri, P., Doksum, K. and Samarov, A. (1997). On average derivative quantile regression. *Ann. Statist.*, **25**, 715–744.

[3] Efromovich, S. (1989). On sequential nonparametric estimation of a density. *Theory Probab. Appl.*, **34**, 228–239.

[4] Efromovich, S. (1994). On adaptive estimation of nonlinear functionals. *Statist. Probab. Lett.*, **19**, 57–63.

[5] Efromovich, S. (1995). On sequential nonparametric estimation with guaranteed precision. *Ann. Statist.*, **23**, 1376–1393.

[6] Efromovich, S. (1999). *Nonparametric Curve Estimation: Methods, Theory and Applications*. Springer-Verlag: New York.

[7] Fama, E.F. and French, K.R. (1996). The CAPM is wanted, dead or alive. *J. Finance*, **51**, 1947–1958.

[8] Hawawini, G. and Keim, D.B. (1999). The cross section of common stock returns: A review of the evidence and some new findings. In *Security Markets Imperfections in Worldwide Equity Markets* (D.B. Kein and W.T. Ziemba, eds.), 3-43. Cambridge Univ. Press: Cambridge.

[9] Hull, J.C. (1999). *Options, Futures and Other Derivatives*, 4th ed. Prentice Hall: Upper Saddle River.

[10] Lo, A.W. (2000). Finance: A selective survey. *J. Amer. Statist. Assoc.*, **95**, 629–635.

[11] Long, T.B.,Jr. (1990). The numeraire portfolio. *J. Fin. Econ.*, **26**, 29–69.

[12] Luenberger, D.G. (1993). A preference foundation for log mean–variance criteria in portfolio choice problems. *J. Econ. Dynam. Control*, **17**, 887–906.

[13] Luenberger, D.G.(1998). *Investment Science.* Oxford Univ. Press: New York.

Address for communication:

SAM EFROMOVICH, Department of Mathematics and Statistics, University of New Mexico, Albuquerque, NM 87131–1141, U.S.A. E-mail: efrom@math.unm.edu

Chapter 11

Interim and Terminal Analyses of Clinical Trials with Failure-Time Endpoints and Related Group Sequential Designs

TZE LEUNG LAI
Stanford University, Palo Alto, U.S.A.

11.1 INTRODUCTION

As pointed out in Wallis (1980), sequential analysis was born in response to demands for more efficient testing of anti-aircraft gunnery during World War II, culminating in Wald's development of the sequential probability ratio test which had an immediate impact on weapons testing. Since the medical community widely accepts hypothesis testing as a means of assessing the reproducibility of the results of an experiment, it was recognized within a few years of Wald's seminal work that the sequential approach might provide a more efficient design of clinical trials to test new medical treatments. Despite substantial methodological development in this direction, sequential analysis received little attention from the biomedical community until the early 1980's following early termination of the *Beta-Blocker Heart Attack Trial* (BHAT). The main reason for this lack of interest was that the fixed sample-size ("sample-size" meaning the number of patients accrued) for a typical trial was too small to allow further reduction by a sequential design while still maintaining reasonable power at the alternatives of interest. On the other hand, BHAT's survival endpoint had to be monitored over a period of 4 years, with all patients accrued within the first 27 months and with periodic reviews of the data by a *Data and Safety Monitoring Board*. The trial was terminated at one of these reviews, 8 months prior to its prescheduled ending. Since interim reviews of the data are usually incorporated in the design and execution of long-term clinical trials at least for the purpose of monitoring safety of the treatments, these trials (of which BHAT is an illuminating example) are particularly suited to group sequential designs.

In Section 11.2 we describe the design of BHAT and the interim analyses of the resulting dataset that led to its early termination. We report some re-analyses of this dataset to illustrate certain statistical issues and methods that are reviewed in Section 11.3. This also provides a roadmap to the literature on the design and analysis of clinical trials with failure-time endpoints and periodic data reviews. Section 11.4 presents some concluding remarks.

11.2 INTERIM AND TERMINAL ANALYSES OF BHAT DATA

The primary objective of BHAT was to determine whether regular, chronic administration of propranolol, a beta-blocker, to patients who had at least one documented *myocardial infarction* (MI) would result in significant reduction in mortality from all causes during the follow-up period. It was designed as a multicenter, double-blind, randomized placebo-controlled trial with a projected total of 4200 eligible patients recruited within 21 days of the onset of hospitalization for MI. The trial was planned to last 4 years, beginning in June 1978 and ending in June 1982, with patient accrual completed within the first 2 years so that all patients could be followed for a period of 2 to 4 years. The sample size calculation was based on a 3-year mortality rate of 18% in the placebo group and a 28% reduction of this rate in the treatment group, with a significance level of 0.05 and 0.9 power using a two-sided logrank test; see BHAT Research Group (1984, p. 388). In addition, periodic reviews of the data were planned to be conducted by a Data and Safety Monitoring Board, roughly once every 6 months beginning at the end of the first year, whose functions were to monitor safety and adverse events and to advise the Steering and Executive Committees on policy issues related to the progress of the trial.

The actual recruitment period was 27 months, within which 3837 patients were accrued from 136 coronary care units in 31 clinical centers, with 1916 patients randomized into the propranolol group and 1921 into the placebo group. Although the recruitment goal of 4200 patients had not been met, the projected power was only slightly reduced to 0.89 as accrual was approximately uniform during the recruitment period.

11.2.1 Analysis by the Data and Safety Monitoring Board

The Data and Safety Monitoring Board arranged meetings at 11, 16, 21, 28, 34 and 40 months to review the data collected so far, before the scheduled end of the trial at 48 months. Besides monitoring safety and adverse events, the Board also examined the standardized logrank statistics (see Section 11.3.1 for details) to examine if propranolol was indeed efficacious. Successive values of these statistics are listed below:

Time (months)	11	16	21	28	34	40
Test statistic	1.68	2.24	2.37	2.30	2.34	2.82

Instead of continuing the trial to its scheduled end at 48 months, the Data and Safety Monitoring Board recommended terminating it in their last meeting because of conclusive evidence in favor of propranolol. Their recommendation was adopted and the trial was terminated on October 2, 1980. It drew immediate attention of the biopharmaceutical community to the benefits of sequential methods, not because it reduced the number of patients but because it shortened a four-year study by 8 months, with positive results for a long-awaited treatment supporting its immediate use.

Note that except for the first interim analysis at 11 months (when there were 16 deaths out of 679 patients receiving propranolol and 25 deaths out of 683 patients receiving placebo), all interim analyses showed normalized logrank statistics exceeding the critical value 1.96

for a single 5% two-sided logrank test. The lack of significance in the first interim analysis seems to be due to the relatively small number of deaths. In comparison, the last interim analysis at 40 months had 135 deaths in the propranolol group of 1916 patients and 183 deaths in the placebo group of 1921 patients. The final report in BHAT Research Group (1982) showed more deaths from both groups (138 and 188) due to additional data that were processed after the interim analysis. The Kaplan-Meier estimates of the respective survival functions at the termination of the trial (BHAT Research Group (1982, p. 1709)) show that the mortality distributions were estimable only up to approximately their 10th percentiles, with the cumulative distribution function for propranolol below that of placebo.

The critical value 1.96 for the standardized logrank statistic only applies to a single interim analysis. To account for repeated testing, the Data and Safety Monitoring Board used an adjustment (which has a critical value 5.46 at the 1st analysis and 2.23 at the 6th analysis) for repeated significance testing with independent and identically distributed normal observations proposed by O'Brien and Fleming (1979). Since logrank statistics (rather than normal observations) were actually used, the Board also appealed to joint asymptotic normality of time-sequential logrank statistics that was established by Tsiatis (1981) shortly before that. See Section 11.3.1.

Actually time-sequential methodology, which was in its infancy at that time, was barely adequate to handle the BHAT data. Moreover, the trial had been designed as a fixed-duration (instead of time-sequential) trial. The Data and Safety Monitoring Board used some informal arguments based on stochastic curtailment (Section 11.3.3) together with a formal group sequential test described in the preceding paragraph to conclude that the propranolol therapy was indeed effective, at the time of the 6th interim analysis. The final report of the study was published by BHAT Research Group (1982), summarizing results of the final analysis of the entire BHAT dataset collected until October 2, 1981, when official patient follow-up was stopped.

11.2.2 Further Analysis of BHAT Data

The Data and Safety Monitoring Board used the O'Brien-Fleming boundary for repeated significance testing. Other group sequential boundaries could be used instead. Siegmund (1985, pp. 133-135) used a modification of Haybittle's (1971) group sequential test, with a critical value 2.65 for the normalized logrank statistics for the first 6 interim analyses and 2.05 for the 7th analysis at the scheduled end of the trial. This test, therefore, again stopped at the 6th interim analysis.

Strictly speaking, the O'Brien-Fleming and Haybittle boundaries for $\{\Sigma_{i=1}^{k} X_i, k \leq K\}$, where the X_i's are i.i.d. normal random variables and K is the prescribed maximum number of interim (including final) analyses, is not applicable to the time-sequential logrank statistics. The reason is a discrepancy between two scales in the data structure, namely, information time and calendar time, as will be explained in Section 11.3. This poses challenging problems at the design stage when one should incorporate in the protocol the boundary that will be used for interim analyses, to ensure validity of the inference. There has been considerable progress toward resolving these problems in the past two decades and Section 11.3.2 provides a brief survey. Gu and Lai (1998, pp. 425-426) used these new developments to "redesign" BHAT, under the assumptions of the original BHAT design and also under the scenario revealed by the actual BHAT data.

BHAT's conclusion was that the hazard ratio of the propranol group to the placebo group was significantly less than 1, assuming a proportional hazards model whose score function is the logrank statistic. What are confidence limits for the logarithm of the hazard ratio, which is the regression parameter β in the Cox regression model? Using a hybrid resampling approach described in Section 11.3.4, we obtain a 90% confidence interval $-0.5 < \beta < -0.08$.

11.3 DESIGN AND ANALYSIS OF GROUP (OR TIME) SEQUENTIAL TRIALS

We first review some important developments in group sequential trials before and after BHAT. Armitage et al. (1969) proposed an alternative to Wald's sequential probability ratio test, called the *repeated significance test* (RST), for testing a normal mean with known variance σ^2 based on successive observations $Z_1, Z_2, \ldots$. The RST of $H_0 : \theta = 0$ stops sampling at stage $T = \inf\{n \leq K : |S_n| \geq a\sigma\sqrt{n}\}$, rejecting H_0 if $T < K$ or if $T = K$ and $|S_K| \geq a\sigma\sqrt{K}$, where $S_n = Z_1 + \ldots + Z_n$ and K

is a prescribed maximum number of observations. Its underlying motivation is to apply the conventional significance test repeatedly until the scheduled end of the trial. Subsequently, Haybittle (1971) proposed the following modification of the RST to increase its power. The stopping rule is still T, but the rejection region is modified to $T < K$ or $|S_K| \geq c\sigma\sqrt{K}$, where $a \geq c$ are so chosen that the overall significance is equal to some prescribed number. Since it is not feasible to examine the data continuously as they accumulate in a double blind, multicenter clinical trial, Pocock (1977) introduced a "group sequential" version of the RST, in which, Z_n represents the sample mean of the nth group (instead of the nth observation) and K represents the maximum number of groups. Instead of the square root boundary, O'Brien and Fleming (1979) proposed to use a constant boundary in the RST.

Although testing for a normal mean appears to be a "toy problem", it can be extended to much more general test statistics by appealing to the central limit theorem. For example, if Z_n is an approximately normal statistic based on m i.i.d. observations in the nth group, then the Pocock or O'Brien-Fleming test can still be used. However, the requirement of equal group sizes here is still an overly stringent restriction. Lan and DeMets (1983) introduced an "error spending" approach to overcome this difficulty. Instead of a normal random walk, they consider a Wiener process $\{W(v), 0 \leq v \leq 1\}$ with stopping rule $T = \inf\{v \in [0, 1] : |W(v)| \geq h(v)\}$, where h is a positive function on $[0, 1]$ such that $P\{T = 0\} = 0$ and $P\{T \leq 1\} = \alpha$, and the infimum of an empty set is considered to be ∞. The *error spending function* is $A(v) = P\{T \leq v\}$, $0 \leq v \leq 1$. Taking v to represent the proportion of information accumulated at time t of interim analysis, $A(v)$ can be interpreted as the amount of Type-I error spent up to time t, with $A(0) = 0$ and $A(1) = \alpha$. In particular, since the asymptotic variance of an asymptotically normal statistic based on n_i i.i.d. observations is proportional to n_i, the proportion of information accumulated at time t_i of interim analysis is $v_i = n_i/n_K$, where n_i is the total number of observations available at time t_i. Hence, Lan and DeMets (1983) proposed to choose $\alpha_j = A(v_j) - A(v_{j-1})$ to determine the stopping boundary b_j $(j = 1, \ldots, K)$ recursively by

$$P_{H_0}\{|W_1| \leq b_1\sqrt{V_1}, \ldots, |W_{j-1}| \leq b_{j-1}\sqrt{V_{j-1}}, |W_j| > b_j\sqrt{V_j}\} = \alpha_j,$$
$$(11.3.1)$$

where W_i denotes the asymptotically normal test statistic at the ith interim analysis and V_i denotes the corresponding variance estimate.

The error spending approach has greatly broadened the scope of applicability of group sequential methods. For example, if one wants to use a constant boundary as in O'Brien and Fleming, one can consider the corresponding continuous-time problem to obtain $A(v)$. The sample sizes at the times of interim analysis need not be specified in advance. Instead, what needs to be specified is the maximum sample size n_K. Lan and DeMets, who had been involved in the BHAT study, were motivated by BHAT to make group sequential designs more flexible. Although it does not require pre-specified "information fractions" at times of interim analysis, the error spending approach requires specification of the terminal information amount, at least up to a proportionality constant. While this is usually not a problem for immediate responses for which, total information is proportional to the sample size, the error spending approach is much harder to implement for time-to-event responses. This is because, for such responses, the terminal information is not proportional to n_K and cannot be known until one carries the trial to its scheduled end, as will be explained below.

11.3.1 Time-Sequential Rank Statistics and Their Asymptotic Distributions

Inspired by the statistical issues raised by BHAT, a comprehensive theory on the asymptotic distributions of time-sequential rank statistics was developed in the 1980's and 1990's. Suppose that a clinical trial for comparing times to failure between two treatment groups X and Y involves n patients who enter the trial serially, are randomly assigned to treatment X or Y, and are then followed until they fail or withdraw from the study or until the study is terminated. Let $T_i' \geq 0$ denote the entry time and $X_i > 0$ the survival time (or time to failure) after entry of the ith subject in treatment group X, and let T_j'' and Y_j denote the entry time and survival time after entry of the jth subject in treatment group Y. Thus the data at calendar time t consist of $(X_i(t), \delta_i'(t)), i = 1, \ldots, n'$, and $(Y_j(t), \delta_j''(t)), j = 1, \ldots, n''$, where

$$X_i(t) = X_i \wedge \xi_i' \wedge (t - T_i')^+, \quad Y_j(t) = Y_j \wedge \xi_j'' \wedge (t - T_j'')^+,$$

$$\delta_i'(t) = I_{\{X_i(t) = X_i\}}, \quad \delta_j''(t) = I_{\{Y_j(t) = Y_j\}},$$

and $\xi_i'(\xi_j'')$ denotes the withdrawal time, possibly infinite, of the ith (j^{th}) subject in treatment group $X(Y)$. At a given calendar time t, one can compute, on the basis of the observed data from the two treatment groups, a rank statistic of the general form considered by Tsiatis (1982):

$$S_n(t) = \sum_{i=1}^{n'} \delta_i'(t) Q_n(t, X_i(t)) \left\{ 1 - \frac{m_{n,t}'(X_i(t))}{m_{n,t}'(X_i(t)) + m_{n,t}''(X_i(t))} \right\}$$

$$- \sum_{j=1}^{n''} \delta_j''(t) Q_n(t, Y_j(t)) \frac{m_{n,t}'(Y_j(t))}{m_{n,t}'(Y_j(t)) + m_{n,t}''(Y_j(t))}, \qquad (11.3.2)$$

where $m_{n,t}'(s) = \sum_{i=1}^{n'} I_{\{X_i(t) \geq s\}}$, $m_{n,t}''(s) = \sum_{j=1}^{n''} I_{\{Y_j(t) \geq s\}}$, and $Q_n(t, s)$ is some weight function satisfying certain measurability assumptions. The case $Q_n \equiv 1$ corresponds to the logrank statistic. Letting $H_{n,t}$ denote a product-limit-type estimator of the common distribution function of the two treatment groups under the null hypothesis, based on $\{(X_i(t), \delta_i(t), Y_j(t), \delta_j(t)) : i \leq n', j \leq n''\}$, Prentice's (1978) generalization of the Wilcoxon statistic is the statistic (11.3.2) with $Q_n(t, s) = 1 - H_{n,t}(s)$. It was extended by Harrington and Fleming (1982) to the case $Q_n(t, s) = (1 - H_{n,t}(s))^\rho$ with $\rho \geq 0$. Let F and G denote the distribution functions of X_i and Y_j, respectively. Assuming the $\{T_i'\}, \{T_j''\}, \{\xi_i'\}, \{\xi_j''\}$ to be i.i.d. sequences, Tsiatis (1982) showed that under the null hypothesis $H_0 : F = G, (S_n(t_1), \ldots, S_n(t_k))/\sqrt{n}$ has a limiting multivariate normal distribution for any k and $0 \leq t_1 < \ldots < t_k$, for a large class of two-sample rank statistics.

Assuming a Lehmann (proportional hazards) family of the form $1 - G(s) = (1 - F(s))^{1-\theta}$, Jones and Whitehead (1979) considered the use of time-sequential logrank statistics $S_n(t)$ to test sequentially over time the one-sided null hypothesis $H_0' : \theta \leq 0$. They suggested plotting $S_n(t)$ versus Mantel's (1966) estimate $V_n(t)$ of the variance of $S_n(t)$ under $F = G$. They argued heuristically that $\{(V_n(t), S_n(t)), t \geq 0\}$ should behave approximately like $\{(v, W(v)), v \geq 0\}$, where $W(v)$ is the standard Wiener process under $\theta = 0$ and is a Wiener process with drift coefficient depending on θ under alternatives near 0. Using this Wiener process approximation, they suggested replacing $(v, W(v))$ in a sequential test for the sign of the drift of a Wiener process by $(V_n(t), S_n(t))$ to construct a corresponding sequential logrank test of H_0'. In particular, they considered the case where the sequential test based on $(v, W(v))$ was an SPRT. Sellke and Siegmund (1983) established weak convergence of $\{S_n(t)/\sqrt{n}, t \geq 0\}$ to a zero-mean Gaussian process with independent increments under $F = G$ and general arrival and withdrawal patterns. In doing so, they provided a rigorous asymptotic justification of the heuristics of Jones and Whitehead (1979) under $H_0 : \theta = 0$. Gu and Lai (1991) later showed that $\{(V_n(t)/n, S_n(t)/\sqrt{n}), t \geq 0\}$ converges weakly to $\{(v, W(v)), v \geq 0\}$ under contiguous proportional hazards alternatives, where $W(v)$ is a Wiener process with $EW(v)/v = c$. This helped provide a rigorous asymptotic justification of the heuristics of Jones and Whitehead (1979) under $H_1 : \theta = c/\sqrt{n}$. Besides Mantel's (1966) estimate, another commonly used estimate (which was actually used by the Data and Safety Monitoring Board of BHAT) of the variance of $S_n(t)$ under $F = G$ is $\#_n(t)/4$, where $\#_n(t)$ is the total number of deaths (from both treatment groups) up to time t (Siegmund (1985), p. 129). Thus, $S_n(t)/\sqrt{\#_n(t)/4}$ is the so-called "standardized logrank statistic" at time t in Section 11.2.1.

For a general weight function of the form $Q_n(t, s) = \psi(H_{n,t}(s))$ in (11.3.2), Gu and Lai (1991) showed that $\{S_n(t)/\sqrt{n}, t \geq 0\}$ converges weakly to a Gaussian process with independent increments and variance function $V(t)$ under the null hypothesis and contiguous alternatives. The mean function of the limiting Gaussian process is 0 under the null hypothesis and is of the form $\mu_g(t)$ under contiguous alternatives that satisfy

$$\int_0^{t^*} \left| \frac{d\Lambda_G}{d\Lambda_F} - 1 \right| d\Lambda_F = O(\frac{1}{\sqrt{n}}), \quad \sqrt{n} \left\{ \frac{d\Lambda_G}{d\Lambda_F}(s) - 1 \right\} \to g(s)$$

as $n \to \infty$, uniformly over closed subintervals of $\{s \in [0, t^*] : F(s) < 1\}$. Here, Λ_F and Λ_G are the cumulative hazard functions of F and G and t^* denotes the time when the trial is supposed to end. In the case of the asymptotically optimal score function $\psi(\cdot) = g(F^{-1}(\cdot))$ for these alternatives, $\mu_g(t) = V(t)$, and consistent estimates $V_n(t)$ of $V(t)$ are given in Gu and Lai (1991, pp. 1420-1421). Therefore the Jones-Whitehead framework can be extended from the logrank score function to a general ψ. In practice, the actual alternatives are unknown and μ_g need not even be monotone when ψ is not optimal for the actual alternatives, such as using logrank statistics for non-proportional hazards alternatives. This means that time-sequential tests based on $S_n(t)$ can achieve *both* savings in study duration and increase in power over the fixed-duration test based on $S_n(t^*)$, as shown by Gu and Lai (1991,1998).

Scharfstein et al. (1997) and Jennison and Turnbull (1997) recently provided proofs of the "folk theorem" that the time-sequential joint distributions of maximum likelihood, or maximum partial likelihood, or semiparametric efficient estimators based on longitudinal data or censored survival data have the following "standard" covariance structure: Let $\hat{\theta}_k$ be the estimator of the true parameter vector θ at the kth interim analysis. Then, under certain regularity conditions, the asymptotic distribution of $(\hat{\theta}_1^T, \ldots, \hat{\theta}_K^T)^T$ is multivariate normal with $\mathrm{Cov}(\hat{\theta}_{k_1}, \hat{\theta}_{k_2}) = I^{-1}(k_2, \theta)$ for all $1 \leq k_1 \leq k_2 \leq K$, where $I(k, \theta)$ is the cumulative information matrix for the assumed parametric/semiparametric model. This is equivalent to the *independent increments* property for the asymptotic distribution of the statistics $\{I(k, \theta)$ $(\hat{\theta}_k - \theta) : 1 \leq k \leq K\}$. Although their results provide a unified theory for linear models with correlated observations, generalized linear mixed models, generalized estimating equations, and proportional hazards regression commonly used in longitudinal studies and survival analysis, the theory does not cover the case where the true model does not belong to the assumed parametric/semiparametric family. An example of such a situation would be when logrank statistics are used for nonproportional hazards alternatives discussed in the preceding paragraph.

It is well-known that tests of treatment effects based on the rank statistics (11.3.2) may lose substantial power when the effects of other covariates are strong. In non-sequential trials, a commonly used method to remedy this when logrank statistics are used is to assume the proportional hazards regression model and to use Cox's partial likelihood approach to adjust for other covariates. Tsiatis et al. (1985) and Gu and Ying (1995) have developed group sequential tests using this approach. A general asymptotic theory for time-sequential methods in proportional hazards regression models with applications to covariate adjustment is given by Bilias et al. (1997). Instead of relying on the proportional hazards model to adjust for concomitant variables, it is useful to have other methods for covariate adjustment, especially in situations where score functions other than logrank are used in (11.3.2) to allow for the possibility of non-proportional hazards alternatives. Lin (1992) and Gu and Lai (1998) developed alternative covariate adjustment methods based on rank estimators and M-estimators in accelerated failure time models and established the associated asymptotics.

11.3.2   Two Time Scales and Some Stopping Rules

As pointed out by Lan and DeMets (1989), there are two time scales in interim analysis of clinical trials with time-to-event endpoints. One is calendar time t and the other is the "information time" $V_n(t)$, which is typically unknown before time t unless restrictive assumptions are made *a priori*. To apply the error spending approach to time-to-event responses, one needs an *a priori* estimate of the null variance of $S_n(t^*)$. Let v_1 be such an estimate. Although the null variance of $S_n(t)$ is expected to be nondecreasing in t under the asymptotic independent increments property, its estimate $V_n(t)$ may not be monotone, and we can redefine $V_n(t_i)$ to be $V_n(t_{i-1})$ if $V_n(t_i) < V_n(t_{i-1})$. Let $A : [0, v_1] \to [0, 1]$ be a nondecreasing function with $A(0) = 0$ and $A(v_1) = \alpha$, which can be viewed as the error spending function of a stopping rule τ, taking values in $[0, v_1]$, of a Wiener process. The repeated significance test whose boundary is generated by $A(\cdot)$ stops at time t_i for $1 \le i < K$ if $V_n(t_i) \ge v_1$ (in which case it rejects $H_0 : F = G$ if $|S_n(t_i)| \ge b_i V_n^{1/2}(t_i)$), or if $V_n(t_i) < v_1$ and $|S_n(t_i)| \ge b_i V_n^{1/2}(t_i)$ (in which case it rejects H_0); it also rejects H_0 if $|S_n(t^*)| \ge b_K V^{1/2}(t^*)$ and stopping has not occurred prior to $t^* = t_K$. Letting $\alpha_j = A(v_1 \wedge V_n(t_j)) - A(V_n(t_{j-1}))$ for $j < K$ and $\alpha_K = \alpha - A(V_n(t_{K-1}))$, the boundary values $b_1, \ldots, b_K$ are defined recursively by (11.3.1) in which, "$\alpha_j = 0$" corresponds to "$b_j = \infty$".

This test has Type-I error probability approximately equal to α, irrespective of the choice of A and the *a priori* estimate v_1. Its power, however, depends on A and v_1. At the design stage, one can compute the power under various scenarios to come up with appropriate choices for A, v_1 and also the score function ψ [with $Q_n(t, s) = \psi(H_{n,t}(s))$] in (11.3.2). Gu and Lai (1999) have developed a power simulation program which can be downloaded from the website http://www.sta.cuhk.edu.hk /minggao/ctrials.htm, and which computes the power of this extension of the Lan-DeMets method to failure-time data. A companion program for determining the sample size to attain prespecified power at a given alternative is also provided.

The requirement that the trial be stopped once $V_n(t)$ exceeds v_1 is a major weakness of the preceding stopping rule. Since one usually does not have sufficient prior information about the underlying survival distributions, the actual accrual rate, the rate of loss to follow-up and noncompliance, v_1 may substantially over- or under-estimate the expected value of $V_n(t^*)$. Scharfstein et al. (1997) and Scharfstein and Tsiatis (1998) have proposed re-estimation procedures during interim analyses to address this difficulty, but re-estimation raises concerns about possible inflation of the Type-I error probability.

Besides the Lan-DeMets-type boundary described above, the power simulation and sample size determination programs of Gu and Lai (1999) also provide the user with two other methods for generating the stopping boundary. One is the method of Slud and Wei (1982) that requires the user to specify positive numbers $\alpha_1, \ldots, \alpha_K$ such that $\Sigma_{j=1}^K \alpha_j = \alpha$, so that the boundary b_j for $|S_n(t_j)|/\sqrt{V_n(t_j)}$ is given by (11.3.1). The other is a Haybittle-type boundary that first chooses b and then determines c by

$$P\left\{|W(V_n(t_K))| \geq c V_n^{1/2}(t_K) \quad \text{or}\right.$$

$$\left.|W(V_n(t_i))| \geq b V_n^{1/2}(t_i) \text{ for some } i < K \,|\, V_n(t_1), \ldots, V_n(t_K)\right\} = \alpha,$$
$$(11.3.3)$$

where $\{W(v), \ v \geq 0\}$ is a standard Brownian motion independent of $\{(X_i, \xi_i', T_i', Y_i, \xi_i'', T_i''), \ i \geq 1\}$. Choosing $b = 2.65$ as in Section 11.2.2 for BHAT again leads to stopping at the 6th interim analysis, but with a somewhat different form of the Haybittle rule than that in Section 11.2.2. Note the difference between (11.3.1) and (11.3.3).

11.3.3 Theory of Group Sequential Boundaries

In the preceding section we have described several classes of stopping boundaries for interim analysis of failure-time data in clinical trials. They involve different levels of complexity for implementation. The simplest approach is that of Slud and Wei (1982), which selects positive numbers $\alpha_1, \ldots, \alpha_K$ that sum to α and determines the boundary by (11.3.1). How the α_j's are chosen, however, plays an important role in its performance. The error spending approach of Lan and De Mets (1983) uses an error spending function of a suitably chosen stopping boundary for testing the drift of a Wiener process to determine the α_j, but it requires an *a priori* estimate of the null variance of $S_n(t^*)$. Updating this estimate during interim analyses as in Scharfstein et al. (1997) and Scharfstein and Tsiatis (1998) adds further to the complexity of the procedure. The Haybittle-type boundary requires the user to specify the threshold b for the first $K - 1$ interim analyses and then determines the terminal threshold c by (11.3.3). In the clinical trial design program developed by Gu and Lai (1999), the default option is to set b such that $P\{|\Sigma_{i=1}^j Z_i| \geq b\sqrt{j} \text{ for some } j \leq k-1\} = \alpha/3$, where $Z_1, Z_2, \ldots$ are independent standard normal random variables. In fact, for any user-specified b, the program first checks whether the above probability is $\leq \alpha/3$ and resets b in the event that this probability exceeds $\alpha/3$.

So far we have focused on two-sided tests of the null hypothesis $F = G$. There are obvious analogs of the above classes of boundaries for the one-sided hypothesis $F \leq G$. Currently, many choices of such boundaries with different levels of complexity are available, but there is little theoretical guidance regarding which one to choose. Since these time-sequential boundaries are essentially extensions (based on Gaussian approximations described in Section 11.3.1) of those used in group sequential tests of $H_0 : \theta = 0$ (or $H_0 : \theta \leq 0$ in the one-sided case) for the mean θ of normal random variables, one may try to look for such guidance from the theory of group sequential tests. However, although there is now a relatively complete theory of fully sequential tests of composite hypotheses in multiparameter exponential families, surprisingly little has been done in developing a corresponding theory for group sequential tests. A major difficulty in extending the theory of fully sequential tests of H_0 based on $Z_1, Z_2, \ldots$ to their group sequential counterparts is the dilemma between good power properties of the test, under the constraint of a prescribed upper bound M on the sample size of the group sequential test, and savings in the expected sample size. The upper bound M on the sample size is typically dictated by the time and financial resources available for the trial and other factors. Note that to test $H_0 : \theta = 0$ versus a simple alternative (or $H_0 : \theta \leq 0$ versus $H_1 : \theta > 0$), the most powerful test (sequential or otherwise) based on no more than M observations is the Neyman-Pearson test with a fixed sample size M, which means no reduction in expected sample size for sequential tests that can take no more than M observations and attain the desired power at a given alternative.

Lai and Shih (2002) recently developed a theory of group sequential tests, based on $Z_1, Z_2, \ldots$, for the parameter θ of an exponential family of densities $e^{\theta z - \psi(\theta)}$ with respect to some σ-finite measure. Let $S_n = Z_1 + \ldots + Z_n$. To test the one-sided hypothesis $H_0 : \theta \leq \theta_0$ at significance level α, suppose no more than M observations are to be taken. The fixed-sample-size test that rejects H_0 if $S_M \geq c_\alpha$ has maximal power at any alternative $\theta > \theta_0$, in particular at the alternative $\theta(M)$ 'implied' by M (in the sense that M can be derived from the assumption that the above fixed-sample-size test has some prescribed power $1 - \tilde{\alpha}$ at $\theta(M)$). Although the protocol of a clinical trial typically justifies its choice of sample-size by stating some conventional level (such as 80% or 90%) at a specified alternative, one often does not have much information prior to a clinical trial to come up with a realistic alternative.

Lai and Shih (2002) began by considering the problem of minimizing $E(T)$ among all group sequential tests with k groups, prespecified group sizes $n_1, n_2 - n_1, \ldots, n_k - n_{k-1}$, and error probabilities not exceeding α and $\widetilde{\alpha}$ at θ_0 and θ_1. This constrained minimization problem can be reduced to a Bayesian optimal stopping problem that can be solved numerically by backward induction. In the case where n_k exceeds the sample size n^* of the Neyman-Pearson test of θ_0 versus θ_1 with error probabilities α and $\widetilde{\alpha}$, the optimal stopping rule is characterized by two boundaries $a_{n_i} < b_{n_i}$ ($i = 1, \ldots, k - 1$) so that stopping occurs if $S_{n_i} \geq b_{n_i}$ (rejecting $H_0 : \theta = \theta_0$) or if $S_{n_i} \leq a_{n_i}$ (accepting H_0). If stopping has not occurred in the first $k - 1$ interim analyses, then H_0 is rejected at the kth analysis if $S_{n_k} \geq c$, where c is so chosen that

$$\sum_{j=1}^{k-1} P_{\theta_0} \left\{ S_{n_j} \geq b_{n_j}, \ a_{n_i} < S_{n_i} < b_{n_i} \quad \text{for} \quad i < j \right\}$$
$$+ P_{\theta_0} \left\{ S_{n_k} \geq c, \ a_{n_i} < S_{n_i} < b_{n_i} \quad \text{for} \quad i < k \right\} = \alpha.$$

Note that although the rejection region in favor of the treatment at each of the k times is one-sided, there is also a lower stopping boundary ('futility boundary') that stops the trial when it becomes futile to demonstrate efficacy of the treatment within the resources allocated to the trial. Lai and Shih (2002) found the following asymptotic approximations to the rejection and futility boundaries as $\alpha + \widetilde{\alpha} \to 0$.

Theorem 11.3.1 *Let $\theta_0 < \theta^* < \theta_1$ be such that $I(\theta^*, \theta_0) = I(\theta^*, \theta_1)$. Let $\alpha + \widetilde{\alpha} \to 0$ such that $\log \alpha \sim \log \widetilde{\alpha}$.*

(i) The sample size n^* of the Neyman-Pearson test of θ_0 versus θ_1 with error probabilities α and $\widetilde{\alpha}$ satisfies $n^* \sim |\log \alpha| / I(\theta^*, \theta_0)$.

(ii) For $L \geq 1$, let $\mathcal{T}_{\alpha, \tilde{\alpha}, L}$ be the class of stopping times T, with possible values $n_1 < \ldots < n_k$ such that $n_k = n^* + L$ and

$$\liminf_{\alpha + \tilde{\alpha} \to 0} (n_i - n_{i-1}) / |\log(\alpha + \tilde{\alpha})| > 0, \tag{11.3.4}$$

which are the stopping times of group sequential tests with error probabilities not exceeding α and $\widetilde{\alpha}$ at θ_0 and θ_1. Then, for given θ and L, there exists $\tau \in \mathcal{T}_{\alpha, \tilde{\alpha}, L}$ that stops sampling when

$$(\theta - \theta_0) S_{n_i} - n_i(\psi(\theta) - \psi(\theta_0)) \geq b \ or \ (\theta - \theta_1) S_{n_i} - n_i(\psi(\theta) - \psi(\theta_1)) \geq \widetilde{b} \tag{11.3.5}$$

for $1 \leq n \leq k - 1$, with $b \sim |\log \alpha| \sim \tilde{b}$, and such that

$$E_\theta(\tau) \sim \inf_{T \in \mathcal{T}_{\alpha,\tilde{\alpha},L}} E_\theta(T) \sim n_\nu + \rho(\theta)(n_{\nu+1} - n_\nu),$$

where $0 \leq \rho(\theta) \leq 1$, $m_{\alpha,\tilde{\alpha}}(\theta) = \min\{|\log \alpha|/I(\theta,\theta_0), |\log \tilde{\alpha}|/I(\theta,\theta_1)\}$ and ν is the smallest $j(\leq k)$ such that $n_j \geq (1 - \epsilon_{\alpha,\tilde{\alpha}})m_{\alpha,\tilde{\alpha}}(\theta)$ for some positive $\epsilon_{\alpha,\tilde{\alpha}} \to 0$.

Under the resource constraint of M on the sample size, it is desirable to adapt to the information on the actual θ gathered during the course of the trial, allowing early stopping at times of interim analysis so that the test has nearly optimal power and expected sample size properties. To achieve these goals in a group sequential test with k groups and group sizes $n_1, n_2 - n_1, \ldots, n_k - n_{k-1}$ so that $n_k = M$, Lai and Shih (2002) use a rejection region of the form $S_{n_k} \geq c$ at the kth analysis, where $c > c_\alpha$ but c does not differ much from c_α. For the first $k-1$ analyses, they adapt the stopping region (11.3.5) to the unknown parameter θ by using the maximum likelihood estimate $\hat{\theta}_{n_i} = (\psi')^{-1}(S_{n_i}/n_i)$ to replace θ. Letting $I(\theta, \lambda)$ denote the Kullback-Leibler information number $E_\theta\{\log[f_\theta(X_i)/f_\lambda(X_i)]\}$, this leads to a stopping region of the form

$$\hat{\theta}_{n_i} > \theta_0 \quad \text{and} \quad n_i I(\hat{\theta}_{n_i}, \theta_0) \geq b, \text{ or} \qquad (11.3.6a)$$

$$\hat{\theta}_{n_i} < \theta(M) \quad \text{and} \quad n_i I(\hat{\theta}_{n_i}, \theta(M)) \geq \tilde{b}, \qquad (11.3.6b)$$

for $1 \leq i \leq k - 1$. If (11.3.6a) holds, reject H_0 upon stopping. If stopping occurs with (11.3.6b), accept H_0. In case stopping does not occur in the first $k - 1$ analyses, reject H_0 if $S_{n_k} \geq c$. The thresholds $b, \tilde{b}$ and c are so chosen that P_{θ_0} (Test rejects H_0) $= \alpha$ and the power of the test at $\theta(M)$ does not differ much from its upper bound $1 - \tilde{\alpha}$. A simple way to choose $b, \tilde{b}$ and c satisfying these properties is as follows: Let $0 < \epsilon < 1/2$ and define $\tilde{b}$ by the equation

$$P\left[\{\hat{\theta}_{n_i} < \theta(M)\} \cap \{n_i I(\hat{\theta}_{n_i}, \theta(M)) \geq \tilde{b}\} \text{ for some } 1 \leq i \leq k - 1\right] = \epsilon\tilde{\alpha},$$

where "P" stands for "$P_{\theta(M)}$". After determining $\tilde{b}$, define b and then

c by the equations

$$\sum_{j=1}^{k-1} P_{\theta_0} \left\{ \widehat{\theta}_{n_j} > \theta_0 \quad \text{and} \quad n_j I(\widehat{\theta}_{n_j}, \theta_0) \geq b, \right.$$

$$n_i I(\widehat{\theta}_{n_i}, \theta_0) 1_{\{\hat{\theta}_{n_i} > \theta_0\}} < b \quad \text{and} \quad n_i I(\widehat{\theta}_{n_i}, \theta(M)) 1_{\{\hat{\theta}_{n_i} < \theta(M)\}} < \tilde{b}$$

$$\left. \text{for} \quad i < j \right\} = \epsilon \alpha,$$

$$P_{\theta_0} \left\{ S_{n_k} \geq c, \; n_i I(\widehat{\theta}_{n_i}, \theta_0) 1_{\{\hat{\theta}_{n_i} > \theta_0\}} < b \quad \text{and} \quad n_i I(\widehat{\theta}_{n_i}, \theta(M)) 1_{\{\hat{\theta}_{n_i} < \theta(M)\}} \right.$$

$$\left. < \tilde{b} \quad \text{for} \quad i < k \right\} = (1 - \epsilon)\alpha.$$

Let $\tilde{\tau}$ be the sample size of the test and $\tilde{\beta}(\theta)$ be its power function. Lai and Shih (2002) proved the following theorem, showing that the test attains the asymptotically minimal value (assuming known θ) of the expected sample size in Theorem 11.3.1 and also has power at $\theta(M)$ comparable to its upper bound $1 - \tilde{\alpha}$.

Theorem 11.3.2 *Let $\alpha + \tilde{\alpha} \to 0$ such that $\log \alpha \sim \log \tilde{\alpha}$.*

  (i) *For every fixed θ, $E_\theta(\tilde{\tau}) \sim n_\nu + \rho(\theta)(n_{\nu+1} - n_\nu)$, where ν and $\rho(\theta)$ are given in Theorem 11.3.1 with $\theta_1 = \theta(M)$.*

  (ii) *$\tilde{\beta}(\theta(M)) \sim 1 - \tilde{\alpha} - \kappa_\epsilon \tilde{\alpha}$, where*

$$\kappa_\epsilon = \epsilon + \left\{ (1 - \epsilon)^{-(\theta(M)-\theta^*)/(\theta^*-\theta_0)} - 1 \right\} \sim \left\{ 1 + (\theta(M) - \theta^*)/(\theta^* - \theta_0) \right\} \epsilon$$

as $\epsilon \to 0$, and $\theta_0 < \theta^* < \theta(M)$ is defined by $I(\theta^*, \theta_0) = I(\theta^*, \theta(M))$.

When the Z_i's are normal with variance $1, \widehat{\theta}_n$ is the sample mean and $I(\theta, \lambda) = (\theta - \lambda)^2/2$. To test $H_0 : \theta \leq 0$, consider the alternative $\theta(M) = \theta_1 > 0$. The futility boundary (11.3.6b) of the preceding group sequential test can be written as $S_{n_i} \leq -(2\tilde{b}n_i)^{1/2} + \theta_1 n_i$, while its rejection boundary is of the form $S_{n_i} \geq (2bn_i)^{1/2}$ for $1 \leq i \leq k - 1$, and $S_{n_k} \geq c$ for $i = k$. Thus, except for the additional futility boundary $S_{n_i} \leq -(2\tilde{b}n_i)^{1/2} + \theta_1 n_i$, the preceding group sequential test is simply the one-sided version of the Haybittle-type test considered in Section 11.3.2, where we have substituted c by $c\sqrt{n_k}$. Lai and Shih (2002) have also extended this group sequential theory to two-sided tests of $\theta = \theta_0$, with and without futility boundaries, and derived the Haybittle-type test of Section 11.3.2 as a special case (for normal Z_i's) of these two-sided tests without futility boundaries. This asymptotic theory and the simulation studies in Lai and Shih (2002) show that the Haybittle-type tests have nearly optimal power and expected sample size properties over a wide range of alternatives, subject to prescribed constraints on the Type-I error probability and maximum sample size. The general form of these Haybittle-type tests in the context of one-parameter exponential families involves estimating the unknown parameter θ by maximum likelihood during the course of the trial and using it to replace θ in the approximately optimal test (11.3.5) that assumes θ to be known. Moreover, the Haybittle-type tests can be easily modified as in Section 11.3.2 for interim analysis, based on time-sequential rank statistics, of clinical trials with failure-time endpoints.

11.3.4 Stochastic Curtailment

The Data and Safety Monitoring Board of BHAT, which was designed as a fixed-duration (instead of time-sequential) trial, actually did not monitor the trial with a stopping boundary. Instead, it used the *conditional power*, which is the conditional probability of rejecting the null hypothesis at the scheduled end of the trial given the current data, along with some speculation about the future data. See DeMets et al. (1984). This concept, which was developed for monitoring the trial, was later described in Lan et al. (1982). The setting assumed was that of a Wiener process $W(v)$, $0 \leq v \leq 1$, with drift coefficient μ. Consider the one-sided fixed-sample-size test of $H_0 : \mu = 0$ versus $H_1 : \mu = \mu_1$ (> 0) based on $W(1)$ with Type-I error probability α and Type-II error probability $\tilde{\alpha}$. Since the conditional distribution of $W(1)$ given $\{W(v), v \leq s\}$ is normal with mean $W(s) + \mu(1 - s)$ and variance $1 - s$, the conditional power at μ given $\{W(v), v \leq s\}$ is

$$\beta_s(\mu) = 1 - \Phi\left((1 - s)^{-1/2}\{\Phi^{-1}(1 - \alpha) - W(s) - \mu(1 - s)\}\right), \quad (11.3.7)$$

where Φ is the standard normal distribution function. Suppose one stops the trial to reject H_0 when $\beta_s(0) > \rho$, and stops to reject H_1 when $\beta_s(\mu_1) < 1 - \tilde{\rho}$ for some $\rho, \tilde{\rho}$ near 1. Then Lan et al. (1982) showed that the Type-I error probability of the test is $\leq \alpha/\rho$, while the Type-II error probability is $\leq \tilde{\alpha}/\tilde{\rho}$. They also appealed to the central limit theorem in extending this argument to asymptotically normal statistics. For BHAT, at the 6th interim analysis, the conditional power under the null trend was found to range from 0.8 (for 120 additional deaths) to 0.94 (for 60 additional deaths and to be 0.89 for the projected number of 80 additional deaths. The Data and Safety Monitoring Board, therefore, concluded that the nominal Type-I error probability would not be inflated by much if the test should be stochastically curtailed in this manner.

A general theory of stochastic curtailment for censored survival data was recently developed by Lin et al. (1999), who also showed an application (namely, monitoring a colon cancer trial). Previous applications of stochastic curtailment to monitoring time-to-event trials include Andersen (1987) on an alcoholic cirrhosis clinical trial and Pawitan and Hallstrom (1990) on the *Cardiac Arrhythmia Suppression Trial*. A Bayesian approach to stochastic curtailment has been developed by Spiegelhalter et al. (1986) using predictive power instead of conditional power.

11.3.5 Confidence Intervals in Group Sequential Designs

In sequentially designed experiments, the sample size is not fixed in advance but is a random variable that depends on the data collected so far. This creates bias in parameter estimation and introduces substantial difficulties in constructing valid confidence intervals. For samples of fixed sizes, an important methodology for bias estimation and construction of confidence intervals without distributional assumptions is the bootstrap method introduced by Efron (1987). It is therefore natural to extend the bootstrap method to sequential sampling. However, Chuang and Lai (1998) showed that bootstrap intervals for a population mean μ following group sequential tests have inaccurate coverage probabilities, even when the variance is known (say, equal to 1). In fact, letting $\bar{Z}_n = S_n/n$ and T be a stopping rule of a group sequential test, $\sqrt{T}(\bar{Z}_T - \mu)$ can no longer be regarded as an approximate pivot since its distribution depends heavily on μ. To get around this difficulty, Chuang and Lai (1998) proposed to "hybridize" the bootstrap and exact methods. To illustrate how it works, consider the simple example where the Z_i's are i.i.d. with unknown mean μ and known variance 1. If the Z_i's are known to be normal, then Rosner and Tsiatis (1988) developed the following method to construct an exact $(1 - 2\alpha)$ level confidence interval for μ from $(T, \bar{Z}_T)$. For each value of μ, one can find by the recursive numerical integration algorithm of Armitage et al. (1969) the quantiles $u_\alpha(\mu)$ and $u_{1-\alpha}(\mu)$ that satisfy

$$P_\mu\{(S_T - \mu T)/\sqrt{T} < u_\alpha(\mu)\} = \alpha = P_\mu\{(S_T - \mu T)/\sqrt{T} > u_{1-\alpha}(\mu)\}. \tag{11.3.8}$$

Hence, the confidence region $\{\mu : u_\alpha(\mu) \leq (S_T - \mu T)/\sqrt{T} \leq u_{1-\alpha}(\mu)\}$ has coverage probability $1 - 2\alpha$. One way of relaxing the assumption of normally distributed Z_i's with mean μ and variance 1 is to assume that $Z_i - \mu$ has some unknown distribution G that has mean 0 and variance 1. After stopping, we can estimate G by the empirical distribution $\widehat{G}_T$ of $(Z_i - \bar{Z}_T)/\widehat{\sigma}_T$, $1 \leq i \leq T$, where $\widehat{\sigma}_T^2 = T^{-1} \sum_{i=1}^{T}(Z_i - \bar{Z}_T)^2$. Let $\epsilon_1, \epsilon_2, \cdots$ be i.i.d. with distribution $\widehat{G}_T$ and let $Z_i' = \mu + \epsilon_i$. Let T' be the stopping rule T applied to $Z_1', Z_2', \cdots$ (instead of $Z_1, Z_2, \cdots$). In analogy with (11.3.8), define the quantiles $\widehat{u}_\alpha(\mu)$ and $\widehat{u}_{1-\alpha}(\mu)$ of the distribution of $(\sum_{i=1}^{T'} \epsilon_i)/\sqrt{T'}$ given $\widehat{G}_T$. An approximate $1 - 2\alpha$ confidence set is

$$\{\mu : \widehat{u}_\alpha(\mu) < \sqrt{T}(\bar{Z}_T - \mu) < \widehat{u}_{1-\alpha}(\mu)\}. \tag{11.3.9}$$

For every fixed μ, the quantiles $\widehat{u}_\alpha(\mu)$ and $\widehat{u}_{1-\alpha}(\mu)$ in (11.3.9) can be computed by simulation. An algorithm is developed to compute (11.3.9) in Chuang and Lai (1998) where the authors also prove the second-order accuracy of (11.3.9) and present simulation studies showing that (11.3.9) compares favorably with the exact confidence interval when the X_i's are normal. The method is also extended in Chuang and Lai (1998) to more complex situations involving nuisance parameters. Chuang and Lai (2000) subsequently developed a relatively complete theory of the hybrid resampling approach. As they pointed out, there are three steps in implementing hybrid resampling. First, one must choose a root $R(\mathbf{Z}, \theta)$, where $\mathbf{Z}$ denotes the vector of observations and θ is the unknown parameter of interest. For example, $R(\mathbf{Z}, \mu) = \sqrt{T}(\bar{Z}_T - \mu)$ in (11.3.9). Secondly, one needs to find a suitable resampling family $\{\widehat{F}_\theta, \theta \in \Theta\}$, where Θ denotes the set of all possible values of θ. Finally, an "implicit" hybrid region of the form $\{\theta : \widehat{u}_\alpha(\theta) < R(\mathbf{Z}, \theta) < \widehat{u}_{1-\alpha}(\theta)\}$ (such as (11.3.9)) has to be inverted into an "explicit" confidence interval.

The hybrid resampling approach developed in Chuang and Lai (1998,2000) assumes equal group sizes. To extend the approach to unequal group sizes, one has to use an ordering method introduced by Siegmund (1978). Suppose that $Z_1, Z_2, \ldots$ are i.i.d. normal random variables with known variance 1 and unknown mean μ, and T is a two-sided stopping rule of the form $T = \inf\{n \in J : S_n \geq b_n$ or $S_n \leq a_n\}$, where $S_n = Z_1 + \ldots + Z_n$ and J is a finite set of positive integers. Siegmund orders the sample space of (T, S_T) as follows: $(t, s) > (t', s')$ whenever (i) $t = t'$ and $s > s'$, or (ii) $t < t'$ and $s \geq b_t$, or (iii) $t > t'$ and $s' \leq a_{t'}$. Let μ_c denote the value of μ for which, $P_\mu\{(T, S_T) \geq (t, s)_{obs}\} = c$, with $(t, s)_{obs}$ denoting the observed value of (T, S_T). Siegmund's confidence interval is $\mu_\alpha \leq \mu \leq \mu_{1-\alpha}$, which has coverage probability $1 - 2\alpha$. Note that this ordering only involves a consideration of possible sample paths that stop (or do not stop if downcrossing of the lower boundary is observed) prior to the observed stopping time. Hence a hybrid resampling version that removes the assumption of normality in Siegmund's method does not require one to generate data beyond the observed stopping time t. See Chuang and Lai (1998) for details and the second-order accuracy of the method. Although the set J considered by Siegmund (1978) is nonrandom, we can still apply it to the case of a random $J = \{n_1, \ldots, n_K\}$ and condition on $(n_1, \ldots, n_K)$ to extend the argument in Chuang and Lai (1998), thereby establishing the second-order accuracy of the hybrid resampling method when the random vector $(n_1, \ldots, n_K)$ is independent of $\{X_i, i \geq 1\}$. See Lai and Li (2002), where a comparative study with an alternative ordering method proposed by Emerson and Fleming (1990) is also given. This alternative ordering method in the latter article is based on normally distributed observations and utilizes the multivariate central limit theorem.

Lai and Li (2002) also generalized this approach (namely, the one that combines sample space ordering with hybrid resampling) to time-sequential survival data in constructing confidence intervals for the logarithm of the hazard ratio of the treatment to the control group. See the last paragraph of Section 11.2.2 for an application to BHAT. Consider the logrank statistic (11.3.2) with $Q_n \equiv 1$ and a group sequential test of $H_0 : F = G$ with stopping rule of the form $\tau = \min\{t_j : |S_n(t_j)| \geq b_j V_n^{1/2}(t_j)\}$, where $t_1 < \ldots < t_K$ denote the calendar times of interim analysis, and $V_n(t)$ is either Mantel's estimate of the null variance of $S_n(t)$ or (total number of deaths up to time t)/4, as used in Section 11.2. The null hypothesis can be rephrased as $H_0 : \theta = 1$ in the Lehmann family $1 - F = (1 - G)^\theta$, or, equivalently, in the proportional hazards model $\Lambda_F = \theta \Lambda_G$, where θ is the hazard-ratio and Λ_F, Λ_G denote the cumulative hazard functions of F and G, respectively. Let P_β denote the probability measure under which $\theta = e^\beta$. Thus, β is the regression parameter in Cox's hazard regression model with covariate that takes the value 1 (representing treatment) or 0 (representing control), for which, $S_n(t)$ is the efficient score statistic. For small β (such that $\sqrt{n}\,\beta \to \mu$), $\{(S_n(t_j)/\sqrt{n}, V_n(t_j)/n) : 1 \le j \le K\}$ converges in distribution to $\{(W(V(t_j)), V(t_j)) : 1 \le j \le K\}$, where $W(\cdot)$ is a Wiener process with drift coefficient μ. This suggests the following ordering of the sample space, which reduces to Siegmund's ordering in the case of confidence intervals for means considered in the preceding paragraph. Let $\Psi_t = S_n(t)/V_n(t)$. Order the sample space of (τ, Ψ_τ) by the following relation:

$$(\tau_1, \Psi_{\tau_1}^{(1)}) < (\tau_2, \Psi_{\tau_2}^{(2)}) \quad \text{if and only if} \quad \Psi_{\tau_1 \wedge \tau_2}^{(1)} < \Psi_{\tau_1 \wedge \tau_2}^{(2)}. \tag{11.3.10}$$

As in the normal mean case, let $p(\beta) = P_\beta\{(\tau, \Psi_\tau) > (\tau, \Psi_\tau)_{\text{obs}}\}$. Then $\{\beta : \alpha < p(\beta) < 1 - \alpha\}$ is a confidence set for β with coverage probability $1 - 2\alpha$. Even if the baseline distribution G should the known, the probability $p(\beta)$ has to be evaluated by simulation. In practice G is unknown and we can replace it by Breslow's estimate $\widehat{G}$ from all the data at the end of the trial. This suggests replacing $p(\beta)$ by

$$\widehat{p}(\beta) = P\{(\tau^{(\beta)}, \Psi_{\tau^{(\beta)}}^{(\beta)}) > (\tau, \Psi_\tau)_{\text{obs}}\},$$

where the superscript (β) means that the observations are generated by hybrid resampling from the baseline distribution $\widehat{G}$, with β as the hazard ratio. When $\widehat{p}(\beta)$ is an increasing function of β, the hybrid resampling confidence interval $\{\beta : \alpha < \widehat{p}(\beta) < 1 - \alpha\}$ with approximate coverage probability $1 - 2\alpha$ becomes an interval whose endpoints $\underline{\beta} < \bar{\beta}$ are defined by $\widehat{p}(\underline{\beta}) = \alpha, \widehat{p}(\bar{\beta}) = 1 - \alpha$. Details on the implementation of the procedure are given in Lai and Li (2002), where simulation studies and analytic results show that the confidence intervals thus constructed have coverage probabilities close to nominal values.

11.4 CONCLUDING REMARKS

A large literature on group sequential methods for clinical trials has emerged after BHAT, and there have been many new developments in methodology and applications. See the recent monographs by Jennison and Turnbull (2000) and Whitehead (1997) and the references therein. A number of long-standing problems concerning calendar time versus information time in the design and analysis of time-sequential trials with failure-time endpoints have been recently resolved. Unlike the period when BHAT was being monitored, there is now a relatively complete methodology to design and perform interim and final analyses of such trials. Since interim reviews of the data are usually incorporated in the design and execution of long-term clinical trials at least for the purpose of monitoring safety of the treatments, these trials are particularly suited to group sequential designs. Data and Safety Monitoring Boards are now routinely set up for such trials, and major advances in information technology have made it possible to process all the data available just before the Data and Safety Monitoring Board meets, in contrast to the BHAT experience in which a substantial amount of additional data (including 8 more deaths) remained to be processed after termination of the trial (see Section 11.2.1).

REFERENCES

[1] Andersen, P.K. (1987). Conditional power calculations as an aid in the decision whether to continue a clinical trial. *Contr. Clin. Trials*, **8**, 67-74.

[2] Armitage, P., McPherson, C.K. and Rowe, B.C. (1969). Repeated significance tests on accumulating data. *J. Roy. Statist. Soc., Ser. A*, **132**, 235-244.

[3] Beta-Blocker Heart Attack Trial Research Group (1982). A randomized trial of propranolol in patients with acute myocardial infarction. *J. Amer. Med. Assoc.*, **147**, 1707-1714.

[4] Beta-Blocker Heart Attack Trial Research Group (1984). Beta-Blocker Heart Attack Trial: Design, methods and baseline results. *Contr. Clin. Trials*, **5**, 382-437.

[5] Bilias, Y., Gu, M.G. and Ying, Z. (1997). Towards a general asymptotic theory for Cox model with staggered entry. *Ann. Statist.*, **25**, 662-682.

[6] Chuang, C.S. and Lai, T.L. (1998). Resampling methods for confidence intervals in group sequential trials. *Biometrika*, **85**, 317-352.

[7] Chuang, C.S. and Lai, T.L. (2000). Hybrid resampling methods for confidence intervals (with discussions). *Statist. Sinica*, **10**, 1-50.

[8] DeMets, D.L., Hardy, R., Freedman, L.M. and Lan, G.K.K. (1984). Statistical aspects of early termination in the Beta-Blocker Heart Attack Trial. *Contr. Clin. Trials*, **5**, 362-372.

[9] Efron, B. (1987). Better bootstrap confidence intervals (with discussions). *J. Amer. Statist. Assoc.*, **82**, 171-200.

[10] Emerson, S.S. and Fleming, T.R. (1990). Parameter estimation following group sequential hypothesis testing. *Biometrika*, **77**, 875-892.

[11] Gu, M.G. and Lai, T.L. (1991). Weak convergence of time-sequential rank statistics with applications to sequential testing in clinical trials. *Ann. Statist.*, **19**, 1403-1433.

[12] Gu, M.G. and Lai, T.L. (1998). Repeated significance testing with censored rank statistics in interim analysis of clinical trials. *Statist. Sinica*, **8**, 411-423.

[13] Gu, M.G. and Lai, T.L. (1999). Determination of power and sample size in the design of clinical trials with failure-time endpoints and interim analyses. *Contr. Clin. Trials*, **20**, 423-438.

[14] Gu, M.G. and Ying, Z. (1995). Group sequential methods for survival data using partial likelihood score processes with covariate adjustment. *Statist. Sinica*, **5**, 793-804.

[15] Harrington, D.P. and Fleming, T.R. (1982). A class of rank test procedures for censored survival data. *Biometrika*, **69**, 553-566.

[16] Haybittle, J.L. (1971). Repeated assessments of results in clinical trials of cancer treatment. *Brit. J. Radiol.*, **44**, 793-797.

[17] Jennison, C. and Turnbull, B.W. (1997). Group sequential analysis incorporating covariate information. *J. Amer. Statist. Assoc.*, **92**,

1330-1341.

[18] Jennison, C. and Turnbull, B.W. (2000). *Group Sequential Methods with Applications to Clinical Trials.* Chapman & Hall/CRC: New York.

[19] Jones, D. and Whitehead, J. (1979). Sequential forms of logrank and modified logrank tests for censored data. *Biometrika,* **66**, 105-113.

[20] Lai, T.L. and Li, W. (2002). Confidence intervals in group sequential trials with random group sizes and applications to survival analysis. *Tech. Rep., Dept. of Statist., Stanford Univ.*, Stanford, California.

[21] Lai, T.L. and Shih, M.C. (2002). Power, sample size and adaptation considerations in the design of group sequential trials. *Tech. Rep., Dept. of Statist., Stanford Univ.*, Stanford, California.

[22] Lan, K.K.G. and DeMets, D.L. (1983). Discrete sequential boundaries for clinical trials. *Biometrika,* **70**, 659-663.

[23] Lan, K.K.G. and DeMets, D.L. (1989). Group sequential procedures: calendar versus information time. *Statist. Med.,* **8**, 1191-1198.

[24] Lan, K.K.G., Simon, R. and Halperin, M. (1982). Stochastically curtailed tests in long-term clinical trials. *Sequential Anal.,* **1**, 209-219.

[25] Lin, D.Y. (1992). Sequential logrank tests adjusting for covariates with accelerated life model. *Biometrika,* **79**, 523-529.

[26] Lin, D.Y., Yao, Q. and Ying, Z. (1999). A general theory on stochastic curtailment for censored survival data. *J. Amer. Statist. Assoc.,* **94**, 510-521.

[27] Mantel, N. (1966). Evaluation of survival data and two new rank order statistics arising in their consideration. *Canc. Chemo. Rep.,* **50**, 163-170.

[28] O'Brien, P.C. and Fleming, T.R. (1979). A multiple testing procedure for clinical trials. *Biometrics,* **35**, 549-556.

[29] Pawitan, Y. and Hallstrom, A. (1990). Statistical interim monitoring of the Cardiac Arrhythmia Suppression Trial. *Statist. Med.*, **9**, 1081-1090.

[30] Pocock, S.J. (1977). Group sequential methods in the design and analysis of clinical trials. *Biometrika*, **64**, 191-199.

[31] Prentice, R.L. (1978). Linear rank statistics with right censored data. *Biometrics*, **65**, 167-179.

[32] Rosner, G.L. and Tsiatis, A.A. (1988). Exact confidence intervals following a group sequential trial: A comparison of methods. *Biometrika*, **75**, 723-729.

[33] Scharfstein, D.O. and Tsiatis, A.A. (1998). The use of simulation and bootstrap in information-based group sequential studies. *Statist. Med.*, **17**, 75-87.

[34] Scharfstein, D.O., Tsiatis, A.A. and Robins, J.M. (1997). Semiparametric efficiency and its implication on the design and analysis of group sequential studies. *J. Amer. Statist. Assoc.*, **92**, 1342-1350.

[35] Sellke, T. and Siegmund, D. (1983). Sequential analysis of the proportional hazards model. *Biometrika*, **70**, 315-326.

[36] Siegmund, D. (1978). Estimation following sequential tests. *Biometrika*, **65**, 341-349.

[37] Siegmund, D. (1985). *Sequential Analysis: Tests and Confidence Intervals*. Springer-Verlag: New York.

[38] Slud, E.V. and Wei, L.J. (1982). Two-sample repeated significance tests based on the modified Wilcoxon statistic. *J. Amer. Statist. Assoc.*, **77**, 862-868.

[39] Spiegelhalter, D.J., Freedman, L.S. and Blackburn, P.R. (1986). Monitoring clinical trials: Conditional or predictive power? *Contr. Clin. Trials*, **7**, 8-17.

[40] Tsiatis, A.A. (1981). The asymptotic joint distribution of the efficient scores test for the proportional hazards model calculated over time. *Biometrika*, **68**, 311-315.

[41] Tsiatis, A.A. (1982). Repeated significance testing for a general class of statistics used in censored survival analysis. *J. Amer. Statist. Assoc.*, **77**, 855-861.

[42] Tsiatis, A.A., Rosner, G.L. and Tritchler, D.L. (1985). Group sequential tests with censored survival data adjusting for covariates. *Biometrika*, **72**, 365-373.

[43] Wallis, A.W. (1980). The Statistical Research Group. *J. Amer. Statist. Assoc.*, **75**, 320-334.

[44] Whitehead, J. (1997). *The Design and Analysis of Sequential Clinical Trials*, 2nd ed. Wiley: Chichester.

Address for communication:

TZE LEUNG LAI, Department of Statistics, Stanford University, Stanford, CA 94305-4065, U.S.A. E-mail: lait@stat.stanford.edu

Chapter 12

Applications of Sequential Tests to Target Tracking by Multiple Models

X. RONG LI
University of New Orleans, New Orleans, USA

TUMULESH K. S. SOLANKY
University of New Orleans, New Orleans, USA

12.1 INTRODUCTION

Multiple-model (MM) method is a powerful, robust, and adaptive approach for solving numerous practical problems. It has many engineering applications, ranging from target tracking to fault detection and identification, and from biomedical signal processing to modern navigation systems.

In the MM approach for solving the problem of estimating a deterministic or random process given a related random (observation) process, a bank of elemental filters [1] operates in parallel at every time, each based on an individual model; the overall estimate is obtained by a certain combination (fusion) of estimates from the individual filters. The

---

[1] A filter estimates a process in the case where data are made available sequentially.

use of multiple models can achieve performance usually significantly superior to that of the best (single-model-based) elemental filter, similar to the superiority in modeling power of a mixture density to a single best density.

The MM method has three generations. In the first generation, which can be called autonomous MM, the elemental filters operate completely independently, and the power of this generation stems from its superior "decision" rule of fusing outputs of elemental filters. These elemental filters cooperate with one another in the second generation, known as cooperating MM, on top of the superior fusion rule. While inheriting the first two generations' strength, the third generation, known as *variable-structure* MM (VSMM), emphasizes the use of a variant set of models that adapts to the environment (as in Li and Bar-Shalom (1996)) while the first two generations have a fixed set of models. The second and third generations are particularly popular for target tracking. For a survey of the MM approach to estimation, the reader is referred to Li (1996). As for target tracking, an informative reference would be Li and Jilkov (2003). The VSMM approach provides state-of-the-art solutions to a number of complex engineering problems, in particular, target tracking, and is promising for many other applications. An easily accessible account of VSMM approach in the context of target tracking is given in Li (2000a).

The most important and rather difficult task in the VSMM approach is *model-set adaptation* (MSA) — determination of the model set in real time based on the data available sequentially. MSA is usually decomposed into two sub-problems: Model-set activation (or generation) and model-set termination. In general, while the former determines (that is, activates or generates) a family of candidate model-sets $M_1, \ldots, M_N$, the latter selects from the candidates $M_1, \ldots, M_N$ the smallest model-set having the largest probability of including the true model of the process being estimated. Typical examples of MSA problem are

Problem 1: Is it better to delete a subset M_1 from the current model-set M?

Problem 2: Given the current model-set M, which subset is better, M_1 or M_2?

Problem 3: Is it better to add *one* of the sets $M_1, \ldots, M_N$ to the current set M?

Problem 4: Is it better to delete *one* of the sets $M_1, \ldots, M_N$ from the current set M?

Problem 5: Is it better to add *some* of the sets $M_1, \ldots, M_N$ to the current set M?

Problem 6: Is it better to delete *some* of the sets $M_1, \ldots, M_N$ from the current set M?

Another important task when applying the MM approach is the design of model-sets. This is needed for all three generations of the MM methods. For more information about this difficult problem, the reader is referred to Li (2002) and Li et al. (2002, item [23]). We mention only a typical subproblem here:

Problem 7: Choose L (known) best sets from the family of candidate sets $M_1, \ldots, M_N$.

Problem 7 differs from Problems 3, 4, 5, and 6 in that it does not involve a special model-set M.

Solutions to these problems are useful for virtually all applications of MM estimation. For example, in maneuvering target tracking, the greatest challenge arises from a lack of knowledge about the motion models of the target being tracked, although quite often we do know that it belongs to a family of candidate model-sets. Another common application of the MM method is in the detection and identification of a fault in a system or process, where different types of faults are represented by different model-sets.

In general, given a set of possible models at some stage and based on sequentially available data, we are interested in addressing various queries related to the current set of models, defined in Problems 1–7. Evidently, these problems can be formulated as a statistical decision problem, particularly as a problem of testing statistical hypotheses. Here, sequential methods are preferred primarily for the following reasons.

- Observations are available sequentially.

- A sequential test usually arrives at a decision substantially faster than a non-sequential test when they are subjected to the same decision error rates. This is crucial because the model-set needs to adapt itself as quickly as possible and the pace at which it adapts amounts to the (average) sample size needed for the test. For example, Wald's *sequential probability ratio test* (SPRT) is the fastest among all tests with the same given decision error rates if the true distribution is really one of the two assumed distributions. Even in the worst case, when the true distribution lies between the two assumed ones, the SPRT usually requires a smaller sample size than a non-sequential test, provided that the allowed Type-I error probability is not too small, which is the case for virtually all MSA problems.

- Using an SPRT-based sequential test, appropriate thresholds for the test statistics can be determined (approximately) without a knowledge of the distribution of the observations, while for a non-sequential test, the optimal thresholds depend on the underlying distribution and are hard to come by.

- A sequential test does not need to determine the sample size in advance, while a non-sequential test does, which is not an easy job in the context of MSA.

However, these hypothesis testing problems are challenging in at least the following aspects.

- Candidate hypotheses (model-sets) are not necessarily disjoint. It is difficult to obtain the corresponding regions of acceptance, rejection, and continuation in an optimal fashion. Such testing problems are usually considered ill-posed and have been rarely studied.

- Since MSA has to be done as quickly as possible even at the risk of having high decision error rates, the sample size cannot be large (it is usually smaller than five). For this reason, tests with nice large-sample (asymptotic) properties are not necessarily good choices.

- Most problems involve multiple hypotheses, each being composite, and the parameters that characterize each hypothesis are usually multi-dimensional.

- The sequential observations are usually correlated and almost never independent or identically distributed.

12.2   AN EXAMPLE

In order to illustrate an application of the MM method, we consider the problem of describing the motion of an aircraft. This is necessary for tracking maneuvering aircrafts in, for example, an air traffic control system. The data used in this example were generated by a performance evaluator. More specifically, they were generated by a system that describes the measurements of the positions of an aircraft that may move in a variety of motion patterns over some time period. For example, the variety of motion patterns for an aircraft could include a constant velocity model, a constant-turn model or some other model. The actual pattern, but not its distribution, is known to the evaluator, and is reflected indirectly in the data. Before we formally look at the data, in Section 12.3, we formulate and develop the statistical methodologies to solve the seven problems outlined in Section 12.1. These methodologies are then used in Section 12.4 to analyze the aircraft-data and their performance is evaluated. The data could be downloaded from the website http://ece.engr.uno.edu/Li.

Most target tracking techniques are based on Kalman filtering, which requires knowledge of target motion and its observation system. The MM method has been used in target tracking (see Li and Jilkov (2003)) to describe not only many other types of target motion (for example, missiles, ground vehicles, and ships; see Li and Jilkov (2000,2001a)), but also measurement systems (Li and Jilkov (2001b)), as well as statistics of motion noise and measurement errors (Li and Bar-Shalom (1994)).

Another major area of application of the MM method is fault detection and identification, where different sensor-, actuator-, or system-faults are represented by different models, as in Watanabe (1992), Basseville and Nikiforov (1993), and Zhang and Li (1998).

Many other applications of the MM method can be found in the survey by Li (1996) and in the numerous references cited therein, including piecewise linearization of nonlinear systems, time-invariant partitioning of time-varying systems, adaptive control, guidance and navigation, adaptive filtering and identification, flight control, seismic signal processing, biomedical signal processing, and geodesy.

12.3   FORMULATION AND STATISTICAL METHODS

Let s be the time-invariant unknown *true mode* in effect during the time period over which, a test of hypotheses is performed. In the hypothesis testing terminology, s is the unknown parameter on which, hypotheses are tested. For convenience, let m be a generic hypothesized value of

s; it represents the parameter of a generic *model* in the MM approach. A set of hypothesized values of s is denoted by M and referred to as a *model-set* in the MM approach for obvious reasons. Since we only deal with the MM approach here, we assume throughout that model-sets have finitely many elements. However, the results presented here are valid for many other hypothesis testing scenarios, including those with many other classes of model-sets. Finally, denote by S the unknown set of possible values of the true mode s.

It has been shown in Li (1996) that the optimal model-set for the MM approach is $M = S$. The performance of MM estimators deteriorates if either extra models are used $(M \supset S)$ or some models are missing $(M \subset S)$. The deterioration worsens as M and S become more mismatched. Given the same degree of mismatch, however, the case of missing models is usually worse than the case of having extra models.

With these effects, given a collection of not necessarily disjoint model-sets $M_1, \ldots, M_N$, various definitions of the best model-set have been proposed in Li (2002) and Li et al. (2002) for a variety of purposes. For example, it may be defined as the one with the smallest cardinality among the model-sets with the largest probability[2] of including the true mode s. In other words, the best among $M_1, \ldots, M_N$ is the set M_j with cardinality $|M_j| = \min_{q \in Q} |M_q|$, where

$$P\{s \in M_q\} = \max_{i \in \{1,\ldots,N\}} P\{s \in M_i\}, \quad \forall q \in Q \subset \{1, \ldots, N\}. \quad (12.3.1)$$

The use of this definition would direct us to a probability-based model-set adaptation. In this article, however, we present an alternative approach based on sequential hypothesis testing.

12.3.1   Sequential Solutions for Problems Involving Two Model-Sets

Problem formulation

Consider **Problem 2** first. It is formulated naturally as:

$$H_1 : s \in M_1 \quad \text{vs.} \quad H_2 : s \in M_2 \quad (12.3.2)$$

and the continuation region corresponds to M. Such a hypothesis testing problem is unconventional and extremely difficult in general because

---

[2]Prior or posterior, depending on whether data are available.

the sets M_1 and M_2 may have common elements. A partinent question is: which model-set should be chosen if the true mode is in the intersection? To the authors' knowledge, there are no general results available for such non-disjoint hypothesis testing problems. Actually, if such a hypothesis testing problem is a proper formulation of a problem of detecting abrupt changes, then the change is not detectable by the definition of detectability based on Kullback information. See, for example, Basseville and Nikiforov (1993). Fortunately, the particular problem here can be solved optimally, as presented below.

Since both hypotheses H_1 and H_2 are composite, the requirements for the Type-I and Type-II errors are usually specified as

$$\max_{m \in M_1} P\{\text{``}H_2\text{''}|s = m\} \leq \alpha, \quad \max_{m \in M_2} P\{\text{``}H_1\text{''}|s = m\} \leq \beta \qquad (12.3.3)$$

or

$$\max_{m \in (M_1 - M_2)} P\{\text{``}H_2\text{''}|s = m\} \leq \alpha, \quad \max_{m \in (M_2 - M_1)} P\{\text{``}H_1\text{''}|s = m\} \leq \beta.$$
$$(12.3.4)$$

Such requirements make a decision quite safe, but perhaps too safe to lose a good deal of efficiency in the sense of requiring a sample size that is too large. Following Wald (1947), the requirements are replaced by considering an weighted average of the Type-I error probabilities, $P\{\text{``}H_2\text{''}|s = m\}$, over the region $A = \{s \in M_1\}$ and the Type-II error probabilities, $P\{\text{``}H_1\text{''}|s = m\}$, over the region $R = \{s \in M_2\}$ with the following weight functions

$$\int_A dW_a(s) = 1 \quad \text{and} \quad \int_R dW_r(s) = 1. \qquad (12.3.5)$$

Then, the question is how to select the weight functions $dW_a(s)$ and $dW_r(s)$. This is not trivial in general, though Wald (1947) suggested a number of ways. For MSA, however, a very natural and intuitively appealing choice is:

$$dW_a(s) = P\{s = m|s \in M_1\}, \quad dW_r(s) = P\{s = m|s \in M_2\} \qquad (12.3.6)$$

This choice satisfies the normalization requirements (12.3.5). As such, we propose in effect the use of the following modified requirements on the *expected* error probabilities:

$$P\{\text{``}H_2\text{''}|s \in M_1\} \leq \alpha, \quad \text{and} \quad P\{\text{``}H_1\text{''}|s \in M_2\} \leq \beta, \qquad (12.3.7)$$

where

$$P\{\text{``}H_p\text{''}|s \in M_q\} = \sum_{m \in M_q} P\{\text{``}H_p\text{''}|s = m\}P\{s = m|s \in M_q\}.$$

We do so because $P\{s = m|s \in M_q\}$, known as model probability, is available in an MM estimator. See Li (1996).

Model-set probability and likelihood

Let z_k be the observation at time k and z^k be the σ-algebra generated by the observation sequence $\{z_\kappa\}_{\kappa \leq k}$ through time k. Let $\tilde{z}_k = z_k - E[z_k|z^{k-1}]$ be the observation residual at time k, that is, the part of z_k which is "unpredictable" from the past. It is available (approximately) from an MM estimator. Let $\tilde{z}^k = \{\tilde{z}_\kappa\}_{\kappa \leq k}$ be the sequence of the observation residuals through time k. Note that the observation sequence $\{z_k\}$ itself is highly correlated, but the residual sequence $\{\tilde{z}_k\}$ is de-correlated. In the Gaussian case, $\{\tilde{z}_k\}$ is independent. In the general case, $\{\tilde{z}_k\}$ is weakly correlated because the computed $E[z_k|z^{k-1}]$ is approximate.

Since the task is to decide on a model-set, the probabilities and likelihoods of a model-set are naturally of major interest.

The *marginal likelihood* of a model-set M_j at time k is the sum of the probabilities $P\{s = m|s \in M_j, z^{k-1}\}$ multiplied by the marginal likelihoods [3] $p[\tilde{z}_k|s = m, z^{k-1}]$, over all the *models* in M_j:

$$
\begin{aligned}
L_k^{M_j} &\equiv p[\tilde{z}_k|s \in M_j, z^{k-1}] \\
&= \sum_{m \in M_j} p[\tilde{z}_k|s = m, z^{k-1}]P\{s = m|s \in M_j, z^{k-1}\}.
\end{aligned}
$$

(12.3.8)

Both $p[\tilde{z}_k|s = m, z^{k-1}]$ and $P\{s = m|s \in M_j, z^{k-1}\}$ are available from the MM algorithm approximately.

The *joint* likelihood of the model-set M_j is defined as $L_{M_j}^k = p[\tilde{z}^k|s \in M_j]$. Note that a subscript k and a superscript k are used for quantities at k and through k, respectively. The *joint* likelihood ratio $\Lambda^k = L_{M_1}^k/L_{M_2}^k$ of *model-set* M_1 to M_2 can often[4] be approximated

---

[3] The likelihood defined here differs from the usual one $p[z_k|s = m, z^{k-1}]$. Likewise for the joint likelihood.

[4] For example, this is exact if $\log(\Lambda^k)$ sequence is a random walk.

for computation by a product of model-set marginal likelihood ratios:

$$\Lambda^k = \prod_{k_0 \le \kappa \le k} \frac{L_\kappa^{M_1}}{L_\kappa^{M_2}} \tag{12.3.9}$$

where k_0 is the test starting time. This simplifies computation greatly. It is important for MSA, where the observation sequence itself is usually not independent. However, the sequence of marginal likelihood ratios defined here is usually approximately independent, since the observation residual sequence is at most weakly correlated and the likelihood ratio is approximately normally distributed under some mild conditions. See, for example, Ross (1987).

The (posterior) probability that the true mode is in a model-set M_j at time k is defined as

$$\mu_k^{M_j} = P\{s \in M_j | s \in \mathbf{M}_k, z^k\} = \sum_{m \in M_j} P\{s = m | s \in \mathbf{M}_k, z^k\},$$
$$\tag{12.3.10}$$

which is the sum of the probabilities of all models in M_j. Here $\mathbf{M}_k$ is the total model-set in effect at time k, which includes M_j as a subset and is problem dependent. The model probability $P\{s = m | s \in \mathbf{M}_k, z^k\}$ for each model m is available (approximately) from an MM estimator.

Optimal sequential tests

Theorem 12.3.1 *Model-Set Sequential Likelihood Ratio Test (MS-SLRT): For Problem 2 with (12.3.2) and the modified requirements (12.3.7), the following SPRT-based test is optimal (that is, uniformly most efficient):*

- *Choose M_1 if $\Lambda^k \ge B$*

- *Choose M_2 if $\Lambda^k \le A$*

- *Use M and continue to test with more observations if $A < \Lambda^k < B$*

where A and B are two positive constants to be determined.

Instead of choosing the conditional and, thus, distinct model probability $P\{s = m_i | s \in M_l\}$ as the weight function, as in (12.3.6), unconditional and, thus, common model probabilities may also be chosen for Problem 2:

$$dW_a(s) = P\{s = m | s \in M\}, \qquad \forall m \in M_1;$$
$$dW_r(s) = P\{s = m | s \in M\}, \qquad \forall m \in M_2. \tag{12.3.11}$$

Then, we have the following test based on the ratio of model-set probabilities, rather than likelihoods.

Theorem 12.3.2 *Model-Set Sequential Probability Ratio Test (MS-SPRT): For Problem 2 with (12.3.2) and the specified expected error probabilities*

$$P\{\text{``}H_2\text{''}, s \in M_1 | s \in M\} \leq \alpha', \quad P\{\text{``}H_1\text{''}, s \in M_2 | s \in M\} \leq \beta', \tag{12.3.12}$$

the following SPRT-based test is optimal (that is, uniformly most efficient):

- *Choose M_1 if $P^k \geq B'$*

- *Choose M_2 if $P^k \leq A'$*

- *Use M and continue to test with more observations if $A' < P^k < B'$*

where A' and B' are two positive constants to be determined and $P^k = \mu_k^{M_1} / \mu_k^{M_2}$ is the model-set probability ratio.

Consider now **Problem 1**. We formulate it as:

Delete M_1 if $H_1 : s \in M_1$ is rejected;

Keep M_1 (that is, adopt M) if $\bar{H}_1 : s \in \overline{M_1}$ is rejected;

Use M if evidence is inconclusive;

where $\overline{M_1} = M - M_1$. Why do we use M (rather than M_1) if $\bar{H}_1$ is rejected? The reason is that the question is whether to delete M_1 from M or not (that is, $\overline{M_1}$ or $M-M_1$ is not an option at all). There are practical reasons why M should be used while the true mode is indeed unlikely to be in $\overline{M_1}$. One such reason is to capture future mode jumps, and another is to facilitate future model adaptation and initialization of new models and filters, apart from accounting for the possibility of a decision error. The price paid for this practice in the context of VSMM estimation is usually low. It usually leads to some extra computation rather than significant performance deterioration even if M_1 is actually better than M.

With this formulation, it is seen that Problem 1 is solved by the above theorems with $M_2 = \overline{M_1}$, since they are valid for arbitrary $M_1 \subset M$ and $M_2 \subset M$ and the theorems have nothing to do with the actions taken after a decision is made.

Discussions

The tests outlined in Theorems 12.3.1 and 12.3.2 are actually SPRT's using the model-set joint likelihood ratio and the probability ratio, respectively. $P\{s = m | s \in M_l, z^{k-1}\}$ is nothing but the probability at time k using data z^{k-1} of mode $s = m$ in model-set M_l. These probabilities are available from the MM estimator based on the model-set M_l. This is a unique characteristic of MSA and, thus, the key to the usefulness of the above theorems. It is this characteristic that enables the technique of weight function to convert a composite hypothesis-testing problem into a simple one with a clear physical justification. These tests also have the additional advantage that virtually no extra computation is needed to carry them out. Note also that the test in Theorem 12.3.2 is more compatible with the definition of the best model-set in (12.3.1).

The use of the weight functions (12.3.6) or (12.3.11) converts the composite and possibly non-disjoint hypothesis-testing problem into a simple one. As a result, any overlap between model sets M_1 and M_2 poses no problem after this conversion. Here, the model-set likelihood (or probability) acts exactly in the same way as the likelihood for a simple hypothesis-testing problem. This is the key to Theorems 12.3.1 and 12.3.2. They are optimal in the sense of having the quickest MSA for the non-disjoint and composite hypothesis-testing problems.

In practice, the constants A and B can be treated as design parameters, obtained from the tuning of an MSA algorithm. Their relationships with the decision error probabilities are given by Wald's approximation:

$$A = \frac{\beta}{1 - \alpha}, \qquad B = \frac{1 - \beta}{\alpha} \qquad (12.3.13)$$

which are exact if Λ^k (or P^k) can only be either in (A, B) or on the boundaries A or B (that is, there is no *overshoot* or excess over the boundaries) and are very accurate if the amount of overshoot is small. In case there is an overshoot, the above values of A and B are more conservative than the optimal choices.

The weight functions used in Theorem 12.3.1 satisfy the normalization requirements (12.3.5), but those in Theorem 12.3.2 do not. However, the latter can be justified as follows. To satisfy the normalization requirements, the following weight functions could be used instead:

$$dW_a(s) \;\; = \;\; P\{s = m | s \in M\}/c_a$$

$$dW_r(s) = P\{s = m | s \in M\}/c_r$$

where $0 < c_a, c_r \leq 1$ are factors such that the normalization requirements (12.3.5) are satisfied. However,

$$\frac{P\{\text{``}H_2\text{''}, s \in M_1 | s \in M\}}{c_a} \leq \alpha, \quad \frac{P\{\text{``}H_1\text{''}, s \in M_2 | s \in M\}}{c_r} \leq \beta$$

is equivalent to (12.3.12) with

$$\alpha' = c_a \alpha, \qquad \beta' = c_r \beta \qquad (12.3.14)$$

whereas (12.3.7) and (12.3.12) are different. The latter uses the same weight for the two error probabilities under the same system mode, which seems more reasonable than using different weights, as in the case of the former. This is achieved at the price of losing the normalization property. Their allowable error probabilities differ and thus the corresponding optimal tests are different.

Theorems 12.3.1 and 12.3.2 are closely related. The test statistic in Theorem 12.3.1 is the *product* of ratios of model-set *likelihoods* (for the present and the past), while in Theorem 12.3.2, it is the ratio of posterior model-set probabilities *at the present time only*, which depends on the past as well as current information. Specifically, posterior probability ratio = joint likelihood ratio × prior probability ratio. Thus, MS-SPRT is better than MS-SLRT if prior probabilities of the model-sets are available, which is usually the case for MSA. In fact, the normalization requirements imply $c_a = P\{s \in M_1\}, c_r = P\{s \in M_2\}$. However, although (12.3.7) and (12.3.12) are equivalent to (12.3.14), the tests in Theorems 12.3.1 and 12.3.2 are distinct because the likelihood ratio in Theorem 12.3.1 accounts for the factors c_a and c_r, which is not the case for Theorem 12.3.2.

For testing a simple hypothesis $H_0 : \theta = \theta_0$ against a simple alternative $H_1 : \theta = \theta_1$, the optimality of the SPRT does not hold if the true θ takes on a value other than θ_0 or θ_1. In fact, if θ is between θ_0 and θ_1, the SPRT may require a sample size even larger than a non-sequential test. The above two tests may suffer from a similar deficiency when $s \notin M_l$. Note, however, that this deficiency is less serious for MSA since the likelihoods in (12.3.8) are given over a set of points, rather than just a single point. A possible remedy for this deficiency is to consider the Kiefer-Weiss problem (see Ghosh and Sen (1991), and Siegmund (1985)) of *finding a test δ with specified bounds for the error probabilities that* minimizes the maximum expected sample size, $\min\limits_{\delta} \sup\limits_{\theta} E[N|\theta]$, for all values of θ. Unfortunately, a satisfactory solution of this problem is not available, although asymptotically optimal solutions are available. For example, there is a so-called 2-SPRT test of Lorden for distributions in the exponential family (see Huffman (1983)). SPRT is usually very efficient in the case where the error probabilities are not very small even if the true θ is between θ_0 and θ_1. This is important for MSA since it has to be done as quickly as possible and, thus, cannot afford to have a large sample size (that is, small error probabilities).

Another significant drawback of SPRT is that it does not have a guaranteed finite upper bound on its stopping time. In other words, SPRT may keep running for an arbitrarily long time. Since a finite bound on the stopping time is important for MSA, the above theorems can be modified to guarantee such a bound as well as decision error probabilities, following a SPRT-type procedure developed recently in Zhu et al. (2002).

Remark 12.3.1 Note that Problem 2 is somewhat related to a problem one typically considers under Gupta's (1956) subset selection setup, in the general area of *multiple comparisons*. Along the lines of the subset selection approach, one may consider the following version of Problem 2. Given the current model-set M, the goal is to find the subset M^* of M, so that the true model is inside the set M^* with some prespecified probability. However, for the target tracking purpose, the formulation given in Problem 2 is more directly applicable as it allows one to have non-disjoint sets M_1 and M_2. Also, in Problem 2, the subsets M_1 and M_2 are not random, whereas in the subset selection approach, the selected subset M^* of M is random. One may look at Gupta and Huang (1981), Gupta and Panchapakesan (1991), and Mukhopadhyay and Solanky (1994) for subset selection and other common formulations available in the area of multiple comparisons.

12.3.2 Sequential Solutions for Multihypothesis Problems

Problem formulation

Consider **Problem 3** first. It is formulated as:

$$H : s \in M \quad \text{vs.} \quad H_1 : s \in M_1 \quad \cdots \quad \text{vs.} \quad H_N : s \in M_N \quad (12.3.15)$$

with the following action: Add M_i (that is, adopt $M \cup M_i$ instead of M_i) if H_i is accepted; do not add M_i (that is, keep using M) if H is accepted; use M to continue the test if the evidence is inconclusive.

These are multihypothesis testing problems, which are much more difficult and complicated than the binary problems. The first difficulty with multihypothesis problems is the determination of a proper performance criterion. When testing binary hypotheses with a fixed sample, it is customary to seek admissible tests that minimize β for a bounded α since they cannot be minimized simultaneously. Alternatively, we may seek an admissible test that minimizes a linear combination of the error probabilities. Both criteria lead to a likelihood ratio test. For a multihypothesis testing problem with a fixed sample-size, there is no natural way to choose tests by controlling some error probabilities while minimizing others. One may consider minimizing a linear combination of the error probabilities. This, however, does not provide an explicit control of any individual error probability, which is actually quite important in MSA.

Sequential methods are more natural for multihypothesis problems. Although it is generally impossible to minimize expected sample size under all hypotheses while controlling all error probabilities, as the SPRT does for binary problems, tests are available that can control all correct-decision probabilities (for example, Sobel-Wald test) or all error probabilities (for example, the Armitage test, the Simons test, and the Lorden test). One is referred to Ghosh and Sen (1991) and Ghosh et al. (1997). However, most techniques available in the statistical literature are not satisfactory for the multihypothesis testing problems of MSA because they do not take advantage of the peculiarities of MSA. For example, MSA has to be done as quickly as possible even at the cost of moderately higher error probabilities and, thus, tests have to be performed with an extremely small sample size. Consequently, asymptotic optimality (requiring a large sample size) is of limited value here, although it provides some support for the hope that the test is also good with a small sample size. Also, many problems in MSA (such as Problems 3, 4, 5, 6) are only partially invariant under permutations of hypotheses, since the current model-set M should be treated specially but the remaining model-sets should be permutation invariant (that is, labeling of the remaining model-sets is irrelevant), while almost all multihypothesis tests proposed in the statistical literature are permutation invariant.

Solutions

In this section, we present three tests for the multihypothesis testing problems of MSA. They are all based on the MS-SPRT (or MS-SLRT).

We first briefly describe the Lorden's (1976) 2-SPRT, since the first test is based on this. Consider a problem of testing a simple hypothesis $H_1 : p = p_1$ against a simple alternative $H_2 : p = p_2$ based on i.i.d. observations coming from a common density p (with respect to some measure) that is not necessarily p_1 or p_2. In the 2-SPRT, a hypothetical hypothesis $H^* : p = p_0$ is introduced and the original problem is reformulated as the following pair of binary problems: T1 ($H_1 : p = p_1$ vs. $H^* : p = p_0$) and T2 ($H_2 : p = p_2$ vs. $H^* : p = p_0$). The 2-SPRT uses a one-sided SPRT for each problem such that it never accepts H^*. If H_1 is rejected in T1 before H_2 is rejected in T2, then H_1 is rejected; otherwise, H_2 is rejected. A nice property of the 2-SPRT is that when the hypothesis $p = p_0$ is true, the expected sample size is minimized, at least asymptotically, as $\alpha, \beta \to 0$. By selecting the correct p_0, the 2-SPRT provides an asymptotically optimal solution to the above-mentioned Kiefer-Weiss problem while controlling the error probabilities.

Multiple Model-Set Sequential Likelihood Ratio Test (*MMS-SLRT*): A solution of Problem 3 with (12.3.15) is the following test:

S1. Perform N one-sided MS-SLRTs simultaneously for N pairs of hypotheses $(H : s \in M$ vs. $H_1 : s \in M_1), \ldots, (H : s \in M$ vs. $H_N : s \in M_N)$. These tests are *one-sided* in the sense that H is never rejected, which is ensured by using thresholds B_i and $A = -\infty$. This step ends when only one of the hypotheses $H_1, H_2, \ldots, H_N$ is not rejected yet:

  – Reject all M_i for which $\Lambda_i^k = L_M^k / L_{M_i}^k \geq B_i$;

  – Continue to test for the remaining pairs until only one of the hypotheses $H_1, H_2, \ldots, H_N$ is not rejected.

Specifically, let K be the smallest sample size (time) by which some $(N - 1)$ of the N alternative hypotheses are rejected by the one-sided MS-SLRTs

$$k_i = \min \left\{ k : \Lambda_i^k \geq B_i, i = 1, \ldots, N, i \neq j \right\}$$
$$K = \min \left\{ k : k \geq k_i, i = 1, \ldots, N, i \neq j \right\}.$$

Then, the test accepts H_j if $\Lambda_j^K < B_j$, where $B_1, \ldots, B_N \geq 1$ are chosen such that the Type-I and Type-II error probabilities for all binary problems are α and β, respectively. If several hypotheses are rejected at last simultaneously, the one with the largest model-set likelihood is accepted.

S2. Perform a (two-sided) MS-SLRT to test $H : s \in M$ vs. $H_j : s \in M_j$, where H_j is the winning hypothesis in Step # 1.

One should note the following important points:

- The current set M is used in the MM estimator until the final decision in Step # 2 is reached.

- This test is only partially invariant under permutations of hypotheses since the current model-set M is treated specially.

- Since the expected sample size is asymptotically minimized in all the one-sided MS-SLRTs of Step # 1 and the MS-SLRT of Step # 2 when $H : s \in M$ is true, it seems reasonable to expect that the above MMS-SLRT has a minimum expected sample size asymptotically. This is important for model-set activation since the system mode should be in (the minimum convex superset of) the current model-set M most of the time, except when it undergoes a jump.

- If MS-SLRTs with B_i of (12.3.13) are used for all binary tests in Step # 1, then the thresholds should all be equal: $B_j = B_i = \frac{\alpha}{1-\beta}$. Alternatively, the threshold suggested in Lorden (1976) may be used.

- The model-set joint likelihood ratio may be replaced by the model-set probability ratio $P_j^k = \frac{P\{s \in M | z^k\}}{P\{s \in M_j | z^k\}}$, leading to a multiple model-set sequential probability ratio test (*MMS-SPRT*), since the latter ratio is indeed a likelihood ratio.

Another solution of Problem 3 with (12.3.15) is the following Multiple-Level Test (*MLT*):

S1 Test all N pairs of hypotheses *separately*:

$$H : s \in M \quad \text{vs.} \quad H_i : s \in M_i$$

Complete all these tests. Let $\mathcal{H}$ be the set of accepted hypotheses from all these tests. Delete H from $\mathcal{H}$ unless $\mathcal{H}$ contains no other hypothesis or H is deemed much more important than the other model-sets.

S2 Let H be the best hypothesis in $\mathcal{H}^5$. Go to Step # 1 to test H against all the other hypotheses in $\mathcal{H}$ pairwise.

Repeat this process until only one hypothesis remains and the corresponding model-set is then adopted. Before the final decision is made, use model-set M provided it is in $\mathcal{H}$ or use the model-set corresponding to the best hypothesis H for the period.

Again, one should make the following important observations:

- Clearly, Step # 1 here is more efficient than the corresponding step in MMS-SLRT or MMS-SPRT, given the same error probabilities. The weakness of this test is the possibility of multiple iterations. In VSMM estimation, the model-sets designed should not have large overlaps and a mode usually does not have a giant jump. These two facts imply that the current model-set usually may be inferior to at most a few other model-sets. Thus, statistically speaking, quite often only one, occasionally two, and rarely more iterations are needed in the MLT. Consequently, this test should be quite efficient for MSA with a proper family of candidate sets.

- The pair of error probabilities α_i and β_i used in the binary tests of Step # 1 is better chosen in such a way that these tests have about the same expected sample size, since the sample size of Step # 1 is equal to the largest sample size needed for the component problems.

- MS-SLRT or MS-SPRT may be used for the binary problems in Step # 1.

We now consider the other problems. **Problem 4** is formulated as:

$$H_0 : s \in M_0 \quad \text{vs.} \quad H_1 : s \in \overline{M_1} \quad \cdots \quad \text{vs.} \quad H_N : s \in \overline{M_N} \quad (12.3.16)$$

---

[5]The best hypothesis is the one that is not rejected in any other binary tests, and is the first hypothesis accepted (if the same error probabilities are used for all binary tests), or the one accepted with the smallest error probability (if the same expected sample size is used). Model-set likelihoods or probabilities are used to break ties if any.

with the following action: Do not delete any model-set M_i (that is, keep using M) if H_0 is accepted; delete M_i if H_i is accepted; use M to continue the test if the evidence is inconclusive. Here $\overline{M_i} = M - M_i$. Several choices of M_0 appear reasonable (for example, $M_0 = \cup_{i=1}^N M_i$) but are not really so. $M_0 = \cap_{i=1}^N M_i$ or $M_0 = (M - \cup_{i=1}^N M_i)$ turns out to be the best for most cases. Consider, for example, $N = 2$ with formulations $F4$ ($H_0 : M_0 = M - (M_1 \cup M_2) = \overline{M_1} \cap \overline{M_2}$) and $F4'$ ($H_0 : M_0' = M_1 \cup M_2$). Let $A = M_1 - M_2, B = M_1 \cap M_2, C = M_2 - M_1, D = M_0$. If $s \in A$, with $F4$, H_2 is true and thus M_2 is deleted, as desired. With $F4'$, both H_0 and H_2 are true and thus M is chosen since evidence would be inconclusive — this is not good. If $s \in B$, with $F4$, none of H_0, H_1, H_2 is true and thus M is chosen. With $F4'$, H_0 is true and thus M is chosen. The situations of $s \in C$ and $s \in A$ are symmetrical. If $s \in D$, with $F4$, H_0, H_1, H_2 are all true and thus M is chosen; with $F4'$, H_1 and H_2 are true and thus M is chosen. Consequently, $F4$ is clearly superior to $F4'$. Using $H_0 : M_0 = M_1 \cap M_2$ leads to the same results as F4 above.

With the above formulation, MMS-SLRT, MMS-SPRT, and MLT are valid for Problem 4 as well, since it is parallel to Problem 3 with the correspondence $M_0 \leftrightarrow M$ and $\overline{M_i} \leftrightarrow M_i$.

Problems 5 and 6 can be solved by repeated use of MMS-SLRT, MMS-SPRT, or MLT, discussed in the context of Problems 3 and 4, respectively. However, a more efficient solution for these problems is the *model-set probability sequential ranking test* (MSP-SRT).

MSP-SRT is a class of tests for dividing a family of model-sets $\{M_i\}_{i=1}^N$ into an *acceptance group* and a *rejection group* and is capable of solving Problems 3–7 efficiently. It reduces to MS-SPRT for Problems 1 and 2. It has multiple versions, all of which have a key *ranking* component: At each time k, rank all N_k of the model-sets $M_1, \ldots, M_N$ that have not yet rejected or accepted as $M_{(1)}, \ldots, M_{(N_k)}$ such that their model-set probabilities $\mu_k^{M_{(i)}} = P\{s \in M_{(i)} | z^k, s \in \mathbf{M}_k\}$ are in a decreasing order:

$$\mu_k^{M_{(1)}} \geq \mu_k^{M_{(2)}} \geq \cdots \geq \mu_k^{M_{(N_k)}},$$

where $\mathbf{M}_k$ is the union of all models not yet rejected, including M if it is present in the problem. Let $\mu_k^M = P\{s \in M | z^k, s \in \mathbf{M}_k\}$ and let $M_{[L]}$ be the model-set at time k such that $\mu_k^{M_{[L]}} = \mu_k^{M_{(L-J)}}$ assuming J model-sets have been accepted. Then, we do the following:

- For Problem 3: Add model-set $M_{(1)}$ if $\frac{\mu_k^{M_{(1)}}}{\mu_k^M} \geq B$ and stop the test; stop considering adding $M_{(j)}, \ldots, M_{(N_k)}$ if $\frac{\mu_k^{M_{(j)}}}{\mu_k^M} \leq A$ and continue for the remaining model-sets.

- For Problem 4: Delete $M_{(N_k)}$ if $\frac{\mu_k^{M_{(N_k)}}}{\mu_k^M} \leq A$ and stop the test; stop considering deleting $M_{(1)}, \ldots, M_{(j)}$ if $\frac{\mu_k^{M_{(j)}}}{\mu_k^M} \geq B$ and continue for the remaining model-sets.

- For Problem 5: Add $M_{(1)}, \ldots, M_{(i)}$ if $\frac{\mu_k^{M_{(i)}}}{\mu_k^M} \geq B$; stop considering adding $M_{(j)}, \ldots, M_{(N_k)}$ if $\frac{\mu_k^{M_{(j)}}}{\mu_k^M} \leq A$; continue for the remaining model-sets.

- For Problem 6: Delete $M_{(i)}, \ldots, M_{(N_k)}$ if $\frac{\mu_k^{M_{(i)}}}{\mu_k^M} \leq A$; stop considering deleting $M_{(1)}, \ldots, M_{(j)}$ if $\frac{\mu_k^{M_{(j)}}}{\mu_k^M} \geq B$; continue for the remaining model-sets.

- For Problem 7: Accept $M_{(1)}, \ldots, M_{(i)}$ if $\frac{\mu_k^{M_{(i)}}}{\mu_k^{M_{[L]}}} \geq B$; reject $M_{(j)}, \ldots$ $M_{(N_k)}$ if $\frac{\mu_k^{M_{(j)}}}{\mu_k^{M_{[L]}}} \leq A$; continue to test until exactly L model-sets have been accepted or not rejected, and thus are chosen.

The model-set probabilities can be replaced by the model-set *joint* likelihoods, resulting in a *model-set likelihood sequential ranking test* (MSL-SRT).

In the above, A and B are design parameters, which control the error probabilities.

Problems 5–7 are thus solved by sequential ranking tests without the need for an explicit formulation.

12.4 AIRCRAFT MOTION EXAMPLE: EVALUATING PERFORMANCE

In this section, we will consider the data described in Section 12.2, to describe the motion of an aircraft. First we detail the process of data generation. Consider the following linear time-invariant system with position-only measurements

$$x_{k+1} = Fx_k + w_k, \qquad z_k = \begin{bmatrix} 1 & 0 & 0 & 0 \\ 0 & 0 & 1 & 0 \end{bmatrix} x_k + v_k$$

with state $x = [x_1, \dot{x}_1, x_2, \dot{x}_2]'$, where process and measurement noise sequences w and v are white, uncorrelated with the initial state x_0, mutually uncorrelated, and have constant means $\bar{w}$, $\bar{v}$ and covariances $Q = 0.004^2 I$, $R = 100^2 I$, respectively. Two generic types of model are considered: nearly *constant-velocity* (CV) model and *constant-turn* (CT) (with a known turn rate) model, given by Li and Bar-Shalom (1993) and Bar-Shalom et al. (2001):

$$F_{CV} = \begin{bmatrix} F_2 & 0 \\ 0 & F_2 \end{bmatrix}, \qquad F_2 = \begin{bmatrix} 1 & T \\ 0 & 1 \end{bmatrix}$$

$$F_{CT} = \begin{bmatrix} 1 & \dfrac{\sin \omega T}{\omega} & 0 & -\dfrac{1-\cos \omega T}{\omega} \\ 0 & \cos \omega T & 0 & -\sin \omega T \\ 0 & \dfrac{1-\cos \omega T}{\omega} & 1 & \dfrac{\sin \omega T}{\omega} \\ 0 & \sin \omega T & 0 & \cos \omega T \end{bmatrix}$$

where $T = 5$ second is the sampling period. Denote by $3°/s$ a CT model with known turn rate $3°/s$. Note that CV corresponds to $0°/s$. These models are commonly used in maneuvering target tracking and, in particular, surveillance for air traffic control, as indicated in Bar-Shalom and Li (1995) and Bar-Shalom et al. (2001).

A performance evaluator was used to evaluate the sequential procedures of the previous sections using 100 observation sequences $\{z_k\}$, where the true mode was known to the evaluator, but not to the sequential procedures. The distribution of the true mode was known to neither the evaluator nor the procedures. The thresholds A and B below were determined by (12.3.13) with $\alpha = \beta$, and thus $AB = 1$. Therefore, only B is indicated. The range of threshold used was $9 \leq B \leq 99$, which corresponds to the Type-I and Type-II error probabilities $\alpha = \beta \in [0.01, 0.1]$ for binary problems. All results are expressed in terms of the discrete time (sample size) k, that is, in multiples of T.

First, consider two model-sets $M_1 = \{0, 3°/s\}$ and $M_2 = \{0, -3°/s\}$. For the 100 observation sequences with two cases ($2°/s$ and $5°/s$, respectively) of the truth, MS-SLRT and MS-SPRT (denoted as "likelihood"

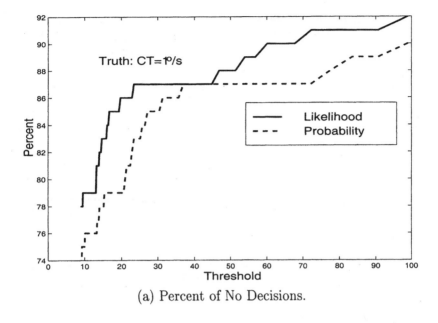

(a) Percent of No Decisions.

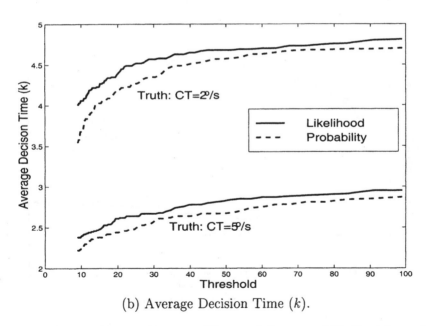

(b) Average Decision Time (k).

Figure 12.4.1: Average Decision Time and Percent of No Decisions of MS-SLRT and MS-SPRT for Two Model-Sets.

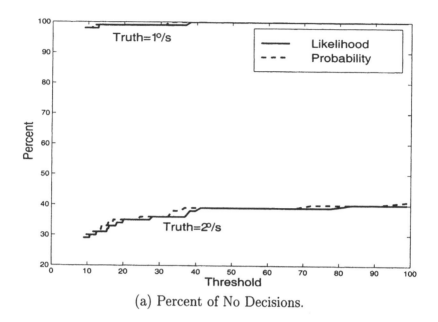

(a) Percent of No Decisions.

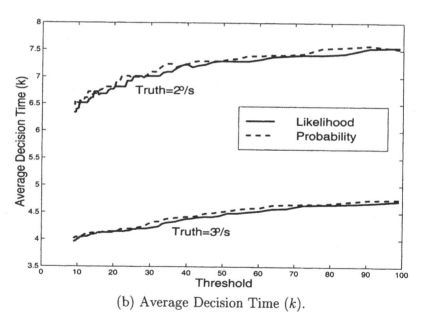

(b) Average Decision Time (k).

Figure 12.4.2: Average Decision Time and Percent of No Decisions of
MMS-SLRT and MMS-SPRT for Three Model-Sets.

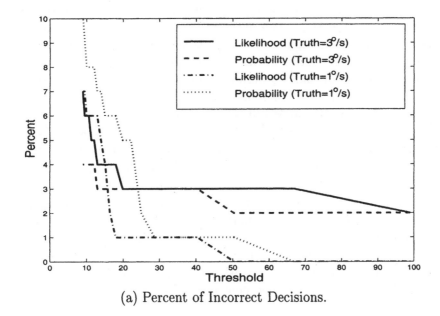

(a) Percent of Incorrect Decisions.

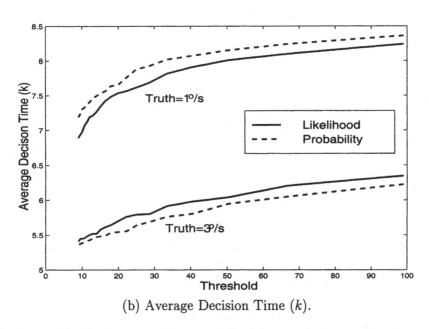

(b) Average Decision Time (k).

Figure 12.4.3: Percent of Incorrect Decisions and Average Decision Time of Sequential Ranking Tests for Five Model-Sets.

and "probability" in the plots, respectively) both led to 100% correct decisions (choose M_1). The average decision times (that is, sample sizes) are shown in Figure 12.4.1(b). When the truth was 1°/s, with a high percentage both MS-SLRT and MS-SPRT refused to make a choice between M_1 and M_2 because they both include the true model (see Figure 12.4.1(a)). The solid and dashed curves correspond to the results of MS-SLRT and MS-SPRT, respectively.

Now, consider three model-sets $M_1 = \{-1°/s, 0, 1°/s\}$, $M_2 = \{-3°/s, -1°/s, 0\}$, $M_3 = \{0, 1°/s, 3°/s\}$ with three cases of the truth, 1°/s, 2°/s, and 3°/s. The results of MMS-SLRT and MMS-SPRT are displayed in Figure 12.4.2.

When the truth was 1°/s, they refused to decide between model-sets M_1 and M_3 because they both included the true model, although M_2 was ruled out. When the truth was 2°/s, M_3 was better than M_1 but their only difference came from that of the models 3°/s and $-1°/s$, which is not large. When the truth was 3°/s, MMS-SLRT and MMS-SPRT led to 100% correct decisions (not shown) and the decision time was shorter.

Finally, consider five model-sets:

$$M_1 = \{-0.5°/s, 0, 0.5°/s\}, M_2 = \{1°/s, 4°/s\}, M_3 = \{-1°/s, -4°/s\},$$

$$M_4 = \{3°/s, 7°/s, 15°/s\}, M_5 = \{-3°/s, -7°/s, -15°/s\}$$

with two cases of the truth: 3°/s and 1°/s. The results for MSP-SRT and MSL-SRT are shown in Figure 12.4.3. In both cases, there was not a single instance of "no decision," the decision error was low, and the decision time was short. When the truth was 3°/s, M_4 rather than M_2 was chosen because M_4 included the true model while M_2 did not (although it "covered" the true model). It is interesting to note that the decision time was shorter when the truth was 3°/s than when it was 1°/s.

The following observations can be made from these evaluations: The percent of decision errors is low and the average decision time (that is, sample size) is short. If the truth lies in an area that is too risky to decide on one model-set, the tests refuse to make a decision in order to avoid a large probability of decision error. Since the probability- and likelihood-ratio tests are optimal under different criteria, one is not superior to the other in general (Figure 12.4.1 is in favor of the former, but Figure 12.4.3 is not).

More information about these tests can be found in Li (2000a,2000b), Li et al. (1999), and Li and Zhang (2000). For example, the MS-SLRT and MS-SPRT are used in the *Model-Group Switching* (MGS) algorithm and verified indirectly by its superior performance. The MMS-SLRT and MMS-SPRT are used in the extended MGS algorithm.

Sequential methodologies are widely used in statistical signal and data processing. They have found many applications in target detection, tracking, and recognition. We mention here a few examples worked out recently. Tartakovsky et al. (2001a,2003) developed a scheme for detection of targets using distributed sensors as an application of sequential procedures. An extension of that work for target recognition as well detection is reported in Tartakovsky et al. (2001b), using invariant sequential procedures. Li et al. (2002) developed an SPRT-type sequential procedure to confirm, maintain, and delete tracks, along with an analysis of the life-span of a track.

12.5 APPENDIX

Proof of Theorem 12.3.1 With the modified requirements (12.3.7), testing the composite hypothesis H_1 against the composite alternative H_2 of (12.3.2) reduces to testing a simple hypothesis against a simple alternative:

$$H_1^* : p = p[\tilde{z}^k | s \in M_1] \quad \text{vs.} \quad H_2^* : p = p[\tilde{z}^k | s \in M_2]. \qquad (12.5.1)$$

This is because the weighted sum of model likelihoods with weight function defined by (12.3.6) turns out to be the expected value of the model likelihood, which is equal to the model-set likelihood: for $l = 1, 2$,

$$p[\tilde{z}^k | s \in M_l] = \sum_{m \in M_l} p[\tilde{z}^k | s = m] P\{s = m | s \in M_l\}.$$

Since the test is actually the SPRT for the above modified simple hypothesis testing problem, the optimality of the test follows from that of the SPRT for this problem.

A regularity condition for the optimality of this test is that the sequence of model-set log-likelihood ratios $\log(\Lambda^k)$ has to be a random walk. ■

Proof of Theorem 12.3.2 This theorem follows essentially from Theorem 12.3.1 by replacing $P\{s = m | s \in M_l\}$ with $P\{s = m | s \in M\}$.

Specifically, with the modified requirements (12.3.12), testing the composite hypothesis H_1 against the composite alternative H_2 of (12.3.2) reduces to testing a simple hypothesis against a simple alternative:

$$H_1^* : p = p[\tilde{z}^k, s \in M_1] \quad \text{vs.} \quad H_2^* : p = p[\tilde{z}^k, s \in M_2]$$

The likelihood ratio of the hypotheses turns out to be the probability ratio of the model-set. Since the test is actually the SPRT for the above modified simple hypothesis-testing problem, the optimality of the test follows from that of the SPRT for this situation. ∎

ACKNOWLEDGMENT

Support for research by ONR grant N00014-00-1-0677, NSF grant ECS-9734285, and NASA/LEQSF grant (2001-4)-01 is greatly appreciated.

REFERENCES

[1] Bar-Shalom, Y. and Li, X.R. (1995). *Multitarget-Multisensor Tracking: Principles and Techniques*. YBS Publishing: Storrs.

[2] Bar-Shalom, Y., Li, X.R. and Kirubarajan, T. (2001). *Estimation with Applications to Tracking and Navigation: Theory, Algorithms, and Software*. Wiley: New York.

[3] Basseville, M. and Nikiforov, I. (1993). *Detection of Abrupt Changes: Theory and Application*. Prentice Hall: Englewood Cliffs.

[4] Ghosh, B.K. and Sen, P.K. (1991). *Handbook of Sequential Analysis*, edited volume. Marcel Dekker: New York.

[5] Ghosh, M., Mukhopadhyay, N. and Sen, P.K. (1997). *Sequential Estimation*. Wiley: New York.

[6] Gupta, S.S. (1956). On a decision rule for a problem in ranking means. *Ph.D. disertation*, Dept. of Statist., Univ. of North Carolina, Chapel Hill.

[7] Gupta, S.S. and Huang, D.Y. (1981). *Multiple Statistical Decision Theory: Recent Developments*. Springer-Verlag: New York.

[8] Gupta, S.S. and Panchapakesan, S. (1991). Sequential ranking and selection procedures. In *Handbook of Sequential Analysis* (B.K. Ghosh and P.K. Sen, eds.), 363-380. Marcel Dekker: New York.

[9] Huffman, M.D. (1983). An efficient approximate solution to the Kiefer-Weiss problem. *Ann. Statist.*, 11, 306-316.

[10] Li, X.R. (1996). Hybrid estimation techniques. In *Control and Dynamic Systems: Advances in Theory and Applications* (C.T. Leondes, ed.), 213-287. Academic Press: New York.

[11] Li, X.R. (2000a). Engineer's guide to variable-structure multiple-model estimation for tracking. In *Multitarget-Multisensor Tracking: Applications and Advances* (Y. Bar-Shalom and D.W. Blair, eds.), 499-567. Artech House: Boston.

[12] Li, X.R. (2000b). Multiple-model estimation with variable structure - Part II: Model-set adaptation. *IEEE Trans. Auto. Control*, 45, 2047-2060.

[13] Li., X.R. (2002). Model-set design, choice, and comparison for multiple-model estimation—Part I: General results. In *Proc. 2002 Int. Conf. Inform. Fusion*, 26-33. Annapolis, Maryland.

[14] Li, X.R. and Bar-Shalom, Y. (1993). Design of an interacting multiple model algorithm for air traffic control tracking. *IEEE Trans. Control Sys. Tech.* (special issue on Air Traffic Control), 1, 186-194.

[15] Li, X.R. and Bar-Shalom, Y. (1994). A recursive multiple model approach to noise identification. *IEEE Trans. Aerosp. Elec. Sys.*, 30, 671-684.

[16] Li, X.R. and Bar-Shalom, Y. (1996). Multiple-model estimation with variable structure. *IEEE Trans. Auto. Control*, 41, 478-493.

[17] Li, X.R. and Jilkov, V.P. (2000). A survey of maneuvering target tracking: Dynamic models. In *Proc. 2000 SPIE Conf. on Signal and Data Proc. of Small Targets*, 4048, 212-235. Orlando, Florida.

[18] Li, X.R. and Jilkov, V.P. (2001a). A survey of maneuvering target

tracking—Part II: Ballistic target models. In *Proc. 2000 SPIE Conf. on Signal and Data Proc. of Small Targets*, **4473**, 559-581. San Diego, California.

[19] Li, X.R. and Jilkov, V.P. (2001b). A survey of maneuvering target tracking—Part III: Measurement models. In *Proc. 2000 SPIE Conf. on Signal and Data Processing of Small Targets*, **4473**, 423–446. San Diego, California.

[20] Li, X.R. and Jilkov, V.P. (2003). A survey of maneuvering target tracking—Part V: Multiple-model methods. In *Proc. 2003 SPIE Conf. on Signal and Data Processing of Small Targets*, **5204**. San Diego, California.

[21] Li, X.R., Li, N. and Jilkov, V.P. (2002). An SPRT-based sequential approach to track confirmation and termination. In *Proc. 2002 Int. Conf. Inform. Fusion*, 951-958.

[22] Li, X.R. and Zhang, Y.M. (2000) Multiple-model estimation with variable structure—Part V: Likely-model set algorithm. *IEEE Trans. Aerosp. Elec. Sys.*, **36**, 448–466.

[23] Li, X.R., Zhao, Z.-L., Zhang, P. and He, C. (2002). Model-set design, choice and comparison for multiple-model estimation—Part II: Examples. In *Proc. 2002 Int. Conf. Inform. Fusion*, 1347-1354. Annapolis, Maryland.

[24] Li, X.R., Zhi, X.R. and Zhang, Y.M. (1999). Multiple-model estimation with variable structure—Part III: Model-group switching algorithm. *IEEE Trans. Aerosp. Elec. Sys.*, **35**, 225–241.

[25] Lorden, G. (1976). 2-SPRT's and the modified Kiefer-Weiss problem of minimizing an expected sample size. *Ann. Statist.*, **4**, 281–291.

[26] Mukhopadhyay, N. and Solanky, T.K.S. (1994). *Multistage Selection and Ranking Procedures: Second-Order Asymptotics*. Marcel Dekker: New York.

[27] Ross, W.H. (1987). The expectation of the likelihood ratio criterion. *Int. Statist. Rev.*, **55**, 315–329.

[28] Siegmund, D. (1985). *Sequential Analysis.* Springer-Verlag: New York.

[29] Tartakovsky, A.G., Li, X.R. and Yaralov, G. (2001a). Sequential detection of targets in distributed systems. In *Proc. 2001 SPIE Conf. on Signal Processing, Sensor Fusion, and Target Recognition,* **4380**, 501–513. Orlando, Florida.

[30] Tartakovsky, A.G., Li, X.R. and Yaralov, G. (2001b). Invariant sequential detection and recognition of targets in distributed systems. In *Proc. 2001 Int. Conf. on Inform. Fusion.* Montreal, Quebec, Canada, WeA3.11–WeA3.18.

[31] Tartakovsky, A.G. and Li, X.R. and Yaralov, G. (2003). Sequential detection of targets in multichannel systems, *IEEE Trans. Inform. Theory,* **49**, 425-445.

[32] Wald, A. (1947). *Sequential Analysis.* Wiley: New York.

[33] Watanabe, K. (1992). *Adaptive Estimation and Control: Paritioning Approach.* Prentice Hall: New York.

[34] Zhang, Y.M. and Li, X.R. (1998). Detection and diagnosis of sensor and actuator failures using IMM estimator. *IEEE Trans. Aerosp. Elec. Sys.,* **34**, 1293–1312.

[35] Zhu, Y.M., Zhang, K.S. and Li, X.R. (2002). An SPRT-type procedure with finite upper bound on stopping time. Submitted for publication.

Addresses for communication:

X. RONG LI, Department of Electrical Engineering, Information and Systems Laboratory, University of New Orleans, New Orleans, LA 70148, U.S.A. E-mail: xli@uno.edu
TUMULESH K.S. SOLANKY, Department of Mathematics, University of New Orleans, New Orleans, LA 70148, U.S.A. E-mail: tsolanky @uno.edu

Chapter 13

A Sequential Procedure That Controls Size and Power In a Multiple Comparison Problem

WEI LIU
University of Southampton, Southampton, U.K.

13.1   INTRODUCTION

A medical study was to be designed to compare the potency of three cardiac substances. In the study, a suitable dilution of one of the three substances was slowly infused into an anesthetized guinea pig and the dosage at which the pig died was recorded. The main research goal was to determine whether significant differences existed among the three substances in terms of potency measured as the dosage at death. The guinea pigs were quite homogeneous in weight and age, and the laboratory enviroment and measurement procedures could reasonably be assumed to be identical. So it was decided that a completely randomized design could be used to collect data. It was also decided that a potency difference of $\delta = 5$ or larger between any two subtances should be detected with a probability of at least $\beta = 0.80$, since a potency difference of $\delta = 5$ was regarded as substantial. From this information, it was clear that the problem consisted of pairwise comparisons among three treatments in a standard one-way analysis of variance layout. So, Tukey's procedure for pairwise comparisons seemed to be appropriate. It was agreed that the Type-I *familywise error* (FWE) rate should be controlled at about $\alpha = 0.05$. The difficulty was to determine how many observations should be required from each of the three treatments, since there was little information on the variances of the observations. The variance, however, was assumed to be the same for all three treatments.

The problem, therefore, could be viewed as follows. Let $X_{i1}, X_{i2}, \cdots$ denote i.i.d. observations from the ith treatment with common distribution $N(\mu_i, \sigma^2)$ where both μ_i and σ^2 are unknown constants, $i = 1, 2, 3$. An $\alpha = 0.05$ multiple test was sought for testing $H_{ij}^0 : \mu_i = \mu_j$ against $H_{ij}^a : \mu_i \neq \mu_j$, $1 \leq i \neq j \leq 3$. The test was required to have a probability of at least 80% of rejecting all the false null hypotheses H_{ij}^0 that satisfied $|\mu_i - \mu_j| \geq \delta (= 5)$.

Had the common variance of the three treatments, σ^2, been known, one possible procedure would be as follows: Take n observations from each treatment, and let $\bar{X}_i(n)$ denote the sample mean of the observations from the ith treatment, $i = 1, 2, 3$. Using Tukey's simultaneous confidence intervals for pairwise comparisons, one would

$$\text{reject } H_{ij}^0 \Leftrightarrow \sqrt{n}|\bar{X}_i(n) - \bar{X}_j(n)|/\sigma > q_3^\alpha, \ 1 \leq i \neq j \leq 3, \ (13.1.1)$$

where q_3^α is the upper α-point of the range distribution with three treatments. Hochberg and Tamhane (1987) and Hsu (1996) provide an excellent review on multiple comparison problems and procedures. From the tables in Miller (1981), one finds $q_3^{0.05} = 3.314$. Upon rejecting H_{ij}^0, one can also make the directional decision $\mu_i > \mu_j$ or $\mu_i < \mu_j$ according to whether $\bar{X}_i(n) > \bar{X}_j(n)$ or $\bar{X}_i(n) < \bar{X}_j(n)$. Since Tukey's procedure is based on a set of $(1 - \alpha)$-level simultaneous confidence intervals, the probability of wrongly rejecting a true H_{ij}^0 or making a wrong directional decision (for example, a decision of $\mu_i > \mu_j$ while in fact $\mu_i < \mu_j$) is no more than α. In order to ensure that this procedure also has the required power property, namely,

$$\inf_{\mu_i \in R} P\{\text{all false } H_{ij}^0 \text{ with } |\mu_i - \mu_j| \geq \delta \text{ are rejected}\} \geq \beta, \ (13.1.2)$$

the sample size n should be suitably chosen. Note that the power function above is clearly bounded below by

$$B(\mu_1, \mu_2, \mu_3) \ = \ P\{\text{all false } H_{ij}^0 \text{ with } |\mu_i - \mu_j| \geq \delta \text{ are rejected}$$
$$\text{with a correct directional decision}\}$$

Also, it is established in Liu (1996) that

$$\inf_{\mu_i \in R} B(\mu_1, \mu_2, \mu_3) = \int_{-\infty}^{\infty} \phi(x)\Phi(-x - q_3^\alpha + \frac{\sqrt{n}\delta}{\sigma})\Phi(x - q_3^\alpha + \frac{\sqrt{n}\delta}{\sigma})dx$$

where ϕ and Φ respectively denote the p.d.f. and c.d.f. of a standard normal distribution. The sample size n can, therefore, be chosen so that the integral above is equal to $\beta = 0.80$. From Table 3 of Liu (1996), one finds $\sqrt{n}\delta/\sigma = 5.123$ and so $n = n_0$ where

$$n_0 = (5.123\sigma/\delta)^2. \tag{13.1.3}$$

In summary, if we knew the variance σ^2, then a sample of size n_0 would be taken from each treatment. After the data were collected, one would

$$\text{reject } H_{ij}^0 \Leftrightarrow |\bar{X}_i(n_0) - \bar{X}_j(n_0)| > \frac{q_3^\alpha \delta}{5.123}, \ 1 \le i \ne j \le 3, \tag{13.1.4}$$

by using $\sigma/\sqrt{n_0} = \delta/5.123$ in (13.1.1). Directional decisions could be made accordingly as before. Of course, the actual sample size must be an integer, and can be chosen as $\langle n_0 \rangle + 1$, where $\langle x \rangle$ denotes the largest integer less than x.

The problem was, of course, that little was known about the variance σ^2 and so, the sample size formula in (13.1.3) could not be used directly. The next section provides a three-stage sampling procedure that was applied to overcome this stumbling block. It determines the sample size adaptively by mimicking the formula (13.1.3).

13.2 A THREE-STAGE PROCEDURE

Although the formula (13.1.3) could not be used to determine the sample size directly, it was needed to estimate the sample size adaptively by estimating σ^2 on the basis of the available observations. It was decided that $m_0 = 10$ observations should be taken initially from each of the three treatments, since it was felt that in order for the study to be credible, a sample of at least ten observations from each treatment was essential. On the other hand, the initial sample size m_0 should not be too large to avoid being larger than n_0 by a wide margin. So, $m_0 = 10$ observations were collected from each of the three treatments and they are given in Table 13.2.1. Based on these observations, σ^2 was estimated in the usual way by

$$\sigma_{m_0}^2 = \frac{1}{3(m_0 - 1)} \sum_{i=1}^{3} \sum_{j=1}^{m_0} (X_{ij} - \bar{X}_i(m_0))^2 = 13.25.$$

Now, the final sample size could be estimated by replacing σ^2 with $\sigma_{m_0}^2$ in (13.1.3) and, in this case, the estimate was given by

$$n_1 \equiv \langle (5.123/\delta)^2 \sigma_{m_0}^2 \rangle + 1 = 14.$$

One possible way forward was to take $n_1 - m_0$ observations, the shortfall between the estimated total sample size and the number of observations already taken, from each treatment. This was the idea behind the famous two-stage procedure proposed by Stein (1945). The potential pitfall of a two-stage approach is that the estimate of σ^2 and hence, that of the total sample size may not be very accurate. The reason is that they are based on only the first-stage samples which are often small in size. We adopted an alternative *three-stage* approach, the idea behind which was first put forward by Hall (1981). After taking the first-stage sample from each treatment and calculating the estimate of the total sample size, a second-stage sample was taken so that the observations in the first and second stages made up a sizable fraction (say, 0.9) of the estimated total sample size. All the observations in the first and second stages were then used to re-estimate σ^2 and hence n_0. Finally, a third sample was taken from each treatment to make up the shortfall between the new estimate of n_0 and the total number of observations already taken in the first two stages.

More precisely, a second sample of size $M_1 - m_0$ was taken from each treatment where

$$M_1 = \max \left\{ \langle 0.9(5.123/\delta)^2 \sigma_{m_0}^2 \rangle + 1, m_0 \right\} = \max\{13, \ 10\} = 13.$$

Note that $\langle 0.9(5.123/\delta)^2 \sigma_{m_0}^2 \rangle + 1$ is about 90% of the estimated total sample size of n_0 based on the first samples. The observations in the second-stage samples are also given in Table 13.2.1. The estimation of σ^2 based on the first-stage and second-stage samples was given by

$$\sigma_{M_1}^2 = \frac{1}{3(M_1 - 1)} \sum_{i=1}^{3} \sum_{j=1}^{M_1} (X_{ij} - \bar{X}_i(M_1))^2 = 13.28,$$

and the new estimate of n_0 was given by

$$\langle (5.123/\delta)^2 \sigma_{M_1}^2 \rangle + 1 = 14$$

A final sample of size $M_2 - M_1 = 1$ was therefore taken from each treatment, where

$$M_2 = \max \left\{ \langle (5.123/\delta)^2 \sigma_{M_1}^2 \rangle + 1, M_1 \right\} = \max\{14,\ 13\} = 14$$

The observation(s) in the third-stage samples is (are) also given in Table 13.2.1. Note that if $M_1 = m_0$, then no observations would have been taken in the second stage, and the third-stage samples would be taken directly. Similarly, if $M_2 = M_1$, then no observations would be taken in the third stage, and M_1 would be the final sample size from each treatment. From the observations collected, our inference on H_{ij}^0 was to mimic that in (13.1.4):

$$\text{reject } H_{ij}^0 \ \Leftrightarrow \ |\bar{X}_i(M_2) - \bar{X}_j(M_2)| > q_3^\alpha\, \delta/5.123 = 3.234,$$

and infer $\mu_i > \mu_j$ if H_{ij}^0 is rejected and $\bar{X}_i(M_2) > \bar{X}_j(M_2)$,

$1 \leq i \neq j \leq 3$. Since

$$\bar{X}_1(M_2) - \bar{X}_2(M_2) = -0.45,$$
$$\bar{X}_1(M_2) - \bar{X}_3(M_2) = -7.21,$$
$$\bar{X}_2(M_2) - \bar{X}_3(M_2) = -6.76,$$

H_{12}^0 was not rejected, H_{13}^0 was rejected and the inference "$\mu_3 > \mu_1$" was made, and H_{23}^0 was rejected and the inference "$\mu_3 > \mu_2$" was made. Hence the potency difference between the first two treatments is not significant, while the potency of the third treatment is significantly higher than those of the first two treatments.

Note, however, that this procedure controls only approximately the Type-I FWE rate at $\alpha = 5\%$ and the power requirement at $\beta = 80\%$, which is based on large-sample results that can be established in a way similar to Hall (1981) and Liu (1997a). Before this procedure was applied to the problem, a simulation study was carried out to assess its small-sample performance on the control of Type-I FWE rate and power. The next section presents the details of the simulation study.

Table 13.2.1: The Dataset from the Study

	Tr.# 1	Tr.# 2	Tr.# 3	Estimated σ^2
1st stage	20	20	32	
	20	30	26	
	19	13	28	
	21	24	29	
	17	18	24	
	20	22	28	$\sigma^2_{m_0} = 13.25$
	20	22	26	
	15	20	27	
	26	15	31	
	15	24	26	
2nd stage	21	16	23	
	25	17	26	$\sigma^2_{M_1} = 13.28$
	26	19	26	
3rd stage	20	23	27	$\sigma^2_{M_2} = 12.47$
Sample means	20.36	20.81	27.57	

13.3   A VALIDATION STUDY VIA SIMULATION

This simulation study was carried out before the procedure was actually applied in order to make sure that the procedure indeed satisfies the Type-I FWE rate and power requirements closely. Of course, the properties of the procedure, such as the Type-I FWE rate, the power, and the average sample size, depend on the variance σ^2 whose value is rarely known. So, it was necessary to study the procedure for a wide range of σ^2 values. The range of σ^2 values used in our simulation study was $1.5 \leq \sigma \leq 14$, which was believed to contain the true unknown value of σ^2 with a very high chance.

As the procedure attains the maximum Type-I FWE rate at the configuration $\mu_1 = \mu_2 = \mu_3 = 0$, this configuration was used in our simulation study to assess the control of Type-I FWE rate. Although it is not clear at which configuration of (μ_1, μ_2, μ_3) the power function in (13.1.2) attains its minimum, its lower bound $B(\mu_1, \mu_2, \mu_3)$ attains the minimum at $(\mu_1, \mu_2, \mu_3) = (0, \delta, 2\delta)$. So this configuration was used in our simulation study to assess the control of minimum power. Table 13.3.1 presents the results of our simulation study based on 100,000 replications.

From the last row of Table 13.3.1, for example, if the true value of σ is 14.0, then n_0 in (13.1.3) is given by 205.8. In this case, the three-stage procedure requires an average sample size of 217.1 from each treatment. Of course the actual sample size M_2 is a random variable. The standard deviation of M_2 is given by 28.48. The maximum Type-I FWE rate is 0.045, which is less than $\alpha(= 0.05)$. The minimum power is no less than 0.812, which is larger than the desired $\beta(= 0.80)$. When σ is small (for example, $\sigma = 1.5$), the procedure required no observations beyond the first stage. In this case, since the first-stage sample of 10 observations from each treatment is more than what is required, the Type-I FWE rate is 0.000, much smaller than $\alpha(= 0.05)$, while the minimum power is no less than 0.991, much larger than $\beta(= 0.80)$.

Accross the range of σ values in Table 13.3.1, the smallest value of the minimum power is 0.796. So, the procedure satisfies the power requirement very closely. On the other hand the procedure may have a Type-I FWE rate as large as 0.06 if the true value of σ happens to be 4.0. Indeed the data in Table 13.2.1 suggests that the true value of σ is around 3.5 since $\sigma_{M_2}^2 = 12.47$. Nevertheless, it was felt that a 0.06 Type-I FWE rate was acceptable for the study under consideration.

Based on the results of this simulation study, the procedure was considered to be satisfactory in the control of both the Type-I FWE rate and power, and therefore was adopted for use in the study.

13.4 CONCLUDING REMARKS

In this paper, we have considered a three-stage procedure that controls both the Type-I FWE rate and the power for multiple comparisons among three treatments. This procedure is simple to use, but can be improved in several directions, however.

Table 13.3.1: Simulation Results of the Sequential Procedure

σ	n_0	$E(M_2)$	std	FWE rate	Power
1.5	2.4	10.0	0.00	0.000	0.991
2.0	4.2	10.0	0.01	0.001	0.952
2.5	6.6	10.1	0.29	0.011	0.885
3.0	9.4	10.9	1.41	0.035	0.826
3.5	12.9	13.3	2.89	0.055	0.800
4.0	16.8	17.1	4.02	0.060	0.796
4.5	21.3	21.8	4.84	0.057	0.798
5.0	26.2	27.1	5.60	0.057	0.800
5.5	31.8	32.9	6.34	0.053	0.805
6.0	37.8	39.4	7.16	0.052	0.804
6.5	44.4	46.3	8.05	0.050	0.806
7.0	51.4	53.8	8.97	0.050	0.805
7.5	59.1	61.9	9.96	0.049	0.806
8.0	67.2	70.5	11.04	0.047	0.809
10.0	105.0	110.5	15.84	0.047	0.807
14.0	205.8	217.1	28.48	0.045	0.812

Firstly, one can fine-tune the size of the third-stage sample and the critical values of the procedure in order to keep the Type-I FWE rate and the power closer to the desired α and β values. This can be achieved by using the second-order large-sample approximations, a technique that was used in Liu (1997a), for example.

Secondly, note that this three-stage procedure is based on the simultaneous confidence interval procedure of Tukey for pairwise comparisons. For the purpose of multiple hypotheses testing, the simultaneous confidence interval procedure can be improved upon by using stepwise multiple tests. For instance, Liu (1996) proposed some step-down and step-up tests for pairwise comparisons among three treatments. Sequential procedures can be easily developed based on these stepwise tests for pairwise comparisons among three treatments. Since these stepwise tests are uniformly more powerful than the simultaneous confidence interval procedure for the purpose of multiple testing, the corresponding sequential procedures should require smaller sample-sizes than the three-stage procedure developed here for controlling the same α and β values.

Thirdly, a three-stage sequential procedure can in general be improved upon in terms of its sample size by using other sequential sampling schemes. For example, one could use the fully sequential sampling of Anscombe (1953), Robbins (1959) and Chow and Robbins (1965), the general k-sample scheme of Liu (1997b), and the hybrid sampling scheme of Liu (1997c). Ghosh et al. (1997) provide an excellent review on several sequential sampling schemes and their applications in various statistical problems. Nikoukar (1996) considers some fully sequential sampling procedures for pairwise comparisons among several treatments which control both the Type-I FWE rate and the power. He also discusses second-order fine-tuning of those procedures. But these improved sampling schemes require more frequent re-estimation of σ^2 and, hence, are more cumbersome to use than the three-stage scheme.

Finally, it is assumed in this paper that the variances of the three treatments are equal. Without the equal variance assumption, Bishop and Dudewicz (1978) proposed a two-sample procedure that controls the size and power (in terms of a non-centrality parameter) for the overall test of equality of several treatment effects. But this procedure runs the risk of substantial oversampling, as many other two-stage procedures do. Although it is straightforward to devise some three-stage and purely sequential procedures in this case, second-order large-sample results may not be easy to establish, and are not available in the literature to date. This is currently under investigation.

ACKNOWLEDGMENT

I would like to thank two referees for their constructive comments.

REFERENCES

[1] Anscombe, F.J. (1953). Sequential estimation. *J. Roy. Statist. Soc.*, **15**, 1-21.

[2] Bishop, T.A. and Dudewicz, E.J. (1978). Exact analysis of variance with unequal variances: test procedures and tables. *Technometrics*, **20**, 419-430.

[3] Chow, Y.S. and Robbins, H. (1965). On the asymptotic theory of

fixed width confidence intervals for the mean. *Ann. Math. Statist.*, **36**, 457-462.

[4] Ghosh, M., Mukhopadhyay, N. and Sen, P.K. (1997). *Sequential Estimation*. Wiley: New York.

[5] Hall, P. (1981). Asymptotic theory of triple sampling for sequential estimation of a mean. *Ann. Math. Statist.*, **9**, 1229-1238.

[6] Hochberg, Y. and Tamhane, A.C. (1987). *Multiple Comparisons Procedures*. Wiley: New York.

[7] Hsu, J.C. (1996). *Multiple Comparisons: Theory and Methods*. Chapman & Hall: New York.

[8] Liu, W. (1996). On some single-stage, step-down and step-up procedures for comparing three normal means. *Comp. Statist. Data Anal.*, **21**, 215-227.

[9] Liu, W. (1997a). On some sample size formulae for controlling both size and power in clinical trials. *J. Roy. Statist. Soc., Ser. D*, **46**, 239-251.

[10] Liu, W. (1997b). A k-stage sequential sampling procedure for estimation of normal mean. *J. Statist. Plan. Inf.*, **65**, 109-127.

[11] Liu, W. (1997c). Improving the fully sequential sampling scheme of Anscombe-Chow-Robbins. *Ann. Statist.*, **25**, 2164-2171.

[12] Miller, R. G.,Jr. (1981). *Simultaneous Statistical Inference*, 2nd ed. Springer-Verlag: New York.

[13] Nikoukar, M. (1996). On fixed-width simultaneous confidence intervals for multiple comparisons and related problems. *Ph.D. Thesis*, Dept. of Math., Univ. of Southampton, Southampton.

[14] Robbins, H. (1959). Sequential estimation of the mean of a normal population. In *Probab. and Statist.*, *H. Cramér vol.* (U. Grenander, ed.), 235-245. Almquist and Wiksell: Stockholm.

[15] Stein, C. (1945). A two-sample test for a linear hypothesis whose power is independent of the variance. *Ann. Math. Statist.*, **16**, 243-258.

Address for communication:

WEI LIU, Department of Mathematics, University of Southampton, Southampton, SO17 1BJ, U.K. E-mail: W.Liu@maths.soton.ac.uk

Chapter 14

How Many Simulations Should One Run?

NITIS MUKHOPADHYAY
University of Connecticut, Storrs, U.S.A.

GREG CICCONETTI
Muhlenberg College, Allentown, U.S.A.

14.1 INTRODUCTION: ONE-SAMPLE T-TEST

We consider a well-known statistical methodology for tests of hypotheses. Suppose that we have a fixed number of independent observations $X_1, ..., X_m$ from a single population with a continuous probability density function (p.d.f.) $f(x)$ and the distribution function (d.f.) $F(x)$. We assume that this population has a finite mean μ and a finite variance $\sigma^2 (> 0)$. In order to test a null hypothesis $H_0 : \mu = \mu_0$ against a two-sided alternative hypothesis $H_1 : \mu \neq \mu_0$ at a given size or level $\alpha, 0 < \alpha < 1$, a customary two-sided test

would reject H_0 in favor of H_1 if and only if $\left| \frac{\sqrt{m}(\overline{X} - \mu_0)}{S} \right| > t_{m-1, \alpha/2}$

$$(14.1.1)$$

where $\overline{X} \equiv \overline{X}_m = m^{-1} \Sigma_{i=1}^m X_i$ is the sample mean, $S^2 \equiv S_m^2 = (m-1)^{-1} \Sigma_{i=1}^m (X_i - \overline{X}_m)^2$ is the sample variance, and $t_{m-1, \alpha/2}$ is the upper $50\alpha\%$ point of the Student's t distribution with $m-1$ degrees of freedom. If the population happens to be normal, that is if

$$f(x) = \tfrac{1}{\sigma\sqrt{2\pi}} \exp\left\{-\tfrac{(x-\mu)^2}{2\sigma^2}\right\}, \ -\infty < x < \infty, \qquad (14.1.2)$$

then the size of the test procedure (14.1.1) will be exactly α. In statistical literature, this is a widely used and recommended test procedure to decide between H_0 versus H_1 even when f is not given by (14.1.2). When f is not given by (14.1.2), that is when the normality of the population distribution can not be justified from practical perspectives, then many standard books and manuals on statistical methodologies recommend that the *sample size m* should preferably be "large", customarily $m \geq 30$. The wisdom behind this comes from the following well-known result:

Under H_0, the test statistic $\frac{\sqrt{m}(\overline{X}_m-\mu_0)}{S_m} \xrightarrow{\mathcal{L}} N(0,1)$ as $m \to \infty$. In other words, the distribution of $\frac{\sqrt{m}(\overline{X}_m-\mu_0)}{S_m}$ can be approximated by the standard normal distribution if the sample size m is "large". Then, the percentage point $t_{m-1,\alpha/2}$ may be replaced in (14.1.1) by $z_{\alpha/2}$, the upper $\tfrac{1}{2}\alpha$ point of the standard normal distribution.

(14.1.3)

Since a population distribution is rarely exactly normal in real applications, practitioners often invoke the idea summarized in (14.1.3) and then follow up with the test procedure (14.1.1). But, then, the size or level of the resulting test may no longer be α.

Now, one may like to address the following pertinent question: If the test procedure (14.1.1) is used in the case of a *specific* non-normal distribution f with a fixed m, how different will the true size of this test be from the nominal target α? One may try to obtain some practically useful guidelines derived from computer *simulation* runs with a few specifically chosen f. An obvious limitation of any computer *simulation*, however, is that one cannot include all possible non-normal f in this investigation.

First, we formulate the problem, rationalize a loss function, and then suggest a criterion to measure the "goodness" of the proposed estimator of size associated with the test (14.1.1). We apply it in the case of a normal and three *specific* non-normal population distributions (Section 14.2). Then, the relevance of an appropriate purely sequential simulation methodology is described (Section 14.3). In the context of this problem, we argue that a natural two-stage simulation methodology can also be pursued and its description is provided (Section 14.4).

These two methodologies are followed by numerical examples and analyses (Section 14.5) where we find clear evidence that the two-stage simulation strategy operationally outperforms its purely sequential counterpart.

We emphasize that for the specific problem on hand, neither the loss function nor these proposed sequential and two-stage simulation methodologies can be explicitly found in the literature. Also, the derivation of some of the associated asymptotics do not appear to be routine and hence, these are appended at the end of the chapter (Section 14.6).

We end this section with a passing remark. As we began reviewing the literature on "simulation" within the areas of computer science and numerical analysis, we quickly discovered that phrases such as *simulation, repetition, replication, run, simulation size, batch* are interchangeably used without discretion. Unfortunately, these "taken-for-granted" terminologies do create a confusing maze from which it becomes quite a challenge to figure out exactly *what* part of an experiment is being "simulated". In one situation, simply generating $m = 10$ observations from a known population distribution is called a *simulation*. In yet another situation, generating $m = 10$ observations from a known population distribution a fixed number of times, say $n = 100$, is called a *simulation*. In the second situation, is the "simulation size" 10 or 100? From this point onward, we will try to be careful in our usage of various terminologies.

14.2   FORMULATION, LOSS AND RISK FUNCTIONS, AND OBJECTIVE

Even with the help of a sophisticated computer, it will be physically impossible to investigate the performance of the statistical methodology (14.1.1) for every single *non-normal* population distribution f. In this investigation, we include one normal and three *specific* non-normal population distributions. Let $\Phi(.)$ denote the standard normal d.f., that is

$$\Phi(x) = \int_{y=-\infty}^{x} \phi(y)dy, \phi(x) = \tfrac{1}{\sqrt{2\pi}} \exp(-\tfrac{x^2}{2}), -\infty < x < \infty,$$
(14.2.1)

and we consider the following fixed choices for F.

1. $F(x) \equiv F_1(x) = \Phi\left(\frac{x-5}{\sqrt{2}}\right)$, that is the population distribution is $N(5, 2)$;

2. $F(x) \equiv F_2(x) = 0.75\Phi\left(\frac{x-5}{1}\right) + 0.25\Phi\left(\frac{x-5}{\sqrt{5}}\right)$, that is the population distribution is a mixture of 75% $N(5, 1)$ and 25% $N(5, 5)$ distributions;

3. $F(x) \equiv F_3(x) = 0.75\Phi\left(\frac{x-5}{\sqrt{4/3}}\right) + 0.25\Phi\left(\frac{x-5}{2}\right)$, that is the population distribution is a mixture of 75% $N(5, \frac{4}{3})$ and 25% $N(5, 4)$ distributions;

4. $F(x) \equiv F_4(x) = 0.75\Phi\left(\frac{x-5}{\sqrt{1/3}}\right) + 0.25\Phi\left(\frac{x-5}{\sqrt{7}}\right)$, that is the population distribution is a mixture of 75% $N(5, \frac{1}{3})$ and 25% $N(5, 7)$ distributions.

$$(14.2.2)$$

Incidentally, the random variables associated with choices #2-#4 are sometimes referred to as *Tukey noises* or *Tukey random variables*.

Our goal is to examine what kind of a key role sequential analysis can play in designing "efficient" experiments with computer simulations. We proceed to highlight our ideas by simulating the one-sample test (14.1.1) when f corresponds to any one of the four choices in (14.2.2). This is just an example! One may examine the one-sample test (14.1.1) by considering an f that corresponds to any other kind of distribution not listed in (14.2.2). This article emphasizes the importance of sequential and multi-stage sampling in any field that may rely upon computer simulations. Obviously, in principle, similar ideas may be used for designing "efficient" computer simulations in general.

A reader may feel tempted to place this investigation within some general context of robust hypothesis testing. It should be clear that our intent is quite different. We certainly do not present a full-blown investigation of robustness properties of one- or two-sample t-test. But, it may not be inappropriate to mention here that many relevant robustness issues associated with both one-sample and two-sample t-tests were addressed by Tukey (1960), Tukey and McLaughlin (1963), Tiku and Singh (1981), Posten et al. (1982), and others.

Let us first explain the particular choices of parameter values in (14.2.2). Suppose that a population p.d.f. is given by

$$f(x) = \tau g(x) + (1 - \tau)h(x), x \in R, 0 < \tau < 1, \qquad (14.2.3)$$

where g and h are also both p.d.f.'s on R. Let μ_g, μ_h denote the means corresponding to the p.d.f.'s g and h respectively. Then, one verifies easily that

$$E_f[X] = \tau\mu_g + (1 - \tau)\mu_h \text{ and } E_f[X^2] = \tau E_g[X^2] + (1 - \tau)E_h[X^2]$$
(14.2.4)

Hence, it checks out easily that all four selected population distributions are such that each corresponding p.d.f. f is continuous, symmetric about $x = 5$, has mean $\mu \equiv \mu_f = 5$ and variance $\sigma^2 \equiv \sigma_f^2 = E_f[X^2] - E_f^2[X] = 2$. The plots of the associated pdf's $f_i(x), i = 1, 2, 3, 4$, are given in Figure 14.2.1.

While the variances of the four distributions in (14.2.2) coincide, we look for other measures that might help in understanding the effect of these distributions on the test (14.1.1). We compare, for example, the tail areas falling beyond the 97.5^{th} percentile of the distribution $F_i, i = 1, 2, 3, 4$.

It should be clear that we have labelled F_2, F_3 and F_4 respectively as "heavier", "lighter", and "heavier" than normal in the sense of having "slower", "faster", and "slower" decreasing tails. In this sense, one also notes that F_4 is heavier than F_2. Now, since the t-test (14.1.1) typically performs worse in the presence of heavy tails and skewness, we might expect a bit more variability in the simulation results for $F = F_2, F_4$.

Table 2.1 : Tail Behavior of the Distributions $F_1 - F_4$

Distribution	$P_F(X > 7.77198)$	Tail Characteristic
F_1 Normal: $P_{F_1}(X > 7.77198) = 0.025$		
F_2	0.028977	Heavier than normal
F_3	0.012386	Lighter than normal
F_4	0.036847	Heavier than normal

We apply the statistical methodology (14.1.1) to test the null hypothesis $H_0 : \mu = 5$ against a two-sided alternative hypothesis $H_1 : \mu \neq 5$ having a preassigned size $\alpha = 0.05$ on four separate occasions, that is when the underlying population distribution is $F_i(x), i = 1, 2, 3, 4$.

Goal: We wish to estimate the Type-I error probability θ_i when the population density f corresponds to $F(x) \equiv F_i(x), i = 1, 2, 3, 4$.

This is accomplished with the help of a computer "simulation" algorithm which is laid out precisely as follows: For the problem on hand, we keep the *sample size* m fixed.

Step #1: First, we fix $m = 5$ and $F(x) \equiv F_1(x)$. We have $t_{4,0.025} = 2.7764$.

Step #2: We generate m observations from this population and obtain the sample mean $\bar{x} = \bar{x}_1$ and standard deviation $s = s_1$ respectively.

Step #3: We evaluate the test statistic $\frac{\sqrt{m}(\bar{x}_1 - 5)}{s_1}$ and check whether $\left| \frac{\sqrt{m}(\bar{x}_1 - 5)}{s_1} \right| > 2.7764$. We reject H_0 when $\left| \frac{\sqrt{m}(\bar{x}_1 - 5)}{s_1} \right| > 2.7764$.

Step #4: We generate and save an indicator variable $u \equiv u_{11}$ which takes the value 1 or 0 according as $\left| \frac{\sqrt{m}(\bar{x}_1 - 5)}{s_1} \right| > 2.7764$ or $\left| \frac{\sqrt{m}(\bar{x}_1 - 5)}{s_1} \right| \leq 2.7764$ respectively.

> When a cycle through Steps #1-#4 is completed, we refer to this as one *pass* or one *simulation*.

But, naturally from one single observed value u_{11}, which is either zero or one, we can not hope to find a reasonable estimate of the Type-I error probability θ_1. We may mention that all standard texts on *simulation* will advocate repeating the previously laid out Steps #1-#4 on n separate occasions. Thus, one would obtain n values $u_{11}, ..., u_{1n}$ for the indicator variable u. The Type-I error probability θ_1 is then customarily estimated by $\bar{u}_{1n} = n^{-1}\Sigma_{j=1}^{n}u_{1j}$, that is the observed sample proportion of rejecting H_0.

Next, we fix $m = 10, 15, 30, 50$ successively, go back to Steps #1-#4 and finish all n passes or simulations to examine the effect of the choice of m on the respective estimate $\bar{u}_{1n}$ for the Type-I error probability θ_1. Then, we can repeat this whole exercise with $m = 5, 10, 15, 30, 50$ and $F(x)$ being successively replaced by $F_i(x)$ in order to estimate θ_i by $\bar{u}_{in}$, the observed sample proportion of rejecting H_0, $i = 2, 3, 4$. But, then, what should be chosen as n? Should we fix $n = 100$ or 500 or 1000 or 10000 or even larger? Customarily, however, one will simply fix $n = 1000$ or 5000 and pass through Steps #1-#4 n times. But, any reasonable way of determining an appropriate n must depend upon how one quantifies the "closeness" between the estimated Type-I error probability $\bar{u}_{in}$ and the unknown $\theta_i, i = 1, 2, 3, 4$. Designing the simulation experiment itself and appropriately determining the number of simulations, namely n, form the core of this article.

In some textbooks, authors have recommended using appropriate sequential experiments for determining the simulation size n. One may refer to Ross (1999, Chapter 7) or Bratley et al. (1988, pp. 29, 40, 76-78, 92, 101) for basic motivations. We remark in passing that Bratley et al. (1988, p. 92) have mentioned the classic sequential paper of Chow and Robbins (1965) a number of times and proposed that n should be determined sequentially depending upon appropriate quantification of the associated "estimation error". Some Chow-Robbins type sequential simulation strategies and their merits relative to fixed-size (that is, when n is fixed in advance) simulation strategies are discussed sporadically in Ross (1999) and somewhat more systematically in Bratley et al. (1988). Neither source offered readers a clear roadmap regarding applications of sequential methods.

14.2.1 The Formulation

In order to formulate the problem properly, we go back to Step #1. We start with a fixed choice of m and a population distribution function $F(x)$. The true unknown Type-I error probability is given by

$$\theta = P_F \left\{ \left| \tfrac{\sqrt{m}(\overline{X} - \mu_0)}{S} \right| > t_{m-1,\alpha/2} \text{ when } \mu = \mu_0 \right\}, 0 < \theta < 1, \quad (14.2.5)$$

and we define an indicator random variable

$$U = \begin{cases} 1 & \text{if } \left| \tfrac{\sqrt{m}(\overline{X} - \mu_0)}{S} \right| > t_{m-1,\alpha/2}, \text{ that is } H_0 \text{ is rejected} \\ 0 & \text{if } \left| \tfrac{\sqrt{m}(\overline{X} - \mu_0)}{S} \right| \leq t_{m-1,\alpha/2}, \text{ that is } H_0 \text{ is not rejected,} \end{cases}$$
$$(14.2.6)$$

which has a Bernoulli(θ) distribution. After Steps #1-#4 have been repeated n times, one obtains n independent Bernoulli(θ) random variables $U_1, ..., U_n$, with their specific realizations denoted by $u_1, ..., u_n$. Now, based on n such simulations, the true but unknown rejection probability θ is customarily estimated by $\overline{U}_n = n^{-1}\Sigma_{i=1}^n U_i$, the sample proportion of rejecting H_0 out of n simulations. It is obvious that $\overline{U}_n$ is an unbiased estimator of θ, and $V_F(\overline{U}_n) = \theta(1 - \theta)/n$ for all fixed n.

14.2.2 A Relative Squared Error Loss Function

We realize that θ is supposedly a small number and thus, one may measure the discrepancy between the estimator $\overline{U}_n$ and θ through the

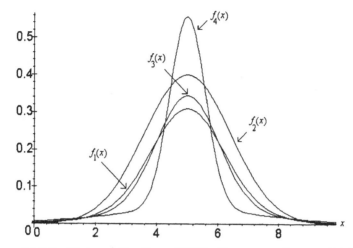

Figure 14.2.1: Plots of the Four PDF's Corresponding to (14.2.3).

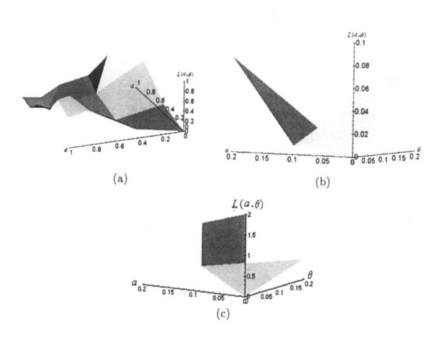

Figure 14.2.2: The Loss Functions Are (a) $L(a, \theta) = (a - \theta)^2$;
(b) $L(a, \theta) = (a - \theta)^2$ Near $(0,0)$; (c) $L(a, \theta) = (\frac{a}{\theta} - 1)^2$ Near $(0,0)$.

following loss function:

$$L_n(\overline{U}_n, \theta) = \left[\frac{\overline{U}_n}{\theta} - 1\right]^2 : \text{"proportionally close" loss.} \quad (14.2.7)$$

This particular loss function makes good sense because when θ is small, we would look at "large" departure between $\frac{\overline{U}_n}{\theta}$ and 1 very seriously. In a situation like this, a traditional squared error loss function, namely $L_n(\overline{U}_n, \theta) = (\overline{U}_n - \theta)^2$, may not be very sensible. When θ is assumed small, we may expect a small value of $\overline{U}_n$. For example, if we have $\overline{U}_n = 0.02, \theta = 0.01$, then the squared error loss is 0.0001 whereas the proportional loss is 1.0. In other words, a squared error loss function will seem to indicate that $\overline{U}_n$ estimates θ quite accurately but, in reality, $\overline{U}_n$ may be several fold larger than θ. The squared error loss is not very sensitive to our specific situation where θ is small, but the proportional loss is. A sharp contrast between a squared error loss function and a proportional loss function, particularly near the origin, is clearly visible from the three plots given in Figure 14.2.2.

In the past, a proportional or a relative squared error loss function has been used by a number of authors in the context of many inference scenarios. Some of the references that come to mind include Nadas (1969), Martinsek (1983,1995), Willson and Folks (1983), Mukhopadhyay and Diaz (1985), and Sriram (1991).

14.2.3 The Objective: Prespecified Proportional Accuracy

The risk function associated with the loss function L_n, defined in (14.2.7), is given by

$$E_F[L_n(\overline{U}_n, \theta)] = E_F\left[\frac{\overline{U}_n}{\theta} - 1\right]^2 = \frac{\theta(1-\theta)}{n\theta^2} = \frac{(1-\theta)}{n\theta}. \quad (14.2.8)$$

Now, one may require that the number of *passes* n be so chosen that the risk of estimation does not exceed a preassigned number $c(> 0)$. In other words, one puts a restriction that

$$\frac{(1-\theta)}{n\theta} \leq c, \text{ which holds if and only if } n \text{ is the smallest}$$
$$\text{integer} \geq c^{-1}\frac{(1-\theta)}{\theta} = n^*(\theta), \text{ say.} \quad (14.2.9)$$

In practice, how should one choose the design constant c? It is essential that c is chosen "small" because then we may expect to claim that on an average $\overline{U}_n$ is in a close proximity of θ. The experimenter may decide that it would be fairly adequate to estimate θ within 10% of the magnitude of θ. In this case, the choice $c = (0.1)^2 = 0.01$ may express the experimenter's feeling quite well. On the other hand, the experimenter may wish to estimate θ within 7% of the magnitude of θ instead. In this case, the choice $c = (0.07)^2 = 0.0049$ may express the experimenter's feeling. Which choice of c should one use in practice? A suitable choice will take into account the required level of accuracy a practical context may demand.

When $\alpha = 0.05$, we may genuinely expect θ to be "around" 0.05. Given this sentiment, an experimenter may then attempt to subjectively quantify the maximum possible error that can be absorbed in practice without causing a significant backlash. Considerations along these lines will guide one to come up with some appropriate number $c(> 0)$ on a case-by-case basis.

Remark 14.2.1 Robbins and Siegmund (1974) gave a sequential estimation procedure for a Bernoulli success probability parameter θ with a loss function $\frac{(\overline{U}_n - \theta)^2}{\theta^2(1-\theta)^2}$. This loss function is appropriate when one feels *a priori* that θ may be close to zero or one. However, this loss function is not very attractive in our situation because we feel that here the parameter θ is close to zero.

14.3   A PURELY SEQUENTIAL METHODOLOGY

In order to estimate θ, supposedly a small percentage, one will run through the Steps #1-#4 at least some minimal number of times, say k_0. We can safely postulate that $0 < \theta < \theta_0(< 1)$ where θ_0 is assumed known. Then, it immediately follows that

$$n^*(\theta) = c^{-1}\frac{(1-\theta)}{\theta} \geq c^{-1}\frac{(1-\theta_0)}{\theta_0}, \qquad (14.3.1)$$

and hence we define the *initial* or *pilot* simulation size

$$k \equiv k(c) = \max\left\{k_0, \left\langle c^{-1}\frac{(1-\theta_0)}{\theta_0}\right\rangle + 1\right\} \text{ where } <a> \text{ is the} \qquad (14.3.2)$$
$$\text{largest integer } < a \text{ and } k_0 \text{ is a known positive integer.}$$

Since θ denotes a small percentage, we may expect only a "small" number of U's to take the value 1 among 100 or more pilot observations.

In practice, if one can afford to work with larger k_0, then one should. But, since the magnitude of $n^* \equiv n^*(\theta)$ remains unknown, we propose the following purely sequential simulation strategy by mimicking the classical techniques developed by Chow and Robbins (1965). We first generate k initial Bernoulli observations $U_1, ..., U_k$ where the pilot sample size k is defined by (14.3.2). Next, we repeat the Steps #1-#4

N times with one additionally generated Bernoulli observation after each pass where $N \equiv N(c)$ is the integer $n(\geq k)$ for which

we observe $n \geq c^{-1} \frac{(1 - \overline{U}_n - n^{-1})}{(\overline{U}_n + n^{-1})}$ for the first time.

Finally, θ is estimated by $\overline{U}_N = N^{-1} \Sigma_{i=1}^N U_i$.

$$(14.3.3)$$

In the stopping rule (14.3.3), we first compare the current simulation size k with the current estimate $\widehat{n_k^*(\theta)} = c^{-1} \frac{(1 - \overline{U}_k - k^{-1})}{(\overline{U}_k + k^{-1})}$ of $n^*(\theta)$. If $k \geq \widehat{n_k^*(\theta)}$, we stop simulations with the pilot observations $U_1, ..., U_k$ and thus estimate θ by $\overline{U}_k$. If $k < \widehat{n_k^*(\theta)}$, then we go back to Steps #1-#4 to generate a new Bernoulli random variable U_{k+1} and compare the current simulation size $k + 1$ with the current estimate $\widehat{n_{k+1}^*(\theta)} = c^{-1} \frac{(1 - \overline{U}_{k+1} - (k+1)^{-1})}{(\overline{U}_{k+1} + (k+1)^{-1})}$ of $n^*(\theta)$. If $k+1 \geq \widehat{n_{k+1}^*(\theta)}$, we stop simulations with the observations $U_1, ..., U_k, U_{k+1}$ and thus estimate θ by $\overline{U}_{k+1}$. But, if $k + 1 < \widehat{n_{k+1}^*(\theta)}$, then we go back to Steps #1-#4 to generate another Bernoulli random variable U_{k+2} and compare the current simulation size $k + 2$ with the current estimate $\widehat{n_{k+2}^*(\theta)} = c^{-1} \frac{(1 - \overline{U}_{k+2} - (k+2)^{-1})}{(\overline{U}_{k+2} + (k+2)^{-1})}$ of $n^*(\theta)$. This process continues until the stated boundary condition in the stopping rule (14.3.3) is satisfied for the very first time.

From the stopping rule (14.3.3), we note that as long as $\overline{U}_n$ is observed zero, the stopping boundary condition can not be satisfied. Now, is there any guarantee that for all fixed c and θ, the sequential procedure will terminate with probability one? Toward that end, we write

$$P_\theta\{N > n\} \leq P_\theta\left\{ n < \tfrac{1}{c} \frac{(1 - \overline{U}_n - n^{-1})}{(\overline{U}_n + n^{-1})} \right\} = P_\theta\{\overline{U}_n < \tfrac{1}{cn+1}\}$$
$$\leq P_\theta\left\{ \left| \frac{\overline{U}_n}{\theta} - 1 \right| > \tfrac{1}{2} \right\},$$

since, for sufficiently large n, the expression $\frac{1}{\theta(cn+1)} - 1$ can be made smaller than $-\frac{1}{2}$. Thus, for sufficiently large n, using Tchebysheff's inequality, we have

$$P_\theta\{N > n\} \le 4E_\theta\left[(\tfrac{\overline{U}_n}{\theta} - 1)^2\right] = \tfrac{4\theta(1-\theta)}{n\theta^2} = \tfrac{4(1-\theta)}{n\theta}.$$

In other words, for all fixed $c(> 0), 0 < \theta < 1$, we obtain

$$0 \le P_\theta(N = \infty) = \lim_{n\to\infty} P_\theta\{N > n\} \le \lim_{n\to\infty} \tfrac{4(1-\theta)}{n\theta} = 0$$
$$\Rightarrow P_\theta(N = \infty) = 0 \text{ so that } P_\theta(N < \infty) = 1.$$

That is, the sequential procedure (14.3.3) does terminate with probability one.

14.4   A TWO-STAGE METHODOLOGY

The motivation is derived from the fact that sampling with larger batches can be carried out quickly and hence one may be able to cut down the operational time significantly compared with purely sequential sampling. More explanations are given Section 14.4.1.

Now, recall the expression of $n^*(\theta)$ from (14.3.1). Again, we define the *initial* or *pilot* simulation size

$$k \equiv k(c) = \max\left\{k_0, \left\langle c^{-1}\tfrac{(1-\theta_0)}{\theta_0}\right\rangle + 1\right\} \text{ where } < a > \text{ is the largest integer } < a \text{ and } k_0 \text{ is a known positive integer.} \qquad (14.4.1)$$

We note that the pilot simulation size k is going to increase as $n^*(\theta)$ increases and, for all practical purposes, k is a lower bound of $n^*(\theta)$. Thus, we may proceed to consider a Stein-type (Stein (1945,1949)) two-stage methodology along the line of Mukhopadhyay and Duggan's (1997) adaptation that was shown to enjoy *second-order* properties in the normal case.

We first generate k initial Bernoulli observations $U_1, ..., U_k$ where the pilot sample size k is defined by (14.4.1). Next, we construct the final simulation size M as follows:

$$M \equiv M(c) = \max\left\{k, \left\langle c^{-1}\tfrac{(1-\overline{U}_k-k^{-1})}{(\overline{U}_k+k^{-1})}\right\rangle + 1\right\}. \qquad (14.4.2)$$

Note that in (14.4.2), we have replaced $\overline{U}_k$ with $\overline{U}_k + k^{-1}$ just to make sure that the ratio $\frac{(1-\overline{U}_k-k^{-1})}{(\overline{U}_k+k^{-1})}$ remains well-defined if it happens at all that $\overline{U}_k = 0$. Now, we explain how this simulation strategy is going to be implemented.

(i) If we find that $M = k$, then we do not generate any additional Bernoulli U's so that the simulation will end simply with the k pilot passes, and then $\overline{U}_k$ will be the final estimator of θ.

(ii) If we find that $M > k$, then we generate additional $M - k$ Bernoulli random variables $U_{k+1}, ..., U_M$ so that the simulation will end with M passes. Then, $\overline{U}_M = M^{-1}\Sigma_{i=1}^{M}U_i$, that is the sample mean of the combined observations $U_1, ..., U_k, U_{k+1}, ..., U_M$, will be the final estimator of θ.

Combining (i) and (ii), we can say that M is the final simulation size and $\overline{U}_M$ is the final estimator of θ. From the stopping rule (14.4.2), we note that for all fixed c and θ, the final simulation size M is finite with probability one.

Such a two-stage procedure for a multivariate normal problem was first developed by Mukhopadhyay (1999). Aoshima (2000) and Aoshima and Mukhopadhyay (1999) developed other multivariate two-stage procedures with *second-order* properties.

14.4.1 The Motivation

One may ask: Why should a practitioner implement the two-stage methodology rather than the earlier purely sequential methodology from (14.3.2)-(14.3.3)? The answer is quite simple indeed. In the sequential strategy (14.3.3), after every pass through Steps #1-#4, one has to check whether n has exceeded $c^{-1}\frac{(1-\overline{U}_n-n^{-1})}{(\overline{U}_n+n^{-1})}$ for the first time, $n \geq k$. We may expect the associated stopping variable N to turn out much larger than k, and then the operational time to check the boundary condition will pile up fast. The two-stage strategy (14.4.1)-(14.4.2) avoids this requirement of checking boundary conditions in its entirety. That is, this latter simulation methodology will terminate faster than the sequential simulation strategy.

In many simulation projects for designing complex systems in computer engineering and information technology, one frequently confronts a bigger system that consists of five or ten such *nested* sub-systems. Now, imagine a complex simulation strategy that consists of three nested sub-simulation strategies. In the first sub-system, suppose that checking the boundary condition takes 0.01 seconds each time and it is completed with 20193 sequential passes beyond k_1, the pilot simulation size. Then, the estimator $W_{N_1}^{(1)}$ of ν_1 thus obtained is used in the second sub-system, and suppose that checking the associated boundary condition takes 0.001 seconds each time and it is completed with 73268 sequential passes beyond k_2, the pilot simulation size. Then, the estimator $W_{N_2}^{(2)}$ of ν_2 thus obtained is used in the third sub-system, and suppose that checking the associated boundary condition takes 0.001 seconds each time and it is completed with 28758 sequential passes beyond k_3, the pilot simulation size. The estimator $W_{N_3}^{(3)}$ of ν_3 thus obtained is then a final output obtained from the whole system. This is just one simulation pass of the full system. Then, the savings in computer operating time alone will be $20193 \times 0.01 + 73268 \times 0.001 + 28758 \times 0.001 = 303.96$ seconds, that is approximately 5.066 minutes, due to implementing the two-stage technique (14.4.1)-(14.4.2) rather than the purely sequential technique (14.3.2)-(14.3.3). The full system, however, may be simulated 100 or 200 times.

When the full system is simulated 100 times, this gives rise to 100 triplets $(N_{1j}, N_{2j}, N_{3j}), j = 1, ..., 100$. Let us denote $T_{ij} = N_{ij} - k_i + 1, i = 1, 2, 3$. such that $T_{1\,\min} = 15642, T_{1\,\max} = 24352, T_{2\,\min} = 68274, T_{2\,\max} = 77852$, and $T_{3\,\min} = 21628, T_{3\,\max} = 32291$ with purely sequential passes beyond the pilot size within a sub-system. Now, if all the purely sequential loops are replaced by the corresponding two-stage simulations within each sub-system, the savings (S) in the computer operating time alone will be approximately between $(0.01 \times 15642 + 0.0001 \times 68274 + 0.001 \times 21628) \times 100$ seconds or 5.14 hours and $(0.01 \times 24352 + 0.0001 \times 77852 + 0.001 \times 32291) \times 100$ seconds or 7.88 hours. In other words, one would have 5.14 hours $\leq S \leq 7.88$ hours. Often experienced hands are hired and paid by the hour to program and test a system, and then this magnitude of savings in computer operation alone will translate into substantial savings in dollars. Therein lies the strength of a two-stage methodology such as the one in (14.4.1).

Remark 14.4.1 The important message we are sending is that the number of simulations should not be arbitrarily fixed. We have put forth this message emphatically with the help of sound arguments and a specific bounded-risk point estimation problem involving a single parameter θ between 0 and 1 that may be close to zero. We believe that the methodologies proposed here could be extended, in principle, for simulation studies of much more complex problems involving (i) estimation of a high-dimensional parameter vector $\boldsymbol{\theta}$, (ii) infinite-dimensional function estimation, or (iii) Markov Chain Monte Carlo investigations where successive iterations are dependent. These remain important and challenging *open problems* since one must start from ground zero by (i) formulating an appropriate loss function and (ii) specifying the kind of "optimization" one may like to pursue on a case-by-case basis. It may, however, be too early to ponder about constructing some general theory and methodology embracing all types of simulations and optimization techniques.

Remark 14.4.2 In the determination of the pilot simulation size k, both sequential and two-stage methodologies exploited the assumption of existence of a known number θ_0 such that $0 < \theta < \theta_0$. Practically in all real-life situation, validity of this assumption would almost be taken for granted. But, no matter how we choose θ_0, it may be possible to think of a distribution f, perhaps pathological in nature, for which this assumption would not hold. In such pathological circumstances, different kinds of sequential and two-stage methodologies for appropriately determining the number of simulations can be developed along the lines of Chow and Robbins (1965) and Mukhopadhyay and Diaz (1985) respectively.

14.5 COMPUTER SIMULATIONS

We have investigated the performance of the sequential methodology (14.3.2)-(14.3.3) and the two-stage methodology (14.4.1)-(14.4.2) on the hypothesis testing procedure (14.1.1) under a variety of scenarios. In particular, we worked under the following configuration combinations: $m = 5, 10, 20, 30, 50$, $k_0 = 100$, $c = 0.1, 0.01, 0.0049, 0.001$ and with the population distribution functions given by $F_i(x)$, $i = 1, 2, 3, 4$ described in (14.2.2). For each of these 240 configurations we are interested in N or M, the stopping simulation size, $\widehat{\alpha}$, the point estimate for the probability of rejecting H_0, and the completion time required by the computer under each configuration. In order to comment on the variability of these recorded values, the entire 240 configurations were repeated 10 times. Since some of the combinations yielded stopping simulation sizes close to $k_0 = 100$, which is supposedly quite small, one complete configuration was run with $k_0 = 200$ to explore how this change might improve estimation of the rejection rate. We will comment on these findings in our concluding remarks.

One may perhaps criticize our decision to stop with only ten complete replications. Once we have a good grasp of how to determine N or M, we can then proceed to come up with the estimate of an appropriate number of complete replications, again sequentially, as the next step in a hierarchical system. Hence, we devote our full energy only on determining N or M with a sequential design first. The variability in N or M and those in the associated estimators of the rejection rates, however, are of secondary interest at this point. We will use ten complete replications as a means of performing simple initial EDA and speak in some generality of possible variations in these estimates. If we wish to gain point estimators for the measures of variation in the future, the suggested ten complete replications may then be utilized as a part of a "pilot simulation" in an appropriately designed two-stage or sequential follow-up study. These sentiments may be reminiscent of determining the number of levels in a hierarchical model — but, one would stop at some step in any hierarchical system!

14.5.1 The Purely Sequential Methodology (14.3.2)-(14.3.3)

Throughout the various configurations, general patterns are quite evident in the sequential procedure (14.3.2)-(14.3.3) and support intuition. Figures 14.5.1-14.5.4 reflect the impact of the distribution, and magnitude of the initial estimate θ_0 on the stopping simulation size N and the rejection rate θ when $c = 0.1, 0.001$, for each of the four distributions $F_i(x), i = 1, 2, 3, 4$ described in (14.2.2). In the Figures 14.5.1-14.5.4, each symbol represents the average obtained from the ten replications for a particular value of the sample size $m = 5, 10, 20, 30, 50$ and the chosen value θ_0. The overall features, trends, and conclusions were very similar when we had fixed $c = 0.01, 0.0049$ and hence, detailed findings are summarized only in the cases where $c = 0.1, 0.001$. Throughout these figures, we use circles, asterisks and crosses to refer to three different values of θ_0 (0.06, 0.10, and 0.30 respectively). For example, in Figure 14.5.1, the circle located in the lower left hand corner corresponds to the average of stopping simulation sizes when $F = F_1, \theta_0 = 0.06, c = 0.1$ and $m = 50$.

This investigation revealed no detectable common pattern for different choices of m and hence we decided to suppress the information concerning the sample size m. In other words, the specific sample size m used in the t-test (14.1.1) had no appreciable impact on either the final simulation size N or the rejection rate θ.

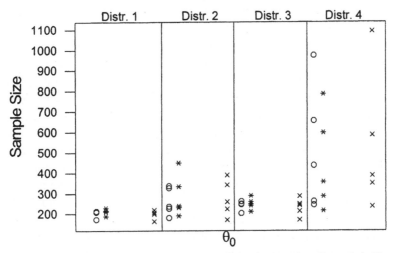

Figure 14.5.1: Variation in Sequential Simulation (or Sample) Size
N by θ_0 for Distributions $F_i, i = 1, 2, 3, 4$, with Risk-Bound $c = 0.1$:
Circles, Asteriks and Crosses Correspond to
$\theta_0 = 0.06, 0.10, 0.30$, Respectively.

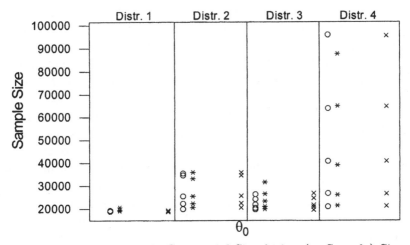

Figure 14.5.2: Variation in Sequential Simulation (or Sample) Size
N by θ_0 for Distributions $F_i, i = 1, 2, 3, 4$, with Risk-Bound $c = 0.001$:
Circles, Asteriks and Crosses Correspond to
$\theta_0 = 0.06, 0.10, 0.30$, Respectively.

From Figures 14.5.1 and 14.5.2 we make the following assertions:
The initial choice of θ_0 does not seem to impact the final simulation size
or its variability. But, reviewing the methodologies (14.3.2)-(14.3.3)
and (14.4.1)-(14.4.2), we are reminded that θ_0 plays a role in determining the pilot simulation size, and hence we were relieved to note that
we were not penalized harshly if we decided to use a conservative choice
of θ_0. We note that the choice of $\theta_0 = 0.06$ was perhaps a bit too risky,
in view of our initial ignorance about the behavior of the rejection rate
θ. The risk-bound c plays its expected role, that is as the risk-bound
decreased, a natural increase in the stopping simulation size could be
seen. Having observed this, it is not surprising to witness an increase
in the spread of N. The choice of the population distribution F also
plays a prominent role in the determination of the stopping simulation
size. Under the ideal normality conditions on F_1 we found the least
variation in the stopping simulation size, followed closely by a similar
pattern under F_3. Recall that F_3 was found to have lighter than normal
tails. Variability increased when we considered F_2 and more so with F_4.
This is exactly as we had expected, since each population distribution
F_2 and F_4 had a heavier than normal tail (Table 14.2.1).

Figures 14.5.3-14.5.4 are concerned with the performance of the rejection rate θ. The comments concerning the stopping simulation size
N seem to have a direct impact on the rejection rates for the four distributions. Again, we note that the sample size m did not lead to any
general pattern and so, information regarding m is suppressed in the
figures. The dotted line at 0.05 marks the target rejection rate under
F_1's normality. The figures suggest that the variation inherent in N
is passed on to the variation of the rejection rate. We note further
that when the risk-bound is $c = 0.1$, the two distributions F_1, F_3 perform quite comparably, while the rejection rates for the distributions
F_2, F_4 have much larger variations. With decreased risk-bound c and,
therefore, with increased stopping simulation size N, we see a much
better picture concerning the rejection rates. The nature of the distributions F_2, F_3, F_4 affects the t-test which seems to perform in a much
more conservative fashion. Moreover, large discrepancies among stopping simulation sizes under $F_i, i = 2, 3, 4$ compared with F_1, indicate
that establishing a common rejection rate for all values of m is perhaps
unattainable.

We should add that heavy tails associated with F_3, F_4 have affected
the rejection rates in the same manner as the stopping simulation sizes.

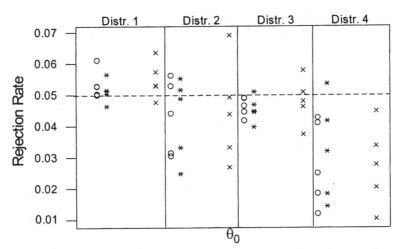

Figure 14.5.3: Variation in Rejection Rate of Sequential Procedure
Due to Simulation Size N by θ_0 for Distributions $F_i, i = 1, 2, 3, 4$,
with Risk-Bound $c = 0.1$: Circles, Asteriks and Crosses
Correspond to $\theta_0 = 0.06, 0.10, 0.30$, Respectively.

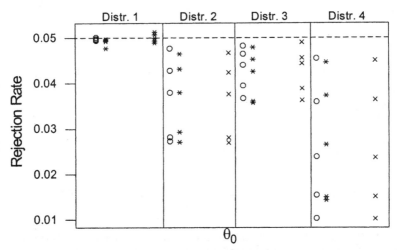

Figure 14.5.4: Variation in Rejection Rate of Sequential Procedure
Due to Simulation Size N by θ_0 for Distributions $F_i, i = 1, 2, 3, 4$,
with Risk-Bound $c = 0.001$: Circles, Asteriks and Crosses
Correspond to $\theta_0 = 0.06, 0.10, 0.30$, Respectively.

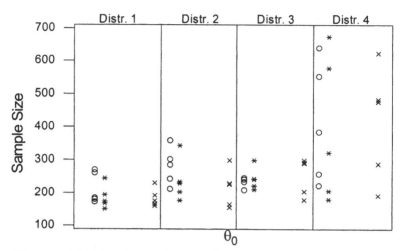

Figure 14.5.5: Variation in Two-Stage Simulation or Sample
Size M by θ_0 for Distributions $F_i, i = 1, 2, 3, 4$, with Risk
$c = 0.1$: Circles, Asteriks and Crosses Correspond to
$\theta_0 = 0.06, 0.10, 0.30$, Respectively.

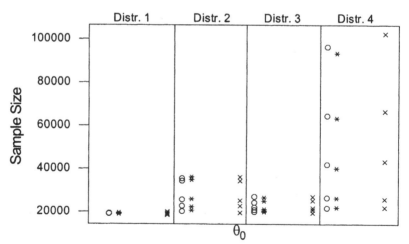

Figure 14.5.6: Variation in Two-Stage Simulation or Sample
Size M by θ_0 for Distributions $F_i, i = 1, 2, 3, 4$, with Risk
$c = 0.001$: Circles, Asteriks and Crosses Correspond to
$\theta_0 = 0.06, 0.10, 0.30$, Respectively.

14.5.2 The Two-Stage Methodology (14.4.1)-(14.4.2)

The two-stage procedure (14.4.1)-(14.4.2) mimics the purely sequential procedure with respect to the behavior of N. We include Figures 14.5.5-14.5.6 for comparison purposes.

Comments concerning the rejection rates based on the two-stage procedure are analogous to the purely sequential procedure. We include Figures 14.5.7-14.5.8 for comparison purposes.

14.5.3 Comparing the Two Methodologies and Conclusions

Our remarks from the previous subsections indicate little difference in the performance of the sequential and the two-stage methodologies as measured by their stopping simulation sizes and rejection rates. On account of such extreme similarities across simulations, our recommendation of one procedure over the other relies heavily on the determination of which procedure runs more efficiently.

Two of the ten configurations ran on a relatively older computer, a Dellx86 Family6 Model 5 Stepping 2 with CPU 448 mHz, 655360 KB of RAM, and only 264 MB free space on a 3.92 GB hard drive. The remaining eight configurations ran on a Dell Workstation PWS530 with CPU 1700 mHz, 1047564 KB RAM, and 33.6 GB free space out of a total hard drive space of 37.2 GB. Simulations were run with the help of S-Plus Version 6.0. Overall time spent for implementing the sequential methodology was 60.14 hours, while 56.46 hours were required for the two-stage study. These times were tallied by having S-Plus call upon the computer's clock time at the beginning and again at the conclusion of each configuration under study. It is worthy to note that in 55.5% of the simulations, the two-stage methodology terminated with a stopping simulation size smaller than that required by the corresponding purely sequential methodology.

However, in 78.875% of the configurations we found that the two-stage methodology completed its task in less time than the corresponding purely sequential methodology. Based on the arguments made earlier in the last part of Section 14.4.1, we fully recommend the use of the two-stage methodology (14.4.1)-(14.4.2) developed in this paper over the sequential procedure (14.3.2)-(14.3.3) for estimating a proportion that is likely to be near zero (or one) via simulation.

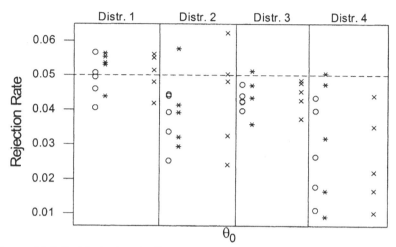

Figure 14.5.7: Variation in Rejection Rate of Two-Stage Procedure
Due to Simulation Size M by θ_0 for Distributions $F_i, i = 1, 2, 3, 4$,
with Risk-Bound $c = 0.1$: Circles, Asteriks and Crosses
Correspond to $\theta_0 = 0.06, 0.10, 0.30$, Respectively.

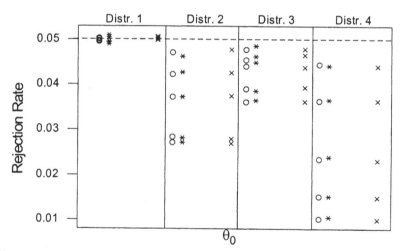

Figure 14.5.8: Variation in Rejection Rate of Two-Stage Procedure
Due to Simulation Size M by θ_0 for Distributions $F_i, i = 1, 2, 3, 4$,
with Risk-Bound $c = 0.001$: Circles, Asteriks and Crosses
Correspond to $\theta_0 = 0.06, 0.10, 0.30$, Respectively.

Throughout the ten repetitions of our 240 configurations, the purely sequential methodology (14.3.2)-(14.3.3) had terminated simulations at the initial value $k_0 = 100$ in 13 out of the 2400 (that is, 0.5417%) configurations, whereas the two-stage methodology (14.4.1)-(14.4.2) terminated simulations analogously in 10 out of the 2400 (that is, 0.4167%) configurations. In all such cases we found that the risk-bound was set at the level $c = 0.1$. This prompted us to increase k_0 to 200 and observe any possible changes. For brevity, however, we ran one repetition of the 240 configurations holding $k_0 = 200$ fixed and found that 7.9% (that is, 19 out of the 240 configurations) of the sequential methodologies and 10.8% (that is, 26 out of the 240 configurations) of the two-stage methodologies terminated with the initial simulation size, $k_0 = 200$. All of these cases occurred when the risk-bound was fixed at the level $c = 0.1$.

One may like to extrapolate what we have just said and consider increasing k_0 even more when $c = 0.1$. But, if that is done, then our proposed stopping rules would terminate at the initial stage with k_0 simulations even more frequently and $\widehat{\theta}$ may not be that great afterall! On the other hand, one must realize that $c = 0.1$ happens to be the largest value of c that we had chosen. For this choice of c and $k_0 = 100$ or 200, one can see that $\left|\widehat{\theta} - \theta\right|$ may be "approximated" by $\sqrt{c}\theta = 0.316\theta$. At this point, rather than worrying too much about the choice of k_0, the experimenter should perhaps do some soul searching and think whether the risk-bound $c = 0.1$ is "good enough" to live with or not. Perhaps, c should be lowered considerably! Statistically designing an experiment is a great idea, but statisticians just cannot take care of every aspect of an experiment in every field of investigation. We would hate to play "God" by pretending to provide "solutions" for all conceivable "statistical conflicts". The fact is that whatever c, θ_0 and k_0 one may decide to fix, very early stopping may occasionally arise. A suitable "resolution" or "balance" may be reached only through teamwork. For example, one may like to try other choices of c, θ_0 and k_0 and critically "examine" differences that may occur in the performance of computer simulations. That ought to guide many experimenters to come close to a practical solution for this type of problems on a case-by-case basis.

From a practical perspective, we would not advocate estimating rejection rates or proportions near zero with a risk-bound as large as 0.1. Hence, we venture to say that our findings indicate very minimal influence of the choice of k_0 on the methodologies provided the risk-bound

c is fixed at a realistically small level.

Since there is some chance that N or M may not go past the pilot phase, an experimenter would need to settle down with a reasonable choice of k_0. The choice of k_0 could be made differently depending upon the substantive field of application. For example, while using computer simulations in biology or electrical engineering, the choice of k_0 would very likely be different. Statistical considerations alone can not settle such issues. But, generally speaking, within a simulation exercise, we recommend that one should really go by the smallest possible k_0 that one can live with. In many situations, k_0 would probably not exceed 100 or 200. Should more observations be needed, one would let the stopping rule (14.3.3) or (14.4.2) determine the appropriate final simulation size. We take solace knowing that estimates have been computed under the bounded risk criterion. Such a statement can not be made when simulations are carried out by haphazardly implementing a fixed number of runs.

| There is truly no substitute for a well-designed simulation study. |

14.6   LARGE-SAMPLE PROPERTIES

From (14.3.3), one can claim easily that $N(c_1) \geq N(c_2)$ with probability one if $0 < c_1 < c_2$. That is, with a smaller choice of c, the resulting simulation size can not go down. Also, from the stopping rules (14.3.3) and (14.4.2), we note immediately that $\min\{N(c), M(c)\} \geq k(c)$ and thus $\lim_{n^*(\theta) \to \infty} \min\{N(c), M(c)\} \geq \lim_{n^*(\theta) \to \infty} k(c) = \infty$ with probability one. So, we can easily claim that $\lim_{n^*(\theta) \to \infty} N(c) = \infty$ and $\lim_{n^*(\theta) \to \infty} M(c) = \infty$ with probability one. Hence, both randomly stopped estimators $\overline{U}_N, \overline{U}_M$ of θ converge to θ with probability one (and hence, in probability) as $n^*(\theta) \to \infty$. One may refer to Hoeffding (1961) or Sproule (1974).

Now, we are in a position to summarize the main large-sample results associated with both the methodologies (14.3.2)-(14.3.3) and (14.4.1)-(14.4.2).

Theorem 14.6.1 *Let $T \equiv T(c)$ be a generic stopping variable where T stands for N (or M) defined in (14.3.3) or (14.4.2).Recall the expression of $n^*(\theta)$ from (14.2.9). Then, as $n^*(\theta) \to \infty$, we conclude:*

(i) $T/n^*(\theta) \to 1$ in probability (P_θ);

(ii) $E_\theta[T/n^*(\theta)] \to 1$;

(iii) $E_\theta[L_T/c] \to 1$.

for all fixed $0 < \theta < 1$.

Its proof in the case of the purely sequential and two-stage methodologies are sketched in Sections 14.6.1 and 14.6.2 respectively. The properties (i)-(ii) reinforce the feeling that we can expect the final simulation size N or M to be very close to the unknown fixed sample size $n^*(\theta)$. The property (iii) states that the purely sequential risk $E[L_N]$ or the two-stage risk $E[L_M]$ for estimating θ by $\overline{U}_N$ or $\overline{U}_M$ respectively may be expected to be close to the set target c.

14.6.1   The Purely Sequential Methodology (14.3.2)-(14.3.3): Proof of Theorem 14.6.1

From (14.3.3), observe the basic inequality

$$c^{-1}\frac{(1-\overline{U}_N-N^{-1})}{(\overline{U}_N+N^{-1})} \leq N < c^{-1}\frac{(1-\overline{U}_{N-1}-(N-1)^{-1})}{(\overline{U}_{N-1}+(N-1)^{-1})} + (k-1)I(N=k)$$

$$\Rightarrow \frac{(1-\overline{U}_N-N^{-1})}{(\overline{U}_N+N^{-1})}\frac{\theta}{1-\theta} \leq \frac{N}{n^*(\theta)} < \frac{(1-\overline{U}_{N-1}-(N-1)^{-1})}{(\overline{U}_{N-1}+(N-1)^{-1})}\frac{\theta}{1-\theta} + \frac{k-1}{n^*(\theta)}I(N=k).$$

$$(14.6.1)$$

We noted earlier that $\lim_{n^*(\theta)\to\infty} N = \infty$ with probability one. So, the randomly stopped estimator $\overline{U}_N$ of θ converges to θ with probability one (and hence, in probability) as $n^*(\theta) \to \infty$. One may refer to Hoeffding (1961), Khan (1969) or Sproule (1974). Also, one notes that $\frac{k-1}{n^*(\theta)}$ is $O(1)$. Thus, part (i) will follow from (14.6.1) if we verify that

$$I(N=k) \overset{P}{\to} 0 \text{ as } n^*(\theta) \to \infty. \qquad (14.6.2)$$

Now, let us pretend that $n^*(\theta)$ goes to infinity in such a way that $c^{-1}\frac{1-\theta_0}{\theta_0}$ remains a positive integer and that $n^*(\theta)$ is large enough so that $k = c^{-1}\frac{1-\theta_0}{\theta_0}$. Thus, for all fixed $0 < \theta < 1$, we can write

$$P_\theta\{N=k\} \leq P_\theta\left\{k \geq c^{-1}\frac{(1-\overline{U}_k-k^{-1})}{(\overline{U}_k+k^{-1})}\right\} = P_\theta\{\overline{U}_k \geq \theta_0\}$$

and hence for any fixed $p(> 0)$, we obtain

$$P_\theta\{N = k\} \le P_\theta[|\overline{U}_k - \theta| \ge \theta - \theta_0] \le (\theta - \theta_0)^{-2p} E_\theta[|\overline{U}_k - \theta|^{2p}]$$
$$= O(k^{-p}) = O([n^*(\theta)]^{-p}),$$
(14.6.3)

with the help of the Markov inequality. Next, with an arbitrary but fixed $\varepsilon > 0$, we can apply the Markov inequality one more time and express

$$P_\theta\{I(N = k) > \varepsilon\} \le \varepsilon^{-1} E_\theta\{I(N = k)\} = \varepsilon^{-1} P_\theta\{N = k\} \to 0$$
$$\text{as } n^*(\theta) \to \infty, \text{ in view of (14.6.3).}$$
(14.6.4)

The last equation verifies (14.6.2) and hence part (i) holds.

Next, we turn to part (ii). For this, we closely follow Woodroofe (1982, Chapter 4) and Ghosh et al. (1997, Section 2.9) to rewrite (14.3.3) as follows:

$N \equiv N(c)$ is the integer $n(\ge k)$ for which we observe
$ng(\overline{U}_n + n^{-1}) \ge c^{-1}$ for the first time where $g(x) = \frac{x}{1-x}$. (14.6.5)

Let us denote $S_n = \Sigma_{i=1}^n \left\{ \frac{\theta}{1-\theta} + \frac{1}{(1-\theta)^2}(U_i - \theta) \right\}, n \ge 1$. But, note that $g'(x) = (1-x)^{-2}$ and $g''(x) = (1-x)^{-3}$ so that we have

$$ng(\overline{U}_n + n^{-1}) = n\left(\frac{\theta}{1-\theta} + n^{-1}\right) + n\frac{1}{(1-\theta)^2}(\overline{U}_n + n^{-1} - \theta) +$$

$$\frac{n(\overline{U}_n + n^{-1} - \theta)^2}{(1-\nu_n)^3} \Leftrightarrow ng(\overline{U}_n + n^{-1}) = S_n + \xi_n, \qquad (14.6.6)$$

with $\xi_n = \frac{n(\overline{U}_n + n^{-1} - \theta)^2}{(1-\nu_n)^3} + \left\{ 1 + \frac{1}{(1-\theta)^2} \right\}$,

where the suitable random variable ν_n is such that it lies between $\min\{\overline{U}_n, \theta\}$ and $\max\{\overline{U}_n, \theta\}$ for large enough n.

Surely, ξ_n is *slowly changing* (Ghosh et al. (1997, p. 59) or Woodroofe (1982, p. 41)). Since $n^{-1}\xi_n$ converges to zero with probability one, we claim (Lemma 2.9.1 from Ghosh et al. (1997, p. 59) or (4.1) from Woodroofe (1982, p. 41)) that $\max_{1 \le k \le n} n^{-1} |\xi_k|$ also converges to zero with probability one as $n \to \infty$. Thus, the two sufficient conditions in the Theorem 2.9.3 of Ghosh et al. (1997, p. 62) or Theorem 4.4 in

Woodroofe (1982, p. 46) hold and hence part (ii) immediately follows.

Now, we are in a position to verify part (iii). Let us denote $Q_N = \Sigma_{i=1}^{N} U_i$, $W_N = \frac{(Q_N - N\theta)}{\sqrt{\theta(1-\theta)n^*(\theta)}}$, and observe that the sequential risk function $E_\theta[L_N]$ can be rewritten as

$$E_\theta\left[\frac{(Q_N - N\theta)^2}{\theta^2 N^2}\right] = E_\theta\left[\frac{(Q_N - N\theta)^2}{\theta^2(n^*(\theta))^2}\right] + E_\theta\left[\frac{(Q_N - N\theta)^2}{\theta^2(n^*(\theta))^2}\left\{\frac{(n^*(\theta))^2}{N^2} - 1\right\}\right].$$
(14.6.7)

Then, using Wald's second lemma (Theorem 2.4.5, Ghosh et al. (1997, p. 27)) and part (ii), we have

$$E_\theta[W_N^2] = \frac{E_\theta(N)\theta(1-\theta)}{\theta(1-\theta)n^*(\theta)} = \frac{E_\theta(N)}{n^*(\theta)} \to 1 \text{ as } n^*(\theta) \to \infty. \quad (14.6.8)$$

Next, using Anscombe's (1952) central limit theorem we can claim that W_N converges in distribution to a standard normal variable as $n^*(\theta) \to \infty$. Now, we can combine this result with (14.6.8) to conclude that W_N^2 is then uniformly integrable.

Then, from (14.6.7) we have

$$E_\theta[c^{-1}L_N] = \frac{\theta n^*(\theta)}{(1-\theta)}E_\theta[L_N] = E_\theta[W_N^2] + E_\theta\left[W_N^2\left\{\frac{(n^*(\theta))^2}{N^2} - 1\right\}\right]. \quad (14.6.9)$$

But, since $N \geq k(c)$, observe that for sufficiently large $n^*(\theta)$, we can write

$$W_N^2\left\{\frac{(n^*(\theta))^2}{N^2} - 1\right\} \leq W_N^2\left\{\frac{(n^*(\theta))^2}{k(c)^2} - 1\right\} = a_{\theta,\theta_0}W_N^2. \quad (14.6.10)$$

Here, a_{θ,θ_0} is a positive number that does not involve c. Now, since W_N^2 is uniformly integrable, it follows from (14.6.10) that $W_N^2\left\{\frac{(n^*(\theta))^2}{N^2} - 1\right\}$ is also uniformly integrable. But, we know that W_N^2 converges in distribution to a χ_1^2 random variable as $n^*(\theta) \to \infty$ and thus combining with part (i), we can claim that $E_\theta\left[W_N^2\left\{\frac{(n^*(\theta))^2}{N^2} - 1\right\}\right] \to 0$ as $n^*(\theta) \to \infty$. At this point, one simply combines (14.6.8)-(14.6.9) to complete the proof of part (iii). ∎

14.6.2   The Two-Stage Methodology (14.4.1)-(14.4.2): Proof of Theorem 14.6.1

From (14.4.2), observe the basic inequality

$$c^{-1}\frac{(1-\overline{U}_k-k^{-1})}{(\overline{U}_k+k^{-1})} \le M < c^{-1}\frac{(1-\overline{U}_k-k^{-1})}{(\overline{U}_k+k^{-1})} + kI(M=k)$$

$$\Rightarrow \frac{(1-\overline{U}_k-k^{-1})}{(\overline{U}_k+k^{-1})}\frac{\theta}{1-\theta} \le \frac{M}{n^*(\theta)} < \frac{(1-\overline{U}_k-k^{-1})}{(\overline{U}_k+k^{-1})}\frac{\theta}{1-\theta} + \frac{k}{n^*(\theta)}I(M=k).$$

$$(14.6.11)$$

We noted earlier that $\lim_{n^*(\theta)\to\infty} k = \infty$, but $k = O(n^*(\theta))$. Using the weak law of large numbers it follows that the estimator $\overline{U}_k$ of θ converges to θ in probability as $n^*(\theta) \to \infty$. Also, essentially the earlier proof of (14.6.2) can be cited here to claim that $I(M=k) \xrightarrow{P} 0$ as $n^*(\theta) \to \infty$. Thus, part (i) will follow from (14.6.11).

Next, we turn to part (ii). Along the line of (14.6.3), one can easily verify that $P_\theta\{M=k\} = O([n^*(\theta)]^{-p})$ for any fixed $p(> 0)$. Also, Fatou's lemma and part (i) together will let us conclude that

$$\liminf_{n^*(\theta)\to\infty} E_\theta[\frac{M}{n^*(\theta)}] \ge E_\theta[\liminf_{n^*(\theta)\to\infty} \frac{M}{n^*(\theta)}] = 1.$$

Thus, in view of the upper bound given by (14.6.11), part (ii) will immediately follow if we can verify the following result:

$$\lim_{n^*(\theta)\to\infty} E_\theta\left[\frac{1}{\overline{U}_k+k^{-1}}\right] = \frac{1}{\theta} \text{ for every fixed } 0 < \theta < 1. \qquad (14.6.12)$$

Now, with some appropriate random variable ψ_k between θ and $\overline{U}_k + k^{-1}$, we can express

$$E_\theta\left[\frac{1}{\overline{U}_k+k^{-1}}\right]$$

$$= \frac{1}{\theta+k^{-1}} - \frac{1}{\theta^2}E_\theta\left[(\overline{U}_k - \theta + k^{-1})\right] + E_\theta\left[(\overline{U}_k - \theta + k^{-1})^2\frac{1}{\psi_k^3}\right]$$

$$= \frac{1}{\theta+k^{-1}} - \frac{1}{k\theta^2} + E_\theta\left[(\overline{U}_k - \theta + k^{-1})^2\frac{1}{\psi_k^3}\right].$$

$$(14.6.13)$$

But, on the set where $\overline{U}_k+k^{-1} > \frac{1}{2}\theta$ is observed, we must have $\psi_k > \frac{1}{2}\theta$ so that with the help of the Cauchy-Schwartz inequality we can write

$$E_\theta\left[(\overline{U}_k - \theta + k^{-1})^2\frac{1}{\psi_k^3}I(\overline{U}_k + k^{-1} > \frac{1}{2}\theta)\right]$$

$$\le \frac{8}{\theta^3}E_\theta[(\overline{U}_k - \theta + k^{-1})^4] = o(1).$$

$$(14.6.14)$$

Also, along the line of (14.6.3), one can easily verify that $P_\theta\{\overline{U}_k \leq \frac{1}{2}\theta\} = O(k^{-p})$ for any fixed $p(> 0)$. Thus, for sufficiently large k, on the set where $\overline{U}_k + k^{-1} \leq \frac{1}{2}\theta$ is observed, we must have $\psi_k > \overline{U}_k + k^{-1} \geq k^{-1}$ so that

$$
\begin{aligned}
E_\theta &\left[(\overline{U}_k - \theta + k^{-1})^2 \frac{1}{\psi_k^3} I(\overline{U}_k + k^{-1} \leq \frac{1}{2}\theta) \right] \\
&\leq k^3 E_\theta[(\overline{U}_k - \theta + k^{-1})^2 I(\overline{U}_k + k^{-1} \leq \frac{1}{2}\theta)] \\
&= k^3 E_\theta^{1/2}[(\overline{U}_k - \theta + k^{-1})^4] P_\theta^{1/2}[\overline{U}_k \leq \frac{1}{2}\theta] \qquad (14.6.15) \\
&= k^3 O(k^{-1}) O(k^{-p}) \\
&= o(1), \text{ if } p \text{ is chosen to exceed 2.}
\end{aligned}
$$

At this point, combining (14.6.13)-(14.6.15), we immediately conclude the validity of (14.6.12), which completes the proof of part (ii).

The part (iii) can be verified in exactly the same way it was handled in the sequential case. ■

ACKNOWLEDGMENT

The authors wish to thank Timothy Mills of Rutgers University's Computer Services Department for conversations on comparing computers running at different speeds. Two referees independently pointed out some confusing statements and caught a number of typographical errors in a previous version that needed substantial tightening. Their critical assessments have led to a considerably clearer presentation for which we remain grateful and we take this opportunity to thank them.

REFERENCES

[1] Anscombe, F.J. (1952). Large sample theory of sequential estimation. *Proc. Camb. Philos. Soc.*, **48**, 600-607.

[2] Aoshima, M. (2000). Second-order properties of improved two-stage procedure for a multivariate normal distribution. *Commun. Statist. Theory Meth.*, **29**, 611-622.

[3] Aoshima, M. and Mukhopadhyay, N. (1999). Second-order properties of a two-stage fixed-size confidence region when the covariance

matrix has a structure. *Commun. Statist. Theory Meth.*, **28**, 839-855.

[4] Bratley, P., Fox, B.L. and Schrage, L.E. (1988). *A Guide to Simulation*, 2nd ed. Springer-Verlag: New York.

[5] Chow, Y.S. and Robbins, H. (1965). On the asymptotic theory of fixed-width sequential confidence intervals for the mean. *Ann. Math. Statist.*, **36**, 457-462.

[6] Ghosh, M., Mukhopadhyay, N. and Sen, P.K. (1997). *Sequential Estimation*. Wiley: New York.

[7] Hoeffding, W. (1961). The strong law of large numbers for U-statistics. *Mimeo Ser.# 302, Inst. of Statist., Univ. of North Carolina*, Chapel Hill, North Carolina.

[8] Khan, R.A. (1969). A general method of determining fixed-width confidence intervals. *Ann. Math. Statist.*, **40**, 704-709.

[9] Martinsek, A.T. (1983). Sequential estimation with squared relative error loss. *Bul. Inst. Math. Acad. Sinica*, **11**, 607-623.

[10] Martinsek, A.T. (1995). Estimating a slope parameter in regression with prescribed proportional accuracy. *Statist. Decisions*, **13**, 363-377.

[11] Mukhopadhyay, N. (1999). Second-order properties of a two-stage fixed-size confidence region for the mean vector of a multivariate normal distribution. *J. Multivar. Anal.*, **68**, 250-263.

[12] Mukhopadhyay, N. and Diaz, J. (1985). Two-stage sampling for estimating the mean of a negative binomial distribution. *Sequential Anal.*, **4**, 1-18.

[13] Mukhopadhyay, N. and Duggan, W.T. (1997). Can a two-stage procedure enjoy second-order properties? *Sankhya, Ser. A*, **59**, 435-448.

[14] Nadas, A. (1969). An extension of a theorem of Chow and Robbins on sequential confidence intervals for the mean. *Ann. Math. Statist.*, **40**, 667-671.

[15] Posten, H.O., Yeh, H.C. and Owen, D.B. (1982). Robustness of the two-sample *t*-test under violations of the homogeneity of variance assumption. *Commun. Statist. Theory Meth.*, **11**, 109-126.

[16] Robbins, H. and Siegmund, D. (1974). Sequential estimation of p in Bernoulli trials. In *Studies in Probab. Statist., E.J.G. Pitman Volume* (E.J. Williams, ed.), 103-107. Jerusalem Academic Press: Jerusalem.

[17] Ross, S.M. (1999). *Simulation*, 2nd ed. American Press: San Diego.

[18] Sproule, R.N. (1974). Asymptotic properties of U-statistics. *Trans. Amer. Math. Soc.*, **199**, 55-64.

[19] Sriram, T.N. (1991). Second order approximation to the risk of a sequential procedure measured under squared relative error loss. *Statist. Decisions*, **9**, 375-392.

[20] Stein, C. (1945). A two sample test for a linear hypothesis whose power is independent of the variance. *Ann. Math. Statist.*, **16**, 243-258.

[21] Stein, C. (1949). Some problems in sequential estimation (abstract). *Econometrica*, **17**, 77-78.

[22] Tiku, M.L. and Singh, M. (1981). Robust test for means when population variances are unequal. *Commun. Statist. Theory Meth.*, **10**, 2143-2159.

[23] Tukey, J.W. (1960). A survey of sampling from contaminated distributions. In *Contributions to Probability and Statistics* (I. Olkin, ed.). Stanford University Press: Stanford.

[24] Tukey, J.W. and McLaughlin, D.H. (1963). Less vulnerable confidence and significance procedures for location based on a single sample: Trimming/Winsorization. *Sankhya*, **25**, 331-352.

[25] Willson, L.J. and Folks, J.L. (1983). Sequential estimation of the mean of the negative binomial distribution. *Sequential Anal.*, **2**, 55-70.

[26] Woodroofe, M. (1982). *Nonlinear Renewal Theory in Sequential*

Analysis. SIAM: Philadelphia.

Addresses for communication:

NITIS MUKHOPADHYAY, Department of Statistics, UBox 4120, University of Connecticut, Storrs, CT 06269-4120, U.S.A. E-mail: mukhop @uconnvm.uconn.edu
GREG CICCONETTI, Department of Mathematical Sciences, Muhlenberg College, Allentown, PA 18104, U.S.A. E-mail: cicconet@muhl enberg.edu

Chapter 15

Sequential Estimation in the Agricultural Sciences

MADHURI S. MULEKAR
University of South Alabama, Mobile, U.S.A.

LINDA J. YOUNG
University of Florida, Gainesville, U.S.A.

15.1   INTRODUCTION

In agriculture, efforts are made to maximize the net revenue. Pest populations, such as insects and weeds, reduce yield, but control of these pests can be expensive. In general, control is only recommended when the pest population exceeds the economic threshold, the point at which the anticipated loss in revenue from pests exceeds the cost of control. The economic threshold often changes with the growth stage of the crop. In general, as the crop matures, the plants can tolerate more and more pests. The economic threshold may be affected by related environmental factors. For example, if beneficial insects such as lady beetles are present, they may be able to provide sufficient control of pest populations, making chemical control unnecessary. An understanding of the population dynamics of each species (crop, pest, beneficial) in the field is needed before optimal use of pesticides can occur. One element in the study of population dynamics is precise estimation of the density of pest and beneficial species. Sequential sampling has proven to be a useful tool in this setting. In this article, a review of sequential estimation procedures used most commonly in agriculture is given.

Rojás (1964) was the first to introduce an agricultural application of sequential estimation. He considered sequential estimation of the density of soil insects with a specified coefficient of variation of the sample mean, but the statistical properties of the method were not investigated. Here the term "density" refers to the average number of insects in a sampling unit. This differs from the traditional statistical use of the term as an abbreviation for a probability density function. In the remainder of this article, the biological meaning of "density" will be assumed all the time and "probability density function" will be used for the statistical meaning of the word.

Karandinos (1976) summarized sequential estimation procedures for the binomial, negative binomial, and Poisson distributions using three different measures of precision: controlling the coefficient of variation D of the sample mean, estimating the mean within h units with a specified level of confidence, and estimating the mean within a certain proportion p of the mean with a specified level of confidence. By recognizing the relationship between the mean and the variance of a negative binomial distribution, Kuno (1969) came up with a stopping rule for a sequential estimation procedure based on the total number T_n of observed pests in n observations, $n = 1, 2, \ldots$, allowing all computations to be made before going to the field. Again, the statistical behavior of the estimation procedure was not explored. Willson (1981) and Willson and Folks (1983) investigated the statistical properties of sequential estimation of the mean of a negative binomial population with a specified coefficient of variation. Willson (1981) studied sequential estimation of the mean of a negative binomial population within h units with a specified level of confidence, and also within a proportion p of the mean with a specified level of confidence. For each of these methods, the stopping rule could be expressed in terms of the total number of species in n observations, $n = 1, 2, \ldots$.

The most commonly used measure of precision in sequential estimation of the mean is the coefficient of variation D of a sample mean. For example, Sorenson et al. (1995) used a sequential egg mass sampling plan for predicting stalk tunneling damage by second generation European corn borer *Ostrinia nubilalis*. Bonato et al. (1995) sequentially estimated the density of two species of phytophagous mites (the cassava green mites *Mononychellus progressivus* and the cotton red mites *Oligonychus gossypii*) on cassava in Africa. Meikle et al. (1998) used sequential plans to determine pest status of two grain store pests, *Prostephanus truncatus* and *Sitophilus zeamais*. Explicit references to applications of other sequential methods in fields such as water quality monitoring and weed control are given by Mukhopadhyay (2002). Often, intuitive approaches to sampling are adopted, and the statistical properties of the associated estimator are not explored. Ease of field implementation is a primary concern. A less than optimal, but easy to obtain estimator, will usually be preferred to an optimal estimator requiring computations in the field. In this article, methods commonly used within agriculture to sequentially estimate the mean with a specified coefficient of variation D will be reviewed, and some new approaches will be considered.

One specific application is given in Section 15.2. The methodology and analysis are described in Section 15.3. Conclusions are given in Section 15.4.

15.2   A SPECIFIC APPLICATION

Pest populations are major threats to agricultural crops. Annually, anywhere from 10% to 20% of crops are lost because of damage due to pests. Control of pests such as insects (or, more generally, arthropods) and weeds is a major environmental, as well as agricultural, concern. Agricultural losses in yield and costs of control have been estimated at $122 billion (USDA (2000a)). Pests destroy an estimated 37 percent of all potential food and fiber crops, despite the widespread use of pesticides in the United States (USDA (2000b)). Losses due to Gypsy moth are approximately $200 million per annum. The Mexican rice borer causes an estimated monetary loss of $10 million to $20 million to the sugarcane industry in the lower Rio Grande Valley. In addition to insects, spread of weeds destroys crops and farmlands. Leafy spurge costs 12.9 million dollars annually in North Dakota alone, and Spotted Knapweed costs 14 million dollars annually in Montana alone. The Bureau of Land Management (USDI-BLM) estimates that infestation of their lands by noxious weeds has increased from 2.5 million acres in 1985 to 9 million acres in 1995. It is estimated that by the year 2010, they will infest 140 million acres in the United States, increasing at a rate of about 20 million acres per year. By then, weed management may be the largest single budget item for the federal land management agencies. Therefore, economical methods of estimating population densities are critical for devising and implementing control methods and studying the effectiveness of such methods (USDA (1998)).

When pesticides were first developed, they were inexpensive and effective. In this setting, farmers tended to use control measures even when population levels were low and not of economic importance. This intensive use of pesticides led to pest species developing resistance to the pesticides. This was first observed in 1914 when San Jose scale (*Quadraspiditus pernicisus* (Comstock)) showed resistance to lime sulfur spray (Metcalf (1955)). Concerns mounted as resistant strains of the house fly (*Músca doméstica L.*) appeared in Sweden and Denmark in 1946, the bedbug (*Cimex lectulrius L.*) in Hawaii in 1947, the mosquitoes (*Cúlex pipiens L.*) in Italy and *Aèdes sollcitan* (Walker) in Florida in 1947, and the human body louse (*Pediculus humànus humànus L.*) in Korea and Japan in 1951 (Brown and Pal (1971)). This resistance persists even after the use of insecticide is discontinued. For example, cotton bollworms and budworms have developed resistance to most commonly used commercial insecticides. As a result, these insects now infest over 75 percent of the United States cotton crop (USDA (2000c)). Additionally, insecticide application to soil causes damage to roots and reduction in soil nutrients, resulting in crop/yield reduction (Felsot et al. (1982)). The impact of pesticides on the environment, especially ground and surface water, is an increasing concern. The observed and potential environmental impacts have led to either restricted use or banning of some pesticides within the United States. All of these concerns led agricultural researchers and farmers to seek better approaches to the control of pest populations. *Integrated pest management* (IPM) was introduced as a means of considering the total agricultural system when making decisions about the use of pesticides. This approach requires an understanding of the population dynamics of the crop, pest species, and beneficials. Precise estimation of the population densities is a critical element in understanding these dynamics, and sequential estimation is a useful tool in this setting.

To illustrate some of the methods, data collected on the number of fleahoppers (*Pseudatomoscelis seriatus* (Reuter)) on cotton, in August 1986, will be used. The data were collected on a 2 hectare field near Chickasha, Oklahoma. It was planted to Paymaster 145 cotton cultivar at 18 pounds of seed per acre in 40-inch rows on June 10, 1986. Cotton fleahoppers are small yellowish-green bugs, about 1/8-inch long, with black specks on the upper surface of the body. Their piercing-sucking mouthparts are used to feed on the leaves and squares of cotton. This feeding can cause extensive loss of small squares during the early fruiting phase of plant development. Various sampling units, including all plants along a meter-row, an individual plant, and a terminal on a plant are typically used in sampling. Here the sampling unit is a terminal on a plant. For the sake of concreteness, the methods described below will be discussed only in the context of adult cotton fleahoppers, but they would apply in an analogous manner to other pest species. Several samples of terminals were inspected. The number of adult cotton fleahoppers on each selected terminal was counted. Since not all terminals were infested with fleahoppers, presence/absence was also noted. Summary statistics for seven samples are reported in Table 15.2.1.

Table 15.2.1: Summary Statistics for Adult Cotton
Fleahopper Sampling

Sampling Date	Sample Size	Sample Mean	Sample Variance	Number of Zeroes
August 4	150	0.687	1.129	86
August 5	200	0.520	0.784	130
August 6	200	0.765	1.075	103
August 7	150	0.767	1.592	79
August 10	150	1.093	2.407	73
August 19	100	0.840	1.348	54
August 20	200	0.925	1.356	94

15.3　METHODOLOGY AND ANALYSIS

An improved understanding of the population dynamics of all species in a field and the interaction among species is an essential component in IPM. Precise estimation of the mean number of insects in a species per sampling unit is key to developing that understanding. The sampling unit may be a leaf, a plant, a square-foot quadrat, a linear-foot of row, or a unit of land-area. Counts of insects from each species within sampling units, such as the terminal of a cotton plant, lead to absolute measures of pest density.

In IPM, discrete distributions are commonly used. For example, using the terminal of a cotton plant as the sampling unit, recording the presence or absence of adult cotton fleahoppers on a plant leads to the binomial (B) distribution. Instead of presence/absence, the number of adult cotton fleahoppers on a terminal may be recorded. Although the Poisson (P) distribution is often the first one considered when modeling count data, it rarely fits biological data. For the Poisson distribution, the mean is equal to the variance. However, in biology, the variance usually exceeds the mean, resulting in over-dispersed count data. The negative binomial (NB) distribution usually provides an adequate fit to these data. Alternatively, the researcher may be unwilling to make any distributional assumptions about the count data, preferring instead a nonparamentric approach. In this section, sequential estimation of the mean with a fixed coefficient of variation D of the sample mean for the nonparametric, binomial, Poisson, and negative binomial cases will be considered. Because the sampling methods are based on the relationship between a mean and a variance, modeled relationships between a mean and a variance may be used to propose other sequential estimation methods. Three such methods will also be discussed.

In the entomological literature, several sampling plans for estimating the population density with a specified coefficient of variation have been proposed. Frequently this is referred to as controlling the *degree of precision* (D) or the *relative variation*. Chandler and Allsopp (1995) used $D = 0.20, 0.25, 0.30$ to sample potato plants for *T. Palmi* infestation. Wang and Shipp (2001) chose $D = 0.15, 0.20, 0.25$, and 0.30 to develop a sequential sampling plan to estimate *Frankliniella accidentalis* (Pergande). Peña and Schaffer (1997) set $D = 0.10, 0.25$ to estimate the citrus leafminer *Phyllocnistis citrella* density on lime. O'Rourke et al. (1998) used $D = 0.25$ and sweep net sampling to estimate aster leafhopper (*Macrosteles quadrilineatus*) in carrots. The parametric and nonparametric approaches to estimating the population mean with a specified coefficient of variation D are used commonly in entomological applications. The stopping rules for these procedures are summarized in Table 15.3.1.

Nonparametric approach: In this approach no assumption is made about the distribution of the sampled population, nor is any specific relation between the population mean and the population variance considered. Rojás (1964) introduced sequential sampling for soil insects. He suggested the following scheme:

Let T_n be the total number of individuals on the first n units sampled. Compute $m_n = T_n/n$ and s_n^2, the mean and the variance, respectively, of the number of individuals counted on the first n units sampled. The *coefficient of variation* (CV), D, of the sample mean is the standard error of the mean divided by the mean:

$$D = \frac{\sqrt{\sigma^2/n}}{\mu}. \qquad (15.3.1)$$

Thus the minimum sample size required to estimate the mean with a CV of D is

$$n = \left(\frac{\sigma}{\mu D}\right)^2. \qquad (15.3.2)$$

Using sample estimates of the parameters in equation (15.3.2) leads to the following intuitive stopping rule for estimating the mean with a CV of D.

Stopping criterion: At each step of sampling, compute $[s_n/(m_n D)]^2$ using the total number of individuals observed on n units, and continue random sampling until

$$N = \text{Smallest integer } n \text{ such that } n \geq \frac{s_n^2}{(m_n D)^2}, \qquad (15.3.3)$$

where N is the random sample size. Once the stopping criterion is met, using the total number of individuals counted on N units, the density at the specified level of precision D is estimated as $\hat{\mu} = T_N/N$.

Extending results of Mukhopadhyay (1978) and Ghosh and Mukhopadhyay (1979), Willson and Folks (1983) showed this procedure to be asymptotically consistent and efficient for distributions with finite means and variances. However, if the underlying distribution is negative binomial, Monte Carlo studies have shown that the estimated density is substantially less precise than desired for regions of the parameter space that are most often observed in studies (Willson (1981)). This procedure requires updating the estimates m_n and s_n^2 after each observation, making it less desirable to field operators than other procedures that do not need such computations.

If the population distribution is known, the relationship between the corresponding mean and variance is also known. Then the stopping rule can be expressed in terms of the cumulative number T_n of individuals counted in n sample units. When sampling stops, the estimate of the density is then T_n/n. Here, sequential estimation for binomial, Poisson, and negative binomial distributions are reviewed.

Table 15.3.1: Stopping Rules for Fixed-Precision Sequential Sampling Plans (Here, "log" Means Natural Logarithm)

Stopping Rule: Continue sampling until
Binomial: $$T_n \geq \frac{n}{nD^2 + 1} \text{ and } 0 < T_n < n$$
Poisson: $$T_n \geq \frac{1}{D^2} \text{ and } n \geq 2$$
Negative binomial: Using UMVUE of μ and relation between μ and σ^2 $$T_n \geq \frac{nk}{D^2 nk - 1} \text{ and } n \geq \frac{1}{D^2 k}$$
Negative binomial: Using UMVUEs of μ and σ^2 $$T_n \geq \frac{nk}{(nk+1)D^2 - 1} \text{ and } n > \frac{1}{D^2 k} - \frac{1}{k}$$
Iwao's patchiness regression: $$T_n \geq \frac{n(\alpha + 1)}{D^2 n - (\beta - 1)} \text{ and } n > \frac{\beta - 1}{D^2}$$
Taylor's power Model: $$T_n \geq \left(\frac{D^2}{a}\right)^{1/(b-2)} n^{(b-1)/(b-2)}$$ that is, $\log(T_n) \geq \dfrac{\log(D^2/a)}{b-2} + \dfrac{b-1}{b-2} \log(n)$
Ecological model: $$T_n \geq \frac{n\beta}{D} \sqrt{MSE\left(1 + \frac{1}{N}\right) + (\log[-\log(1 - p_T)] - \bar{p})^2 s_\beta^2}$$
Modified ecological model: $$T_n \geq C + \frac{n\beta}{D} \sqrt{MSE\left(1 + \frac{1}{N}\right) + (\log[-\log(1 - p_T)] - \bar{p})^2 s_\beta^2}$$

Binomial: Because counting the number of adult cotton fleahoppers on a terminal can be extremely time consuming, especially for large populations, researchers are often interested in observing the presence or absence of adult cotton fleahoppers on each terminal. When the presence (1) or absence (0), and not the actual count of adult cotton fleahoppers, is recorded for each of n randomly selected plants, the binomial distribution provides an appropriate model. Suppose n units are sampled at random and that the probability p ($0 < p < 1$) of a unit being infested is constant. For a binomial distribution, the mean is np and the variance is $np(1 - p)$. The population proportion p of units infested is estimated using its UMVUE, m_n, the proportion of infested sampling units. The variance is estimated by $s^2 = nm_n(1 - m_n)$. A sequential method of estimating the mean with a specified CV of D follows by using these estimates of mean and variance in (15.3.2). The relationship between the mean and the variance permits the stopping rule to be simplified, as shown in Table 15.3.1.

Sequential estimation of the binomial parameter p has been extensively studied, including works by Robbins and Siegmund (1974), Cabilio and Robbins (1975), and Martinsek (1983). The above method is presented because of its widespread use in agriculture. See Karandinos (1976) and Ruesink (1980), for examples.

Poisson: Some information is lost if just the presence/absence and not the actual number of adult cotton fleahoppers on a terminal is recorded. So, count data are recorded quite often. If the adult cotton fleehoppers are randomly distributed in the field so that each fleahopper is equally likely to be on each cotton plant terminal, then the number of adult fleahoppers per terminal will follow a Poisson distribution. Because the mean and variance of a Poisson random variable are equal, the sample mean m is the UMVUE of both. Again, using this estimator in (15.3.2), the stopping rule given in Table 15.3.1 is easily derived. Since this method is rarely used, compared to the other methods presented here, it will not be discussed further.

Negative binomial: The Poisson distribution is rarely observed in nature. Usually, the variance is substantially larger than the mean, leading to over-dispersed or aggregated populations. Several probability distributions have been proposed as models for this over-dispersion. The negative binomial model is the one most widely applied as it often fits biological data well. One reason for this may be that at least 17 different models are known to give rise to this distribution (Boswell and Patil (1970)). For this distribution, the variance is larger than the

mean. Since the mean of a distribution is of primary interest in most
biological sciences, Anscombe (1949) described the negative binomial
probability mass function (p.m.f.) using two parameters, namely, the
mean μ ($\mu > 0$) and the *aggregation parameter* k ($k > 0$, not necessarily
an integer) as

$$P_x = \binom{x+k-1}{k-1} \left(\frac{k}{\mu+k}\right)^k \left(\frac{\mu}{\mu+k}\right)^x, \quad x = 0, 1, 2, \ldots, \quad (15.3.4)$$

where P_x is the probability of observing exactly x individuals in a sampling unit. The parameter k is the measure of *aggregation* or *crowding*
and is usually unknown. A smaller value of k indicates higher degree
of aggregation. The measure of aggregation is related to the mean μ
by $k = \mu^2/(\sigma^2 - \mu)$, where σ^2 is the population variance. As k approaches infinity, the negative binomial p.m.f. approaches that of the
Poisson distribution. The *maximum likelihood estimator* (MLE) of k
has no closed form solution and has not been used in sequential sampling plans. The *method of moments estimator* (MME) of k is

$$\hat{k} = \frac{m^2}{(s^2 - m)} \quad \text{provided that} \quad s^2 > m. \quad (15.3.5)$$

An intuitive estimator of the negative binomial variance is obtained
by substituting estimates of μ and k into the variance-to-mean relationship:

$$\hat{\sigma}^2 = m + \frac{m^2}{\hat{k}} \quad (15.3.6)$$

A sequential procedure developed using the intuitive estimator of
the variance (given by (15.3.6)) and the MME of k (given by (15.3.5))
in relation (15.3.2) reduces to the same stopping rule given by (15.3.3).
This procedure does not perform well in the regions of the parameter
space most often encountered in biology. For example, if $D = 0.25$
and $(\mu, k) = (1, 1)$ and $(5, 5)$, simulations indicate that the actual
coefficient of variation of the mean, D, is 0.42 and 0.32, respectively
(Willson (1981)). Thus, when sampling from the negative binomial
population, knowledge, or a precise estimate, of k allows more precise
estimation of the population density. Based on experience with adult
fleahoppers in cotton, it is believed that k should be 1 or close to 1
(see Young and Willson (1987)). In the work that follows, it will be
assumed that k is one.

Kuno (1969,1972) presented a sequential procedure for estimating the mean of a population with variance of the form $\sigma^2 = a\mu + b\mu^2$, where a and b are known constants. The negative binomial distribution falls in this category with $a = 1$ and $b = 1/k$. A sequential procedure developed using $\hat{\mu} = m_n$ (the UMVUE of the mean) and $am_n + bm_n^2$ as the estimate of σ^2 (coming from the relation between the variance and the mean) results in the stopping rule for the negative binomial distribution given in Table 15.3.1.

Willson and Young (1983) and Willson and Folks (1983) suggested estimating the variance using the UMVUE, $s_n^2 = n(nk + 1)^{-1}(km_n + m_n^2)$, leading to the second negative binomial stopping rule in Table 15.3.1. Simulations indicate that this procedure gives precise estimates of density with desired precision.

The negative binomial procedure results in more precise density estimates when using an estimate of k smaller than the true value. If a range of estimates for k is available, then using a conservative estimate of k is advisable in the density estimation process so that the desired precision is attained. Bliss and Fisher (1953), Bliss and Owen (1958), and Willson et al. (1984) presented different methods for estimating k.

In addition to the nonparametric and parametric methods discussed above, other models for the mean-variance relationship have been proposed. These modeled relationships are then used in (15.3.1) to develop sequential estimation procedures. The statistical properties of these methods have not been investigated. Three such approaches will be considered here: Iwao's patchiness regression, Taylor's power model, and the ecological model relating the mean and the probability of presence. In these three cases, empirical models replace the distributional assumptions of the parametric models.

Iwao's patchiness regression: Lloyd's mean crowding index, m^*, is defined as the "mean number of other individuals in the quadrant per individual". Lloyd (1967) gave

$$m^* = \frac{\sum_{j=1}^{Q} x_j(x_j - 1)}{\sum_{j=1}^{Q} x_j} = \frac{\sum_{j=1}^{Q} x_j^2}{\sum_{j=1}^{Q} x_j} - 1,$$

where Q is the total number of quadrats in the area, x_j is the number of individuals in the jth quadrat, $j = 1, 2, \ldots, Q$. He showed that an approximate relationship between mean crowding (m^*) and the density of the pest population (μ) is given by:

$$m^* = \mu + \left(\frac{\sigma^2}{\mu}\right) - 1 \qquad (15.3.7)$$

Based on empirical studies, Iwao (1968) proposed that mean crowding was linearly related with the density, i.e.

$$m^* = \alpha + \beta\mu. \qquad (15.3.8)$$

If this relationship holds, using Lloyd's index in (15.3.7), the relationship between the variance and the mean can be expressed as

$$\sigma^2 = (\alpha + 1)\mu + (\beta - 1)\mu^2. \qquad (15.3.9)$$

This relationship holds for binomial ($\alpha = -p, \beta = 1$), Poisson ($\alpha = 0, \beta = 1$), and negative binomial ($\alpha = 0, \beta = 1 + 1/k$) distributions. The coefficient β is used as a measure of the aggregation level or the degree of departure from randomness. When the slope β is one, the individuals are assumed to be randomly dispersed in the population.

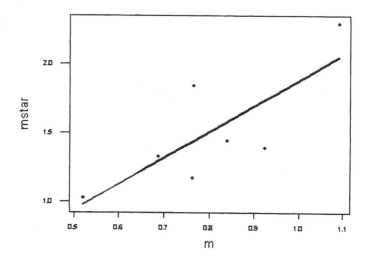

Figure 15.3.1: Fitted Line Plot for m^* Versus m.

To model the mean-variance relationship, datasets are collected on several sampling occasions. The sample mean and sample variance are computed for each dataset. Then, μ is estimated using m, and m^* is estimated as $[m + (s^2/m) - 1]$ for each dataset. Iwao's patchiness regression coefficients α and β are estimated by regressing m^* on m across all sampling occasions. The coefficient of determination (R^2) is used to assess the strength of relationship. The stopping rule is developed using Iwao's patchiness regression coefficients and the relation (15.3.7) in (15.3.2) as shown in Table 15.3.1. It must be noted that no evaluation of the properties of this sequential estimation method has been conducted.

Figure 15.3.1 shows a fitted line plot of mean crowding versus density for the datasets in Table 15.2.1. The relation is linear and the estimated least squares regression equation is $\hat{m}^* = 0.003 + 1.875m$. The R^2 value of 0.61 observed for these data is smaller than those commonly observed. This may be due to the small range in the observed means and the small number of samples used to estimate the relation.

Using the estimated coefficients, the stopping rule would become

$$N = \text{Smallest integer } n \text{ such that}$$

$$T_n \geq \frac{1.003n}{D^2 n - 0.875} \text{ and } n > \frac{0.875}{D^2}.$$

(15.3.10)

Figure 15.3.2 shows the stopping boundaries for the datasets in Table 15.2.1 when different values are specified for D. For adult cotton fleahoppers, $D = 0.2$ is commonly used by researchers for density estimation. For this precision level, the stopping boundary for sample size $n > 21.88$ is $T_n \geq 1.003n/[0.04n - 0.875]$. Using a first sample of size 150 we note that $T_{150} = 103 > 29.4$ (boundary). Therefore the density estimate is $\hat{\mu} = 103/150 = 0.687$ cotton fleahoppers per terminal.

Chandler and Allsop (1995) used Iwao's patchiness regression with $D = 0.10$ and 0.25 to estimate density of adult froghopper *Ecoscarta carnifex* (F.) and associated symptoms on sugarcane crops in northern Queensland, Australia. Sorenson et al. (1995) used decision lines constructed using Iwao's mean crowding-mean method to predict stalk tunneling damage by second-generation European corn borer, *Ostrinia nubilalis* (Hübner) in field corn in eastern North Carolina. Wang and Shipp (2001) used Iwao's fixed-precision sequential sampling plan for estimating *F. occidentalis* adult population density in flowers of greenhouse cucumber at precision levels $D = 0.15(0.05)0.30$.

Taylor's power model: Based on empirical evidence, Taylor (1961, 1984) suggested that the relationship between the population density and the corresponding variance could be modeled by

$$\sigma^2 = a\mu^b.$$

(15.3.11)

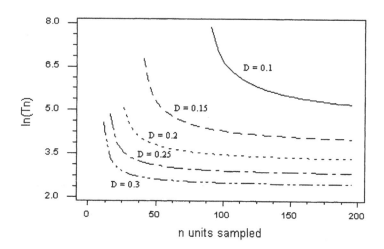

Figure 15.3.2: Boundaries for Sequential Plan Using
Iwao's Patchiness Coefficients. Here, "ln" Means Natural Logarithm.

The coefficient a is interpreted as a sampling factor (Southwood
(1978)), and the coefficient b is a constant measure of dispersion for
a species. A value of $b > 1$ indicates an aggregated dispersion, $b = 1$
indicates a random dispersion, and $b < 1$ indicates an under dispersion.

While a may change with environment, b is considered fixed for any
given species. However, Downing (1986) demonstrated that b varies
with the number of samples taken and the range of means considered.
Also, Taylor et al. (1998) claim that b may be stage-specific for species
and influenced by environmental conditions, exactly as host specificity
in parasitoids is specific yet variable. Thus, Taylor's law has excited
interest as well as controversy.

A sequential procedure for estimating the population mean using
Taylor's power model has been proposed (Ruesink (1980)). Table 15.3.1
gives the stopping rule in terms of Taylor's power model coefficients,
obtained by substituting the variance estimated from Taylor's power
model into (15.3.2) and solving for T_n.

In applications, a logarithmic transformation of the Taylor's power
model,

$$\log(s^2) = \log(a) + b\log(m), \qquad (15.3.12)$$

is routinely used in the estimation process. In practice, data are collected on several different sampling occasions. Then, m and s^2 are
computed for each dataset. The coefficients a and b of Taylor's power

model are estimated by regressing $\log(s^2)$ on $\log(m)$ across all sampling occasions. Typically the coefficient of determination (R^2) is used to assess the strength of the relationship. High values of R^2 are commonly observed in practice. For example, O'Rourke et al. (1998) reported that for Aster leafhopper data, $R^2 = 0.88$. Heinz and Chaney (1995) show the results of applying Taylor's power model to 30 different sets of data from two different time periods and 5 different strata. Inference is based on the assumptions of least-squares regression. Seldom are the assumptions associated with least-squares regression fully met. For example, typically, the variation in the sample mean is ignored in the estimation process. The variation of the sample variance tends to increase with the sample mean. The impacts of these departures from the least-squares regression assumptions are seldom assessed in practice.

Figure 15.3.3 shows the fitted line plot for $\log(s^2)$ versus $\log(m)$ for the data in Table 15.2.1. It shows a fairly linear relationship. The modeled relationship is $\sigma^2 = e^{0.602}\mu^{1.338}$, with $R^2 = 0.800$. The estimated power model coefficients are $a = 1.8258$ and $b = 1.338$. For the data presented here, the stopping rule would take the following form:

$$N = \inf\{n : \log(T_n) \geq 0.399 - 1.51\log(D^2) + 0.511\log(n)\}.$$

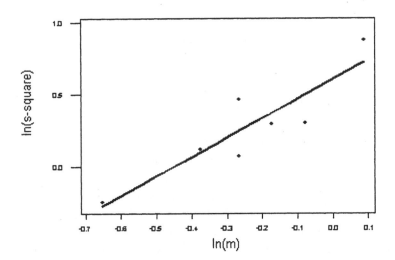

Figure 15.3.3: A Fitted Line Plot of $\log(s^2)$ Versus $\log(m)$. Here, "ln" and "log" Mean Natural Logarithm.

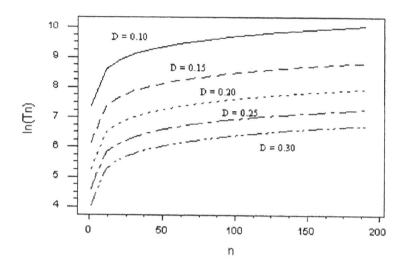

Figure 15.3.4: Boundaries for Plan Using Taylor's Power Model
Estimates. Here, "ln" Means Natural Logarithm.

Figure 15.3.4 shows the stopping boundaries for the data in Table
15.2.1 with different specified values for D. Taylor et al. (1979) showed
that $s^2/m \to 1$ as $m \to 0$. This limiting effect distorts the results of the
power model, particularly for $m < 2$ and $s^2 < 4$ (Taylor and Woiwod
(1982)). For low m values, relation (15.3.12) tends to underestimate
values of b. Perry and Woiwod (1992) suggest removing mean values
below the "Poisson line" given by $m < a^{1/(b-1)}$ to reduce spurious
correlations between the $\log(m)$ and $\log(s^2)$ values. Since all the means
in our data (Table 15.2.1) fall below the Poisson line of 2.073, Taylor's
power model may not be appropriate to use.

Cho et al. (2000) used a fixed-precision-level sampling plan developed using the parameters from Taylor's power model to estimate
population densities within 25% of the mean of immature *Trips palmi*
Karney on fall potato on Cheju Island, Korea. Hangstrum et al. (1997)
applied Taylor's power model and developed a nonlinear variance-mean
relation to study densities of insects in stored-grain data. Hamilton et
al. (1998) applied the sequential estimation method developed by using variance estimates from Taylor's power model for Colorado potato
beetle in eggplant. O'Rourke et al. (1998) used Taylor's power model
for validation of density estimation using a fixed-precision sampling

plan for Aster leafhopper (Homoptera: Cicadellidae) in carrot. They used 260 datasets to estimate Taylor's coefficients and 15 independent datasets to validate the sampling plan. Heinz and Chaney (1995) used it to estimate the population density of *Liriomyza huidobrensis* (Blanchard) larvae in celery collected from commercial fields in the coastal valleys of California.

Ecological model: In the entomological literature, this model is also known as a binomial sequential sampling model, empirical model, or $p_T - m$ model. Counting the number of adult cotton fleahoppers on a terminal of a cotton plant is time consuming, especially when the population is large. This has led researchers to consider presence/absence sampling; however, one cannot avoid some loss of information while doing so. One possible way to regain some of the lost information is to try to establish a relationship between the proportion infested and the population mean. Then, estimation procedures can be based on the proportion infested, and the estimated proportion used to estimate the population density. Kono and Sugino (1958) proposed the following empirical linear model to relate the density μ to p_T, the proportion of units infested at tally threshold T:

$$\log(\mu) = \alpha + \beta \log\left[-\log(1 - p_T)\right], \qquad (15.3.13)$$

where the tally threshold T is the number of individuals present on a sample unit before the unit is classified as "infested". The model parameters α and β are estimated using least-squares regression for a specified tally threshold. See Cho et al. (2000) for an application. As shown by Doran et al. (1995), this model is not strictly linear for some species, and a square root or an arcsine transformation of means or proportions infested is used. As described in the case of the previous two procedures, data are collected from several sampling occasions. From each sampling occasion, the number of units infested with at least T individuals, and the total number of individuals on the sampled units are recorded. Then, the proportion of units infested and the density are estimated from each sample as follows:

$$\hat{p}_T = \frac{\text{Number of units infested for tally threshold } T}{\text{Number of units inspected}}$$

and m is the sample mean number of individuals per sampling unit.

The model parameters can be estimated by regressing $\log(m)$ on $\log\left[-\log\left(1 - \hat{p}_T\right)\right]$. Further improvements might be possible by considering the errors in the variables. Snedecor and Cochran (1980) give an expression for the variance of the predicted value for a given predictor using a linear relation. Applying that result to the empirical linear model used here, we get the following expression for the variance of the density estimate:

$$\sigma_m^2 = \beta^2 \left(MSE \left(1 + M^{-1} \right) + (\log \left[- \log(1 - p_T) \right] - \bar{p})^2 s_\beta^2 \right) \quad (15.3.14)$$

where MSE is the mean squared error from fitting equation (15.3.13), M is the number of pairs of data used to estimate α and β from equation (15.3.13), $\bar{p}$ is the average of $\log \left[- \log(1 - p_T) \right]$ values used in the regression of (15.3.13), s_β^2 is the sample estimate of the variance of β in (15.3.13), and n is the number of units sampled from the population.

Using this estimate of the variance of density in (15.3.2), we get $D = \sigma_m/\mu$. Then, a sequential stopping rule can be given as in Table 15.3.1.

Figure 15.3.5 shows the relation of $\log(m)$ versus $\log \left[- \log(1 - p_T) \right]$ for the data in Table 15.2.1. The estimated relationship is $\widehat{\log(\mu)} = 0.309 + 1.152 \log \left[- \log(1 - p_T) \right]$. Now, a sequential estimation of the binomial parameter could be implemented in this setting. For the above fitted model, $MSE = 0.00989$, $M = 7$, $\bar{p} = -0.4824$, $s_\beta^2 = 0.0468$, and $n = 1150$. Using these statistics in the criterion given in Table 15.3.1, the stopping boundary is computed. At each step of sampling, p_T is observed and the boundary is computed using the following relation:

$$T_n \geq 6624 \sqrt{0.0113 + 0.0468 \left(\log \left[- \log(1 - p_T) \right] + 0.4824 \right)^2}.$$

Figure 15.3.6 shows the boundary computed for $D = 0.2$, and the data collected. As we can see from the plot, after the sixth sample is taken, the total number of fleahoppers exceeds the boundary. Thus, the estimated density is $\hat{\mu} = 723/950 = 0.761$.

In the development of sequential hypothesis tests about a population mean, Schaalje et al. (1991) and Nyrop and Binns (1992) suggested improved methods of estimating the variance. Here we use these improved estimators to develop sequential estimation procedures for population means. Schaalje et al. (1991) proposed a correction (using the mean-variance relation described by Taylor's power model) to modify the approximation of the variance of a sample mean estimated as a function of the proportion of sample units infested. He proposed a correction factor C for the relation (15.3.14), where C takes into account the mean-variance relation described by Taylor's power model. With a

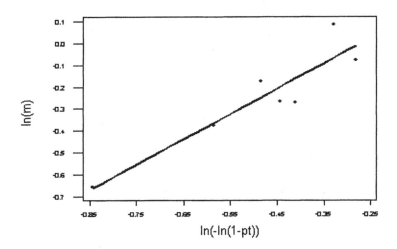

Figure 15.3.5: Fitted Line Model for $\log(m)$ Versus $\log(-\log(1 - p_T))$. Here, "ln" and "log" Mean Natural Logarithm.

correction factor C, (15.3.14) now looks like:

$$\sigma_m^2 = \beta^2 \left(MSE \left(1 + M^{-1}\right) + \left(\log\left[-\log(1 - p_T)\right] - \bar{p}\right)^2 s_\beta^2 + C \right)$$
$$(15.3.15)$$

where

$$C = \frac{\beta^2 \hat{p}_T}{n(1 - \hat{p}_T) \, \log(1 - \hat{p}_T)^2}$$
$$- \frac{1}{n} \exp\left[\log(a) + (b - 2)\left(\alpha + \beta \log\left[-\log(1 - \hat{p}_T)\right]\right)\right].$$

Note that, a and b are estimates obtained from Taylor's power model. Using this modification, the revised stopping rule is shown in Table 15.3.1.

Giles et al. (2000) used the $p_T - m$ linear model to estimate *Schizaphis graminum* intensity using 115 occasions from hard red winter wheat fields in Oklahoma. They also used the sequential probability ratio test developed around p_{ET} (that is, p_T corresponding to economic thresholds) to test $H_0 : p_0 < p_{ET}$ and $H_1 : p_1 > p_{ET}$ for greenbugs on winter wheat. Cho et al. (2000) used the binomial sampling model to estimate the densities of immature *Trips palmi* Karny on fall potato. Doran et al. (1995) used a binomial model for insecticide application to control pests in cabbages grown on Prince Edward Island, Canada.

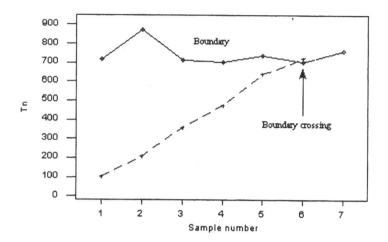

Figure 15.3.6: Boundary for the Ecological Model with Data.

Nyrop and Binns (1992) used Taylor's power model to estimate σ^2 as $s^2 = am^b$, and then estimated the parameter k as $\hat{k} = m^2/(s^2-m) = m/(am^{b-1} - 1)$. They claim that the parameter k estimated this way accounts for changes in k due to changes in the population mean. Using estimated parameters for the empirical model and estimated values of k, they investigated the effect of changing the tally threshold T when using a negative binomial model. They showed that increasing the average number of samples to make a decision will not improve the robustness of a sequential relationship based on the $p_T - m$ relationship. This is because variability in the $p_T - m$ relationship or variability in k overshadows the binomial sampling plan.

15.4   CONCLUDING REMARKS

Generally, sequential methods for density estimation are not used by farmers for decision-making. Sequential estimation of the population mean is commonly employed in agriculture, especially in entomology, by researchers as part of an overall effort to gain a deeper understanding of the population dynamics of pests, and to explore alternative methods for reducing uses of pesticides. Often those collecting the data have little or no experience with statistics. Thus, consideration is always given to ease of implementation. This may lead to the use of less than optimal methods. Many of the methods being used today are intuitive or ad-hoc ones for which, the statistical properties are largely unknown. For example, the modeled relationship between the mean and the variance in Iwao's patchiness regression, Taylor's power model or the ecological model provide a foundation for reasonable sequential estimation rules. Again, the statistical properties of estimation procedures based on these modeled relationships have not been fully explored. It is not known how well these methods perform for moderate values of D such as 0.1, 0.2, or 0.3 — values commonly assumed in practice. Many opportunities exist for providing a firmer theoretical foundation for these methods. Other multistage sampling schemes such as two- and three-stage-sampling are available in the statistical literature and theoretical properties have been developed for them (see, for example, Mukhopadhyay (2002)). But most of them require field computations, something these scientists want to avoid. They may be easily adapted, but they are not being used.

ACKNOWLEDGMENT

We thank Jerry H. Young for the use of his data.

REFERENCES

[1] Anscombe, F.J. (1949). The statistical analysis of insect counts based on the negative binomial distribution. *Biometrics*, **5**, 165-173.

[2] Bliss, C.I. and Fisher, R.A. (1953). Fitting the negative binomial distribution to biological data and note on the efficient fitting of the negative binomial. *Biometrics*, **9**, 176-200.

[3] Bliss, C.I. and Owen, A.R.G. (1958). Negative binomial distributions with a common k. *Biometrika*, **45**, 37-58.

[4] Bonato, O., Baumgartner, J. and Gutierrez, J. (1995). Sampling plans for Mononychellus preogresivus and Oligonychus gossypii (Acari: Tetranychidae) on Cassava in Africa. *J. Econ. Entomol.*, **88**, 1296-1300.

[5] Boswell, M.T. and Patil, G.P. (1970). Chance mechanisms generating the negative binomial distributions. In *Random Counts in Scientific Work, 1: Random Counts in Models and Structures* (G.P. Patil, ed.),

3-22. Pensylvania State Univ. Press: State College.

[6] Brown, A.W.A. and Pal, R. (1971). *Insecticide Resistance in Arthropods.* WHO Monograph Ser. **38**. WHO: Geneva.

[7] Cabilio, P. and Robbins, H. (1975). Sequential estimation of p with squared relative error loss. *Proc. Nat. Acad. Sci., U.S.A.*, **72**, 191-193.

[8] Chandler, K.J. and Allsopp, P.G. (1995). Sampling adults of Eoscarta carnifex (Hemiptera: Cercopidae) and their associated symptoms on sugarcane. *J. Econ. Entomol.*, **88**, 1301-1306.

[9] Cho, K., Kang, S. and Lee, G. (2000). Spatial distribution and sampling plans for Thrips palmi (Thysanoptera: Thripidae) infesting fall potato in Korea. *J. Econ. Entomol.*, **93**, 503-510.

[10] Doran, A.P., Sears, M.K. and Stewart, J.G. (1995). Rvaluation of binomial model for insecticide application to control Lepidopterous pests in cabbage. *J. Econ. Entomol.*, **88**, 302-306.

[11] Downing, J.A. (1986). Spatial heterogeneity: Evolved behaviour or mathematical artefact? *Nature*, **323**, 255-257.

[12] Felsot, A.S., Wilson, J.C., Kuhlman, D.E. and Steffy, K.L. (1982). Rapid dissipation of carbofuran as a limiting factor in corn rootworm control in fields with histories of continuous carbofuran use. *J. Econ. Entomol.*, **75**, 1098-1103.

[13] Ghosh, M. and Mukhopadhyay, N. (1979). Sequential point estimation of the mean when the distribution is unspecified. *Commun. Statist. Theory Meth.*, **8**, 637-652.

[14] Giles, K.L., Royer, T.A., Elliott, N.C. and Kindler, S.D. (2000). Development and validation of binomial sequential sampling plan for the greenbug (Homoptera: Aphididae) infesting winter wheat in the southern plains. *J. Econ. Entomol.*, **93**, 1522-1530.

[15] Hamilton, G.C., Lashomb, J.H., Arpaia, S., Cianese, R. and Mayer, M. (1998). Sequential sampling plans for Colorado potato beetle (Coleo -ptera: Chrysomelidae) in eggplant. *Environ. Entomol.*, **27**, 33-38.

[16] Hangstrum, D.W., Subramanyam, B.H. and Flinn, P.W. (1997). Nonlinearity of a generic variance-mean equation for stored-grain insect sampling data. *Environ. Entomol.*, **26**, 1213-1223.

[17] Heinz, K.M. and Chaney, W.E. (1995) Sampling for Liriomyza huidobrensis (Diptera: Agromyzidae) larvae and damage in celery. *Environ. Entomol.*, **24**, 204-211.

[18] Iwao, S. (1968). A new regression method for analyzing the aggregation pattern of animal populations. *Res. Popul. Ecol.*, **10**, 1-20.

[19] Karandinos, M.G. (1976). Optimum sample size and comments on some published formulae. *Bul. Entomol. Soc. Amer.*, **22**, 417-421.

[20] Kono, T. and Sugino, T. (1958). On the estimation of the density of rice stem borer. *Japanese J. Appl. Entomol. Zool.*, **2**, 184-188.

[21] Kuno, E. (1969). A new method of sequential sampling to obtain the population estimates with a fixed level of accuracy. *Res. Popul. Ecol.*, **11**, 127-136.

[22] Kuno, E. (1972). Some notes on population estimation by sequential sampling. *Res. Popul. Ecol.*, **14**, 58-73.

[23] Lloyd, M. (1967). Mean crowding. *J. Animal Ecol.*, **36**, 1-30.

[24] Martinsek, A.T. (1983). Sequential estimation with squared relative error loss. *Bul. Inst. Math. Acad. Sinica.*, **11**, 607-623.

[25] Meikle, W.C., Markham, R.H., Holst, N., Djomamou, B., Schneider, H. and Vowotor, K.A. (1998). Distribution and sampling of Proste -phanus truncates (Coleptera: Bostrichidae) and Sitophilus zeamais (Coleptera: Curculionidae) in maize stores in Benin. *J. Econ. Entomol.*, **91**, 1366-1374.

[26] Metcalf, R.L. (1955). Physiological basis for insecticide resistence to insecticides. *Physiol. Rev.*, **35**, 192-232.

[27] Mukhopadhyay, N. (1978). Sequential point estimation of the mean

when the distribution is unspecified. *Tech. Rep. #312, Dept. of Theor. Statist.*, *University of Minnesota*, Minneapolis, Minnesota:

[28] Mukhopadhyay, N. (2002). Sequential sampling. In *The Encyclopedia of Environmetrics*, **4** (A.H. Shaarawi and W.W. Piegorsch, eds.), 1983-1988. Wiley: Chichester.

[29] Nyrop, J.P. and Binns, M.R. (1992). Algorithms for computing operating characteristic and average sample number functions for sequential sampling plans based on binomial count models and revised plans for European red mite (Acari: Tetranychidae) on apple. *J. Econ. Entomol.*, **85**, 1253-1273.

[30] O'Rourke, P.K., Burkness, E.C. and Hutchison, W.D. (1998). Development and validation of a fixed-precision sequential sampling plan for aster leafhopper (Homoptera: Cicadellidae) in carrot. *Environ. Entomol.*, **27**, 1463-1468.

[31] Peña, J.E. and Schaffer, B. (1997). Intraplant distribution and sampling of the citrus leafminer (Lepido ptera: Gracillariidae) on lime. *J. Econ. Entomol.*, **90**, 458-464.

[32] Perry, J.N. and Woiwod, I.P. (1992). Fitting Taylor's power law. *Oikos*, **65**, 538-544.

[33] Robbins, H. and Siegmund, D. (1974). Sequential estimation of *p* in Bernoulli trials. In *Studies in Probab. and Statist., Edwin J. G. Pitman vol.* (E.J. Williams, ed.), 103-107. North-Holland: Amsterdam.

[34] Roján, B.A. (1964). La binomial negativa y la estimacion de intensidad de plagas en el suelo. *Fitotecnia Latinamer*, **1**, 27-36.

[35] Ruesink, W.G. (1980). Introduction to sampling theory. In *Sampling Methods in Soybean Entomology* (M. Kogan and D.C. Herzog, eds.), 94-104. Springer-Verlag: New York.

[36] Schaalje, G.B., Butts, R.A. and Lysyk, T.J. (1991). Simulation studies of binomial sampling: A new estimator and density predictor, with special reference to Russian wheat aphid (Homoptera: Aphididae). *J. Econ. Entomol.*, **84**, 140-147.

[37] Snedecor, G.W. and Cochran, W.G. (1980). *Statistical Methods*, 7th ed. Iowa State University Press: Ames.

[38] Sorenson, C.E., Van Duyn, J.W., Kennedy, G.G., Bradley, J.R., Jr., Eckel, C.S. and Fernandez, G.C.J. (1995). Evaluation of a sequential egg mass sampling system for predicting second-generation damage by European corn borer (Lepidoptera: Pyralidae) in field corn in North Carolina. *J. Econ. Entomol.*, **88**, 1316-1323.

[39] Southwood T.R.E. (1978). Assessing and interpreting the spatial distributions of insect population. *Ann. Rev. Entomol.*, **29**, 321-357.

[40] Taylor, L.R. (1961). Aggregation, variance, and the mean. *Nature*, **189**, 732-735.

[41] Taylor, L.R. (1984). *Ecological Methods, with Particular Reference to Insect Population*, 2nd ed. Chapman & Hall: London.

[42] Taylor, L.R. and Woiwod, I.P. (1982). Comparative synoptic dynamics. I. Relationships between inter- and intra-specific spatial and temporal variance/mean population parameters. *J. Animal Ecol.*, **51**, 879-906.

[43] Taylor, L.R., Woiwod, I.P. and Perry, J.N. (1979). The negative binomial as a dynamic ecological model for aggregation, and the density dependence of k. *J. Animal Ecol.*, **48**, 289-304.

[44] Taylor, R.A.J., Lindquist, R.K. and Shipp, J.L. (1998). Variation and consistency in spatial distribution as measured by Taylor's power law. *Environ. Entomol.*, **27**, 191-201.

[45] USDA (1998). Biological control of rangeland weeds. *USDA-ARS 1998 Annual Report*, Project Number 5436-22000-006-00.

[46] USDA (2000a). Biology and collection of natural enemies of orchard and urban insect pests. *USDA-ARS 2000 Annual Report*, Project Number 4012-22000-013-00.

[47] USDA (2000b). Development of pesticide application technologies

for control of pests in field crops. *USDA-ARS 2000 Annual Report*, Project Number 6402-22000-031-00.

[48] USDA (2000c). Crop production. *USDA-ARS National program 2000 Annual Report*.

[49] Wang, K. and Shipp, J.L. (2001). Sequential sampling plans for western thrips (Thysanoptera: Thripidae) on greenhouse cucumbers. *J. Econ. Entomol.*, **94**, 579-585.

[50] Willson, L.J. (1981). Estimation and testing procedures for the parameters of the negative binomial distribution. *Ph.D. Thesis, Dept. of Statist.*, Oklahoma State Univ., Stillwater.

[51] Willson, L.J. and Folks, J.L. (1983). Sequential estimation of the mean of the negative binomial distribution. *Commun. Statist. Sequential. Anal.*, **2**, 55-70.

[52] Willson, L.J., Folks, J.L. and Young, J.H. (1984). Multistage estimation compared with fixed-sample-size estimation of the negative binomial parameter *k*. *Biometrics*, **40**, 109-117.

[53] Willson, L.J. and Young, J.H. (1983). Sequential estimation of insect population densities with a fixed coefficient of variation. *Environ. Entomol.*, **12**, 669-672.

[54] Young, J.H. and Willson, L.J. (1987). Use of Bose-Einstein statistics in population dynamics models of arthropods. *Ecol. Modeling*, **36**, 89-99.

Addresses for communication:

MADHURI S. MULEKAR, Department of Mathematics and Statistics, University of South Alabama, 307 University Boulevard, ILB 325, Mobile, AL 36688-0002, U.S.A. E-mail: mmulekar@jaguar1.usouthal.edu
LINDA J. YOUNG, Department of Biostatistics, College of Medicine, Health Science Center, P.O. Box 100212, University of Florida, Gainesville, FL 32610-0212, U.S.A. E-mail: lyoung@biostat.ufl.edu

Chapter 16

Whither Group-Sequential or Time-Sequential Interim Analysis in Clinical Trials?

PRANAB KUMAR SEN
University of North Carolina, Chapel Hill, U.S.A.

16.1 INTRODUCTION

During the past four decades, *clinical trials* have emerged as a means for quantitative appraisal of various human health related problems. In this evolution, clinical and environmental epidemiology, toxicology, biomedical and clinical sciences, as well as a variety of other public health disciplines, have merged to form an interdisciplinary field of research, invaded by some challenging statistical methodological issues. These clinical trials, generally motivated by the search for an answer to a medical or epidemiological question, vary in their scopes and objectives, and also in their biomedical or bioenvironmental undercurrents. They are quite likely to be engulfed with *medical ethics, cost-benefit* prospects, administrative protocols, and drug-market incentives, so that (statistical) decision making aspects are to be judged from a much broader perspective. The human reaction to drug reception, generally referred to as *drug-response*, is not always easy to assess.

Moreover, it reveals enormous variablity, mainly due to variations in human metabolism, and also due to a number of genetic, familial, and environmental factors which are yet to be fully appraised. Some of these environmental factors work in slow progression and are quite latent in nature. For that reason, precise quantitative formulations, in terms of mathematical models that are customarily used in physical and chemical studies, may not be feasible in clinical trials. Hence, standard biometric methodology may not suffice for clinical trials.

In clinical trials which incorporate *multiple looks* into accumulating experimental or clinical output (datasets), the traditional approach of drawing statistical conclusions following the termination of a study may not be valid. To statistically validate such multiple looks, more complex statistical designs of clinical trials may be necessary and these, in turn, may require considerable modifications of *randomization, replication* and *local control*. Optimal statistical modeling (design) is often a challenging task. As a result, for valid and effective statistical analysis (decision making), a number of issues need to be resolved thoroughly.

The past three decades have witnessed a phenomenal growth of statistical literature in clinical trials. One of the basic features in this respect is the need to have some kind of *statistical monitoring* of these trials. The primary motivation for such surveillance is to have safeguards against toxicity/side-effects when drugs are being used on humans. Medical ethics can also signal some other general concerns. There are time and cost (versus benefit) considerations which advocate surveillance. From a purely statistical standpoint, there are some genuine methodological issues that need to be addressed properly. This task has been accomplished to a certain extent, albeit largely in the form of theory for theory's sake. It might be better to examine how well statistical methodologies derived on paper fit with contemplated real-life applications. It is even more urgent now than in the past, because otherwise we might be lost in the wilderness of *knowledge discovery and data mining* (KDDM) and related purely empirical or exploratory data-analytic approaches.

We are primarily interested in appraising the present status of statistical reasoning in surveillance and *interim analysis* of clinical trials. In the next section, we sketch an outline of the genesis and basic structure of interim analysis in clinical trials. It is in this perspective that the task of surveillance is assessed in Section 16.3. The relevance of (time-)sequential analysis in this perspective is depicted in Section 16.4. The concluding section deals with a general appraisal of some commonly used statistical procedures along with their suitability (from the viewpoints of validity and robustness) in real-life applications.

16.2 INTERIM ANALYSIS: GENESIS AND ANATOMY

Interim analysis typically relates to statistical modeling and analysis of accumulating clinical data to facilitate drawing medical or clinical conclusions with adherence to cost-benefit analysis and appraisal of other therapeutic aspects. In clinical trials involving humans, there may be much less experimental control (compared to conventional biometric as well as dosimetric studies), resulting in a possibly large number of extraneous influencing factors (not all ascribable) that may affect the experimental outcome in a rather complex manner. On top of that, usually medical ethics and other study protocols lead to monitoring of accumulating evidence, either on a periodic basis or sporadically over the tenure of the study. This requires possibly different statistical designs, as well as appropriate statistical tools for both modeling and analysis. With this genesis, interim analysis has only recently been given the much-needed statistical conjugation, although it has been in practice for quite sometime. In its traditional usage with a persistent medical focus, the clinicians (and epidemiologists, to a certain extent) would love to look into their follow-up studies to monitor for possible side-effects or toxicity. However, they were probably unaware that having multiple looks on the accumulating dataset for drawing conclusions (including a possible decision of early termination) might drastically alter the statistical picture regarding the postulated risk of making incorrect clinical decisions. Three decades ago, the (bio-)statisticians came forward rightfully arguing in favor of *statistical interim analysis* that could eliminate some of these impasses.

Peter Armitage and his colleagues should be credited with the early formulation of *sequential medical trials* (Armitage (1975)), and also for addressing the issue of *repeated significance testing* (RST) to adjust the Type-I error rate in interim analysis of certain specific simple studies. Along the same line, *group sequential* (GS) procedures were developed, mostly in the 1970's, to address some of the error-rate issues in interim analysis. However, as we shall see later, in a clinical trial, a follow-up scheme typically arises in a way that vitiates the usual regularity assumptions underlying RST or GS procedures. For example, assumptions regarding independence, normality of errors, linearity of regression (more generally, *generalized linear models* (GLM)) or stationarity of increments (in a stochastic process formulation) may not be tenable. Therefore, the challenge has been to develop novel statistical concepts and methodologies that suit interim analysis in a sound statistical way. Alhough there has been sustained progress along this line, more remains to be accomplished.

One may refer to Sen (1999) for some illustrations of clinical trials wherein interim analysis has played a basic role in the decision making process. These examples are about (i) smoking and health hazards, (ii) artificial sweeteners and health hazards, and (iii) lipids and cardiovascular disease-related hazards. In these cases, the *primary endpoint* is the failure time, or time to the onset of a disease. In other words, one observes a nonnegative random variable whose distribution is positively skewed. Further, under a placebo versus treatment protocol, we may not have valid reasons to take it for granted that the two associated distributions are related to each other by a shift in location or a scale factor only. On top of that, there are usually a large number of explanatory (or auxiliary) variables, and the regression of the failure time on these explanatory variables may not follow a GLM, not to speak of an ordinary linear model. Designs for these clinicical studies, all involving humans, were guided by medical ethics dictating that no subject should receive a treatment which is detected to be inferior or found to have undesirable side-effects. At the same time, these studies were so planned that epidemiologic effects or impacts were clearly understood. A similar feature can be found in a majority of clinical trials that have been conducted during the past three decades. For these reasons, a clinical trial is typically considered only when a good deal of pertinent information has already been collected from laboratory studies, animal experimentation (dosimetry), and therapeutical studies. This way, clinical trials can be categorized into four types, depending on the phase of the information acquisition. These are:

Phase I. Such trials are administered to healthy volunteers to check for possible side-effects and toxicity, and also to determine tolerable dose levels.

Phase II. This relates to further investigation on the safety and dosage aspects, with some consideration of therapeutic efficiency.

Phase III. This relates to a group of well-defined patients on which, a new drug-package is tested and compared clinically with a control/placebo group. In a sense, this constitutes a real clinical trial, whereas Phase I and Phase II trials are regarded as preclinical studies.

Phase IV This relates to monitoring for safety after the (new) drug has gone into usage.

Often Phase III and IV trials are classified on the basis of whether or not the drug has received approval. Large-scale comparative studies are generally labeled as Phase III trials while Phase IV trials refer to genuine post-marketing surveillance.

From the standpoint of bioethics, advocacy groups have voiced concerns about clinical trial-related exploitation in third world countries where affordability of costly drugs is a primary issue. There are allegations that *placebo-controlled trials* (PCT) are unethical when effective therapy is available for the condition being treated or studied, regardless of the condition or consequences of deferring treatments (Temple and Ellenburg (2000)). The WMA (World Medical Association), in its 1997 Helsinki Declaration, documents ethical principles for clinical investigations. In any medical study, every patient (including those in a control group, if any) should be assured of the best proven diagnostic and therapeutic care. This declaration put PCT's under scrutiny, and led to the formulation of *active-controlled equivalence trials* (ACET) which may show that a new therapy is superior (or not inferior) to an existing one. But an ACET may not have all the other characteristics of a PCT. So, while addressing the issues involving interim analysis, one should keep in mind whether a clinical trial is a PCT or an ACET, and whether or not the WMA declarations should be taken into account in the subsequent statistical considerations.

It is clear from the above discussion that the nature of interim analysis could be quite different from one phase to another. In Phase I studies, healthy volunteers may be quite different from the target-patients in terms of their drug-responses as well as metabolism. The main emphasis in an interim analysis would be on early detection of possible toxicity and side-effects. Also, the tolerable dose levels for healthy volunteers may be quite different from those for the target-patients and hence, the dosage aspect merits careful considerations too. Phase II studies dip more into (short- and long-term) safety issues along with dosage recommendations, so that target-patients are under active consideration. Assessment of therapeutic efficacies also calls for careful appraisal. This way, there could be multiple endpoints and more impact of dosimetry (that is, animal studies) on statistical planning and analysis. With possibly multiple objectives, it may be desirable to address these issues in a balanced way. In Phase III trials, a proposed new drug-package is to be compared experimentally with an existing or standard drug (sometimes, with a placebo or no-treatment group), for a well-defined group of patients. In this setup, guided primarily by medical ethics and the possibility of injustice to some specific groups (such as the groups receiving an inferior drug), one must require that the new drug comply with medical regulations as well as toxicity/side-effects documentations acquired from earlier phases. In this sense, Phase III trials are statistically more interesting, and we would like to iterate the interim analysis schemes in this setup.

Our contemplated type of clinical trials typically involves follow-up times. Examples include the remaining life-time of a patient undergoing chemotherapy, or the quality of life of a patient diagnosed with Parkinson's disease. A more notable example might be the impact of high blood-cholesterol levels on cardiovascular diseaes. Typically, in a clinical trial (as was conducted by the National Heart, Lung and Blood Institute (NHLBI) during the period 1972-1984), a group of people known to have high cholesterol levels is to be compared with individuals in a placebo or low-cholesterol group over a period of time, with other demographic factors closely matching. The failure times (that is, the time to occurrence of a heart problem) from the two groups are to be contrasted. From cost and time considerations, a trial can at best be planned for a reasonable period of time (12 years in the above case), so that one is essentially confronted with a two-sample model with possibly multiple end-points, many auxiliary or explanatory variables, and censoring due to the imposed time constraints. In this setup, medical ethics may prompt us to stop a trial at an early stage if by that time there is clear evidence in favor of the treatment group. This is actually for the benefit of the subjects, because all surviving subjects could then be switched over to the better treatment. An early termination may also be provoked if any serious side-effect or toxicity is found to develop in the treatment group. This setup naturally calls for monitoring of a trial (at least on a periodic basis) in order to review the outcomes regularly and take appropriate actions. If such monitoring or surveillance is made purely from operational considerations, no matter how carefully it is done, it will likely cause serious problems with regard to the statistical perspectives. The planning and modeling of the study, as well as effective statistical analysis of the trial outcomes, will be adversely affected. On these accounts, there is a need to incorporate statistical guidelines in designing a surveillance or monitoring scheme, so that interim analysis has a solid foundation. While we relegate the methodological perspectives to the next section, we would like to bring out the principal features of statistically validated interim analysis schemes here. Primarily, there are four important statistical issues relating to interim analysis schemes:

(A) Validation of interim analysis schemes;
(B) Termination of a trial: statistical perspectives;
(C) Drawing valid and efficient statistical conclusions;
(D) Statistical interpretation of dervied conclusions.

In conventional agricultural experiments, statistical planning or design involves randomization which reduces or eliminates bias and also validates the subsequent statistical reasoning. Replication provides a measure of precision for the statistical conclusions that are drawn, while local controls improve the efficacy by further reducing the variability associated with the experiment. In clinical trials, a conventional (unrestricted) randomization scheme may not be adoptable, but usually some restricted randomization schemes are used instead. This may make the statistical formulation of a model and its subsequent analysis more complex. Replication means repetition of an experiment under identical setups. In most clinical trials involving humans, the experiment may not be replicable in an identical fashion. Rather, a reasonable number of subjects are chosen, and efforts are maximized to make use of the available statistical information in the most efficient way. Yet there are many roadblocks such as cost, time and recruitability. The participating individuals are also highly variable in terms of their metabolism and other physiological characteristics, so that any degree of local control seems to be even more difficult to guarantee. Medical ethics, clinical practices, inter-clinic variations, and other demographic factors seem to affect the overall variability among subjects in a clinical trial than the plots in an agricultural study, or the litters in an animal (or dosimetric) study. Further, one may have to reconcile possibly conflicting viewpoints of regulatory agencies and drug-research groups, which are occasionally at odds with each other. One other aspect of clinical trials merits a careful appraisal. In order that statistical conclusions can be drawn from a trial, it is essential that datasets are acquired under reasonable statistical safeguards, so that data management and monitoring are also statistically challenging tasks. All these clearly indicate why an understanding of basic clinical principles by statisticians, as well as an understanding of basic statistical principles by clinicians, is essential for the success of a clinical trial. Finally, one should note that interpretation of accumulating data and formulation of a stopping rule for a trial are predominantly statistical tasks. Therefore, an interim analysis scheme for a clinical trial must be statistically valid, and at the same time, consistent with the general objectives of the study. Generally speaking, statistical formulations are far more complex in interim analyses than in conventional medical studies.

From an administrative viewpoint, in clinical trials ranging over a number of weeks, months or years, a review of the trial data at regular intervals is often a common practice. This way, one ends up with a discrete interim analysis scheme. In other studies, continuous monitoring of a trial may be recommended which gives rise to a continuous interim analysis scheme. As will be made clear in the next section, from statistical perspectives, these two types involve somewhat different approaches. Nevertheless, they also share a common feature: appraisal of accumulating datasets on a follow-up scheme. It is in the context of this follow-up scheme that sequential methodologies are really relevant, though with considerable departure from the classical sequential analysis pioneered by Wald (1947). *Group-sequential procedures* (GSP), *repeated signnificance testing* (RST), and *time-sequential procedures* (TSP) have all been evolving on appropriate interim analysis schemes. Depending on their properties, one may be preferred to another in any particular situation. Much of our discussions in the following sections will revolve around such sequential schemes.

16.3 STATISTICAL SURVEILLANCE

Given the particular nature of interim analysis as outlined in the preceding section, it is of fundamental importance to appraise a clinical trial from statistical perspectives, and in the light of these assessments, to prescribe suitable statistical solutions which are methodologically sound and yet practically adoptable. One may be referred to the work of Peter Armitage and his colleagues in the 1970's where the need for such statistical tools was discussed clearly. Here, we first describe RST schemes. In their formulation, Armitage and his co-workers considered specific statistical problems dealing with parametric models (such as normal and binomial). We consider here only the normal model. The binomial case could be dealt with in a similar manner.

Consider $K(\geq 2)$ independent samples, each of size n, from a normal population with unknown mean μ and known variance. Without any loss of generality, we assume the variance to be 1. Suppose that it is intended to test the null hypothesis $H_0 : \mu = 0$ against $H_1 : \mu \neq 0$. A test statistic based on the sample average is to be computed from the first sample (denoted by T_1, say). If T_1 is significant, the testing procedure is stopped along with the rejection of the null hypothesis. Otherwise, a test statistic (call it T_2) based on the combined first and second samples is computed, and a similar test of significance is carried out. If T_2 is significant, the testing procedure is terminated and the null hypothesis is rejected. Otherwise, one proceeds to the third stage. This procedure is continued to the Kth stage, if no early termination has resulted from the previous steps. The following feature are quite evident:

(A) Although the K samples are themselves independent, the set $\{T_1, T_2, \ldots, T_K\}$ does not consist of independent elements. However, in this specific case, there is some underlying *independent increment* feature that can be used to reformulate the problem to some extent.

(B) If one wants to test the null hypothesis H_0 based on a given significance level (say α^*) at each step, the overall significance level of the RST scheme outlined above can be quite different from (in fact, much larger than) α^*.

Suppose that $C_j, j = 1, \ldots, K$, are the critical values for the test statistics $T_j, j = 1, \ldots, K$. Then, the overall significance level of this RST can be derived as

$$
\begin{aligned}
\alpha &= P\{|T_1| \geq C_1 | H_0\} \\
&+ P\{|T_1| < C_1, |T_2| \geq C_2 | H_0\} + \cdots \\
&+ P\{|T_j| < C_j, j \leq K - 1, |T_k| \geq C_k | H_0\}.
\end{aligned}
\qquad (16.3.1)
$$

Therefore, the crux of the problem is to determine the critical values $C_1, \ldots, C_K$ in such a way that the left hand side is equal to a preassigned number. If $K = 2$, this probability can be evaluated by using the joint bivariate normality of (T_1, T_2), though it depends on the component significance level for the first stage. That way, we may not have a unique choice for the pair $\{C_1, C_2\}$. Moreover, the procedure becomes quite cumbersome if K is larger than 2, and especially so when it is much larger! If we let $T_j = \sqrt{jn}\bar{X}_j$, $j = 1, \ldots, K$, where $\bar{X}_j$ denotes the pooled mean of the first j samples (of pooled size jn), then we may note that under H_0, marginally each T_j has a standard normal distribution. It might be tempting to choose a critical value, say z^*, and to have $C_1 = \cdots = C_K = z^*$, so that the left hand side of (16.3.1) corresponds to a preassigned level α. This is the basic idea of RST so that one really has a repeated significance testing with the same level at each stage. However, analytical computation of z^* may be intractable, and mostly numerical and simulation-based methods have been used to approximate z^*. Pocock (1977) provided a table of z^*-values corresponding to an overall significance level (OASL) α and some selected values of K. We quote them from Sen (1999) in Table 16.3.1.

In a group-sequential testing (GST) scheme, by letting T_j stand for the pooled sample sum for the first j samples, normalized by the scale factor $\sqrt{n}$, one has essentially a truncated group sequential plan with a prefixed upperbound for the sample number, so that the Wald (1947) methodology can be adapted to provide a solution.

Table 16.3.1: Pocock's z^* Values

K	$\alpha = 0.05$	$\alpha = 0.10$
1	1.960	1.645
2	2.178	1.875
3	2.289	1.992
4	2.361	2.067
5	2.413	2.122
10	2.555	2.270

Nevertheless, for this discrete time-point setup with a finite boundary, the computation of the critical values (say, z_o^*) may be equally cumbersome, if not more. Again, numerical methods have mostly been adopted to provide suitable approximations. We refer to O'Brien and Fleming (1979) and Sen (1999) for the following table:

Table 16.3.2: O'Brien and Fleming's z_0^* Values

K	$\alpha = 0.05$	$\alpha = 0.10$
1	1.960	1.645
2	2.797	2.373
3	3.471	2.962
4	4.048	3.466
5	4.562	3.915
10	6.597	5.695

It may be remarked that RST and GST share a common motivation but differ in their test statistics. In RST, normalized sample means are incorporated while in GST, sample sums are used. As such, for $K \geq 2$, their boundaries are not isomorphic. For either scheme, the computation becomes prohibitively laborious as K increases. Moreover, if the subsample sizes are not equal, the computations do not provide correct values. To bring out the relevance of interim analysis, we consider here the following (oversimplified) formulation.

Consider a cohort of n subjects divided into a placebo group and a treatment group of sizes n_1 and n_2 respectively, all entering into the study at a common time-point that we record as 0. These subjects are clinically followed through and as failures occur, their failure-times are recorded along with other relevant information. The accumulating dataset is reviewed at regular calendar time intervals, say, every 3 months. Thus, at the end of each interval, we typically have data on the failures occurring in that interval along with other relevant variables. The numbers of failures occurring in these intervals are random variables (non-negative integers), and on top of that, the ordered failure times are neither independent nor identically distributed (even under a null hypothesis of homogeneity). The very basic assumptions of equal sample sizes and independent increments underlying a group-sequential plan may not be tenable. The next best thing might be the formulation of a stochastic process (for example, a Poisson process) having independent and homogeneous increments. However, since we must start with n subjects and have a follow-up scheme, such a formulation may not always work smoothly. In this respect, a martingale formulation due to Chatterjee and Sen (1973), and elaborated further in Sen (1981b), paves the way for incorporating a suitable sequential methodology in an interim analysis scheme. Therefore, in the next section, we present such a scheme briefly and provide statistical solutions with emphasis on time-sequential methods. One may refer to Sen (1999) for some related discussions.

16.4 EVOLUTION OF TIME-SEQUENTIAL METHODS

Clinical trials generally involve follow-up studies or longitudinal data models and clinical monitoring extends over time until the end point occurs. Thus, there is a natural emphasis on the time factor embedded in such studies. In conventional sequential plans, the responses being spontaneous, independent observations are made sequentially, so that the emphasis is on the *sample number* (SN). The target is to evolve a plan that optimizes the SN. Since the SN is typically random in a sequential plan, its *average* (ASN) is taken as a criterion, and an optimal plan may correspond to minimizing the ASN, subject to having good power properties for a test or good performance characteristics for an estimator. In a clinical trial, however, we might have a predetermined number of subjects who are followed-up over time in our quest for treatment effects, so that the response variable (typically the failure time) is not spontaneous. Thus, in a time-sequential setup, one may have an accumulating dataset relating to the observable failures along with other auxiliary and explanatory variables. As a result, clinical datasets, being updated over time, may lack the property of independent increments. Also, homogeneity of increments in a conventional sequential plan may not be tenable in a time-sequential setup, since the failure time distribution and other constraints may generally lead to heterogeneity and introduce more complexities. Apart from that, while suitable parametric models involving a fixed number of parameters may work well in conventional sequential plans, such models may not be that appropriate in a time- sequential plan where there are often many explanatory variables. This is the main reason why sometimes it becomes difficult to argue in favor of a group-sequential plan based on independent and equally spaced subsamples. As a matter of fact an analogue of SN in a time-sequential setup is the *information time*, not the usual calendar time. If an interim analysis plan is made in advance, it is typically based on calendar time whereas its impact is judged from an information time perspective. This makes it necessary to reconcile the information and calendar times in such a setup. Finally, in a conventional sequential plan, early stopping is allowed in favor of either the null or the alternative hypothesis, depending on the statistical evidence acquired upto that stage. On the other hand, in interim analysis, an early termination of a trial is advocated only when statistical evidence accumulates against the null hypothesis.

Keeping these issues in mind and following Chatterjee and Sen (1973), we consider the following scenario in a discrete time parameter setup: For n subjects invloved in a clinical trial with ordered failure times $t_1, \ldots, t_n$, we consider a sequence of statistics $\{T_{n,k}, 1 \le k \le n\}$ where $T_{n,k}$ captures statistical information upto the time point t_k. Suppose that the $T_{n,k}$'s are so normalized that under the null hypothesis, they have a common mean (assume it to be 0, without loss of generality) and variance $V_{n,k}$ which is nondecreasing in $k(\le n)$. Define a stochastic process $\{W_n(t), 0 \le t \le 1\}$ by letting $k_n(u) = \max\{k : V_{n,k} \le uV_{n,n}\}$ for $u \in (0,1)$, and

$$W_n(t) = \{V_{n,n}\}^{-1/2} T_{n,k_n(t)}, \ t \in (0,1). \tag{16.4.1}$$

Defined in this way, $k_n(t)$ becomes the information time corresponding to the calendar time t. For various nonparametric statistics, the relationship between these two entities be studied explicitly (Sen 1981b). A martingale (array) characterization of the $T_{n,k}$'s for a broad class of statistics (Sen (1981b)), together with the functional central limit theorems for martingales (arrays), leads us to assume that under the null hypothesis

$$W_n \overset{\mathcal{D}}{\Rightarrow} W, \text{ as } n \to \infty, \tag{16.4.2}$$

where $W = \{W(t), t \in (0,1)\}$ is a Brownian motion on the unit interval $(0,1)$. This basic result allows us to formulate a time-sequential plan as follows.

Consider the random functions $W^{+*} = \sup\{W(t) : t \in (0,1)\}$ and $W^* = \sup\{|W(t)| : t \in (0,1)\}$. Then, it is known that for every $\lambda \geq 0$,

$$
\begin{aligned}
P\{W^{+*} \geq \lambda\} &= (2/\pi)^{1/2} \int_\lambda^\infty \exp\{-y^2/2\}dy = 2[1 - \Phi(\lambda)], \\
P\{W^* \geq \lambda\} &= \sum_{k=-\infty}^\infty (-1)^k[\Phi((2k+1)\lambda) - \Phi((2k-1)\lambda)].
\end{aligned}
$$
$$\tag{16.4.3}$$

For $\lambda \geq 1$, the second expression may well be approximated from above by 2 times the first expression. The last two expressions provide critical levels for the one- and two-sided cases which are denoted by C_α^+ and C_α respectively, where α stands for the significance level.

For a one-sided alternative hypothesis problem, we may proceed as follows: Continue the trial as long as $T_{n,k} < C_\alpha^+ V_{n,n}^{1/2}$. If for the first time the inequality is violated at time point t_K, the trial is stopped at that time along with the rejection of the null hypothesis. If, on the other hand, no such $K(\leq n)$ exists, then the null hypothesis is accepted after the completion of the trial. For the two-sided case, we work with the absolute values of $T_{n,k}$ and the critical level C_α. It is also possible to modify the procedure to allow censoring, as well as a fixed duration of time, and we refer to Sen (1981b, Chapter 11) for details. Instead of horizontal boundaries for a Brownian motion, square-root boundaries and a few other types have also been considered in the literature. This approach applies to progressively censored likelihood ratio statistics (Sen (1976)), linear rank statistics (Chatterjee and Sen (1973)), functions of empirical distributions (Sinha and Sen (1982)), and other statistics which are governed by the martingale-based weak invariance principle; For rank statistics, the exact distribution-free property (under the null hypothesis) holds while for others, we may have to be satisfied with the asymptotically distribution-free property. This martingale-based weak invariance principle has also been established for the partial likelihood statistics (Cox (1972)) under the proportional hazards model (Sen (1981a)), although the progressively censored variance function needs to be estimated there from the accumulating data. This might slow down the asymptotic approximations for moderately large sample sizes. Moreover, in many practical applications, it might be a problem to verify whether or not the proportional hazards model is appropriate, and without this verification, the resulting adoption may not lead to a reasonable solution. These developments come under the heading of progressively censored schemes that allow continuous monitoring of a trial without violating the desired overall significance level. In this spirit, it is more like a time-sequential procedure, and we shall now compare it with a group-sequential plan in the context of interim analysis.

Independent and homogeneous increments and only a few "statistical looks" into accumulating data underlie a typical group-sequential plan in interim analysis. It has been pointed out before that an interim analysis scheme is usually adopted on a calendar time basis, while the information time may depend on the underlying failure distribution which is unknown. Thus, most of the simplistic methods mentioned in the previous section may not be entirely justified here. If we revisit the tables from Section 16.3 and compare those cut-off points with the time-sequential cut-off points mentioned in this section, we will observe that for K greater than 8, the critical values are quite close to each other. In other words, if we intend to have ten or more interim analyses on a calendar time basis, the use of a time-sequential plan may avoid the inaccuracries due to differences between the information time and the calendar time. On top of that, a time-sequential plan would not result in a procedure that is too conservative. But if K is as small as 3 or 4, then the critical values adapted from a time-sequential plan may lead to a coservative procedure, albeit there is no guarantee that the tables from Section 16.3 would be fully justified (in view of the difference between calendar time and information time). Realizing the limitations of Pocock's and O'Brien-Fleming's formulae, and motivated by the validity of Brownian motion approximations for stochastic processes arising in clinical trials, Lan and DeMets (1983) came up with a clever idea of a *spending function* for the significance level that can be applied to a discrete parameter case. Their suggested approach includes both Pocock's and O'Brien-Fleming's approaches as special cases. In the Lan-DeMets's formulation too, the time points (unfortunately in the information time scale) $0 < t_1 < \cdots < t_k \leq 1$ are to be specified in advance. However, we have explained earlier that these t_j's would be random when viewed from prespecified time-points (based on a calendar time scale), and this introduces some inaccuracies in a specified spending function. As a matter of fact, in clinical trials, there are usually a number of auxiliary or explanatory variables, in addition to some stochastic covariates. As a result, the information time may not be prespecified at the start of a trial. Rather, it has to be progressively estimated as more and more datasets are available (over time). Therefore, the simple-minded approach of Lan and DeMets (1983) would require some adjustments to suit interim analysis in real-life clinical trials, especially involving multi-center studies. In addition, the notion of a spending function is essentially related to the design of a clinical trial so that it helps maximize the information or power of statistical tests based on the trial data. Unfortunately, unlike Wald's (1947) sequential probability ratio test, the drift function of stochastic processes arising in clinical trials may not be generally linear, and hence, no clinical design can be optimal for all possible alternatives, even in a local sense. Thus, the choice of a spending function needs to be based on some other considerations, and also needs to be modified to suit random points on the information scale. For these reasons, we see again that time-sequential approaches based on continuous monitoring (applicable to discrete time parameter cases as well) are more appealing for interim analysis if the number of such "looks" is not too small.

16.5 CONCLUDING REMARKS

We have noted that dominant clinical factors and medical ethics have a distinct impact on interim analysis where purely statistical methodologies may not be of any profound use. In practice, there is a general tendency to adopt a theoretical model, but only after modifying it through extensive simulations or numerical studies to make it approximately valid and suitable for a particular real-life scenario. In early stages, because of the non-negative nature of the response variables (for example, failure times), parametric models with exponential, Weibull or log-normal distributions were tried. The Armitage approach was only a first step toward more statistically rational solution. But a major question was how to cope with various explanatory variables, especially when the regression model was not expected to have the usual linear (or even a generalized linear) form. While dealing with progressively censored likelihood ratio statistics and some associated quantile processes, general asymptotics were developed in Sen (1976,1979), and they turned out to be useful in a much greater context. Cox (1972) introduced the proportional hazards model which opened a whole new avenue for research and paved the way for countless more publications in the past 30 years. Yet, there are concerns with unreserved adoption of such semiparametric models. In the most simple case, the Cox formulation and the classical log-rank statistics are close to each other and for either one of them, a martingale-based statistical methodology works well. However, staggered entry plans, censoring (which may not be random or noninformative), and multiple end-points have all contributed to the overall complexity of such formulations. A more general statistical plan is needed to suit such complex problems. Some of these plans are discussed in Sen (1981b,1999), and Sinha and Sen (1982). There still remains a lot to be accomplished in this direction. We also refer to Whitehead (1991) for a treatise of sequential methods in clinical trials. Coming back to the use of semiparametric models in interim analysis, the basic assumption of multiplicative intensities may not be tenable in many real-life applications. For this reason, we need to appraise the positive and negative impacts of interim analysis in clinical trials from a broader statistical perspective, in order to achieve a balance between clinical precision and statistical assessments. For a more methodological treatment of time-sequential estimation problems we may refer to Ghosh et al. (1997). They are potentially useful in the hypothesis testing problem as well. In fact, I am not sure whether the familiar regulatory agencies (for example, the Food and Drug Administration in the United States) rely upon statistical appraisals, unless such appraisals are accompanied by solid clinical evidence. This may call for an alternative approach of blending statistical intuition with clinical observations. The journal entitled *Statistics in Medicine* has already acquired some reputation in this direction, though there is a genuine need for statistical surviellance in this general area. Whatever statistical formulation we need to make for interim analysis, it should be design-consistent, and at the same time, adhere to clinical and medical perspectives. In conclusion, we may advocate time-sequential procedures over other sequential plans on grounds of validity and practical convenience.

ACKNOWLEDGMENT

The author is grateful to both reviewers for their most critical reading of the manuscript which helped in eliminating numerous typos and some other statistical obscurities too.

REFERENCES

[1] Armitage, P. (1975). *Sequential Medical Trials*, 2nd ed. Blackwell: Oxford.

[2] Chatterjee, S.K. and Sen, P.K. (1973). Nonparametric testing under progressive censoring. *Cal. Statist. Assoc. Bul.*, **22**, 18-58.

[3] Cox, D.R. (1972). Regression models and life tables (with discussions). *J. Roy. Statist. Soc. Ser. B*, **34**, 187-220.

[4] Ghosh, M., Mukhopadhyay, N. and Sen, P.K. (1997). *Sequential Estimation*. Wiley: New York.

[5] Lan, K.K.G. and DeMets, D.L. (1983). Discrete sequential boundaries for clinical trials. *Biometrika*, **70**, 659-663.

[6] O'Brien, P.C. and Fleming, T.R. (1979). A multiple testing procedure for clinical trials. *Biometrics*, **35**, 549-556.

[7] Pocock, S.J. (1977). Group sequential methods in the design and analysis of clinical trials. *Biometrika*, **64**, 191-199.

[8] Sen, P.K. (1976). Weak convergence of progressively censored likelihood ratio statistics and its role in asymptotic theory of life testing. *Ann. Statist.*, **4**, 1247-1257.

[9] Sen, P.K. (1979). Weak convergence of some quantile processes arising in progressively censored tests. *Ann. Statist.*, **7**, 414-431.

[10] Sen, P.K. (1981a). The Cox regression model, invariance principles for some induced quantile processes and some repeated significance tests. *Ann. Statist.*, **9**, 109-121.

[11] Sen, P.K. (1981b). *Sequential Nonparametrics: Invariance Principles and Statistical Inference.* Wiley: New York.

[12] Sen, P.K. (1999). Multiple comparisons in interim analysis. *J. Statist. Plan. Inf.*, **82**, 5 - 23.

[13] Sinha, A.N. and Sen, P.K. (1982). Tests based on empirical processes for progressive censoring schemes with staggering entry and random withdrawal. *Sankhya, Ser. B*, **44**, 1-18.

[14] Temple, R. and Ellenberg, S.S. (2000). Placebo-controlled trials and active-controlled trials in the evaluation of new treatments, I: Ethical and scientific issues. *Ann. Inter. Med.*, **133**, 455 - 463.

[15] Wald, A. (1947). *Sequential Analysis.* Wiley: New York.

[16] Whitehead, J. (1991). Sequential methods in clinical trials. In *Handbook of Sequential Analysis* (B.K. Ghosh and P.K. Sen, eds.), 593-611. Mercel Dekker: New York.

[17] World Medical Association Declaration of Helsinki (1997). Recommendation guiding physicians in biomedical research involving human subjects. *J. Amer. Med. Assoc.*, **277**, 925 - 926.

Address for communication:

PRANAB K. SEN, Department of Biostatistics, 3105E McGavran-Greenberg, CB# 7420, School of Public Health, University of North Carolina, Chapel Hill, NC 27599-7420, U.S.A. E-mail: pksen@bios.unc. edu

Chapter 17

Change-Point Detection in Multichannel and Distributed Systems

ALEXANDER G. TARTAKOVSKY
University of Southern California, Los Angeles, U.S.A.

VENUGOPAL V. VEERAVALLI
University of Illinois at Urbana-Champaign, Urbana, U.S.A.

17.1   INTRODUCTION

The goal of this paper is to show that recent advances in change-point detection theory, as described in Basseville and Nikiforov (1993), Dragalin (1995,1996), Lai (1995,1998), Pollak (1985,1987), Tartakovsky (2003), and others, can be successfully applied to (i) certain practical problems related to rapid detection of targets in multichannel and multisensor distributed systems, and (ii) the problem of building high-speed anomaly detection systems for early detection of intrusions in large-scale distributed computer networks. We show that the asymptotic theory that has been developed for change-point detection is useful in practical engineering problems too, and it allows for the development of efficient algorithms that are easily implemented. In addition to that, these algorithms have certain optimality properties.

In the standard formulation of a change-point detection problem, one assumes having a sequence of observations whose distribution changes at some unknown instant λ, $\lambda = 1, 2, \ldots$. The goal is to detect this change as soon as possible, subject to false alarm constraints. We are interested in two generalizations of this problem.

The first generalization refers to *multichannel* systems, where there are N sequences of observations whose distributions follow a certain law up to some unknown instant λ, $\lambda = 1, 2, \ldots$. After this instant, one of the populations (and only one) changes its statistical properties. We wish to detect the change as soon as possible after it occurs, subject to constraints on the rate of false alarms. We assume that it is not necessary to indicate which channel has changed. Only the fact that a change occurred is important.

The second generalization corresponds to a situation where the information available for decision-making is distributed (or decentralized). Here, the observations are made at a set of L distributed sensors. The sensors' observations could in general be multichannel, and at the change-point λ, one channel at *each* sensor could change distribution. The sensors send quantized versions of their observations to a *fusion center* where change detection is performed based on the messages from all sensors.

These generalizations are of considerable practical importance and they arise in a variety of applications such as biomedical signal processing, quality control engineering, finance, link failure detection in communication networks, intrusion detection in computer networks, and target detection in surveillance systems. In particular, a typical scenario involves the detection of a target which appears randomly at an unknown time in an N-channel system (for example, infrared, radar, sonar). It is necessary to detect the target "as quickly as possible" while maintaining the false alarm rate at a given level. Another important application area is intrusion detection in distributed high-performance computer networks. Large scale attacks, such as external denial-of-service attacks or internal man-in-the-middle attacks, occur at unknown points in time and should be detected in the early stages by observing abrupt (usually small) changes in the network traffic compared to the "normal" (legitimate) mode. These application areas will be emphasized in Section 17.4.

17.2 MULTICHANNEL CHANGE DETECTION: THEORY

17.2.1 Problem Formulation

We assume that the observed stochastic process $\boldsymbol{X}_k = (X_{1,k}, \ldots, X_{N,k})$ is an N-component process. The component $X_{i,k}$, $k = 1, 2, \ldots$, corresponds to observations obtained from the ith channel of an N-channel system, and all of the channels can be observed simultaneously. Let $\boldsymbol{P}_\infty$ stand for the probability measure when the change does not occur ($\lambda = \infty$), and let $\boldsymbol{P}_{\lambda,i}$ be the probability measure when the change occurs at time λ in the ith channel. Note that if $\lambda = \infty$, then $\boldsymbol{P}_{\infty,i} = \boldsymbol{P}_\infty$ for all i. Further, let $\boldsymbol{E}_\infty$ and $\boldsymbol{E}_{\lambda,i}$ denote the corresponding expectations. A sequential change-point detection procedure is described by a random time τ depending on the observations, which is a *stopping time* (with respect to the family of sigma-algebras $\mathcal{F}_k = \sigma(\boldsymbol{X}_1, \ldots, \boldsymbol{X}_k)$, $k \geqslant 0$) at which it is declared that a change has occurred. Typically, this stopping time (time of alarm) is defined as a first time $k \geqslant 1$ when some statistic exceeds a threshold that controls the rate of false alarms.

Designing the quickest change detection procedures usually involves optimizing the tradeoff between two kinds of performance measures, one being a measure of detection delay, and the other being a measure of the frequency of false alarms. There are two standard mathematical formulations for the optimum tradeoff problem. The first of these is a minimax formulation, due to Lorden (1971) and Pollak (1985), in which the goal is to minimize the worst-case delay, $\text{ES}_i(\tau) = \sup_\lambda \text{ess sup}\, \boldsymbol{E}_{\lambda,i}[(\tau - \lambda + 1)^+ | \boldsymbol{X}_1, \ldots, \boldsymbol{X}_{\lambda-1}]$ or $\text{D}_i(\tau) = \sup_\lambda \boldsymbol{E}_{\lambda,i}(\tau - \lambda | \tau \geqslant \lambda)$, subject to a lower bound on the mean time between false alarms ("ess sup" of a set being the supremum of it except possibly a subset of measure zero). The second is a Bayesian formulation, proposed by Shiryaev (1963,1978), in which the change-point is assumed to have a geometric prior distribution, and the goal is to minimize the expected delay subject to an upper bound on the false alarm probability.

In what follows, we consider a minimax approach with Pollak's measure $\text{D}_i(\tau)$, or with a more general measure

$$\text{D}_i^m(\tau) = \sup_{1 \leqslant \lambda < \infty} \boldsymbol{E}_{\lambda,i}\{(\tau - \lambda)^m \,|\, \tau \geqslant \lambda) \quad \text{for} \quad m > 0, \qquad (17.2.1)$$

which is nothing but the mth moment of the detection delay, in the worst case, assuming that the change occurs in the ith channel. Indeed, if we show that the values of $\text{D}_i^m(\tau)$ are relatively small for all $1 \leqslant i \leqslant N$ and $m > 0$, then we will have shown that the entire distribution $\boldsymbol{P}_{\lambda,i}(\tau - \lambda | \tau \geqslant \lambda)$ is concentrated close to the change-point.

The constraint imposed on the false alarms is that the mean time to false alarm, $\boldsymbol{E}_\infty(\tau)$, should exceed a predefined number $T > 0$. That means we are interested in the class of detection procedures $\boldsymbol{\Delta}(T) = \{\tau : \boldsymbol{E}_\infty(\tau) \geqslant T\}$. It is very difficult, if not impossible, to find an optimal procedure that minimizes the "risk functions" (17.2.1) in the class $\boldsymbol{\Delta}(T)$ for an arbitrary value of T. For this reason, we will consider the asymptotic setting as $T \to \infty$, that is, we are interested in the following problem: Compute

$$\inf_{\tau \in \boldsymbol{\Delta}(T)} \text{D}_i^m(\tau) \text{ as } T \to \infty \text{ for all } i = 1, \ldots, N.$$

When we evaluate the performance of a change detection procedure in particular problems and examples, we will be interested in the *average detection delay* (ADD) and the *false alarm rate* (FAR), which are defined by:

$$\text{ADD}(\tau) = \frac{1}{N} \sum_{i=1}^{N} \text{D}_i(\tau), \quad \text{FAR}(\tau) = \frac{1}{E_\infty(\tau)}. \tag{17.2.2}$$

17.2.2 Detection Procedure and False Alarm Rate

Write $\boldsymbol{X}_i^n = (X_{i,1}, \ldots, X_{i,n})$ and $\boldsymbol{X}^n = (\boldsymbol{X}_1^n, \ldots, \boldsymbol{X}_N^n)$ for the concatenation of the first n observations from the ith channel, and from all N channels, respectively. We suppose that the data in the channels, that is the vectors $\boldsymbol{X}_1^n, \ldots, \boldsymbol{X}_N^n$, are mutually independent. However, in general, we do not assume that the data in a particular channel, say $X_{i,1}, X_{i,2}, \ldots$, are i.i.d. before or after the change.

Let $\boldsymbol{P}_\infty$ stand for the probability measure under which the conditional density of $\boldsymbol{X}_k$ given $\boldsymbol{X}^{k-1} = \boldsymbol{x}^{k-1}$ is

$$f_{0,k}(\boldsymbol{x}_k \mid \boldsymbol{x}^{k-1}) = \prod_{\ell=1}^{N} p_{0,k}(x_{\ell,k} \mid \boldsymbol{x}_\ell^{k-1})$$

for every $k \geqslant 1$. For any $1 \leqslant \lambda < \infty$, we use $\boldsymbol{P}_{\lambda,i}$ to denote the probability measure under which the conditional density of $\boldsymbol{X}_k$ given $\boldsymbol{X}^{k-1} = \boldsymbol{x}^{k-1}$ is

$$f_{i,k,\lambda}(\boldsymbol{x}_k \mid \boldsymbol{x}^{k-1}) = \begin{cases} \prod_{\ell=1}^{N} p_{0,k}(x_{\ell,k} \mid \boldsymbol{x}_\ell^{k-1}) & \text{if } \lambda > k \\ p_{i,k}(x_{i,k} \mid \boldsymbol{x}_i^{k-1}) \prod_{\ell=1, \ell \neq i}^{N} p_{0,k}(x_{\ell,k} \mid \boldsymbol{x}_\ell^{k-1}) & \text{if } \lambda \leqslant k. \end{cases}$$

In other words, if the change occurs in the ith channel at the time $\lambda = n$, the conditional probability density function (p.d.f.) of the ith component changes from $p_{0,n}(x \mid \boldsymbol{x}_i^{n-1})$ to $p_{i,n}(x \mid \boldsymbol{x}_i^{n-1})$.

Next, let $H_{\lambda,i}$ be the hypothesis that the change occurs in the ith channel at time $\lambda \in \{1, 2, \ldots\}$, and let H_∞ be the hypothesis that the change does not occur at all (that is, $\lambda = \infty$). Then the *log-likelihood ratio* (LLR) between the hypotheses $H_{\lambda,i}$ and H_∞ is

$$Z_i^\lambda(n) := \log \frac{d\boldsymbol{P}_{\lambda,i}}{d\boldsymbol{P}_\infty}(\boldsymbol{X}^n) = \sum_{k=\lambda}^{n} \log \frac{p_{i,k}(X_{i,k} \mid \boldsymbol{X}_i^{k-1})}{p_{0,k}(X_{i,k} \mid \boldsymbol{X}_i^{k-1})}, \quad \lambda \leqslant n. \tag{17.2.3}$$

For $i = 1, \ldots, N$, define the statistics $R_i(n) = \sum_{\lambda=1}^{n} e^{Z_i^\lambda(n)}$, and then combine these statistics to form the mixture

$$R(n) = N^{-1}(R_1(n) + \cdots + R_N(n)).$$

The detection procedure $\tau^* = \tau^*(A)$ is defined as

$$\tau^*(A) = \min\{n \geqslant 1 : R(n) \geqslant A\} \quad (\tau^* = \infty \text{ if no such } n \text{ exists}),$$

where A is a positive number (threshold) that is selected so that the FAR is no larger than $1/T$. In other words, the moment of alarm is the first time n such that the statistic $R(n)$ exceeds the threshold A.

Note that the statistic $R_i(n)$ is the Shiryaev-Roberts statistic for detecting a change in the ith channel (see, for example, Pollak (1985,1987)). The detection procedure τ^* is therefore an extension of the Shiryaev-Roberts procedure adapted to detect changes in multichannel systems.

We begin with an examination of the detection procedure τ^* under the hypothesis H_∞. The following lemma establishes a simple lower bound for the average run length to false alarm $\boldsymbol{E}_\infty(\tau^*(A))$ in a general case. In Subsection 17.2.3, this bound will be improved in the i.i.d. case.

Lemma 17.2.1 *For an arbitrary stochastic model and any $A > 0$,*

$$\frac{1}{\text{FAR}} = \boldsymbol{E}_\infty \tau^*(A) \geqslant A. \tag{17.2.4}$$

Proof: Since $\boldsymbol{E}_\infty\{e^{Z_i^n(n)} \mid \boldsymbol{X}^{n-1}\} = 1$, it is easily verified that, for any i, the statistics $R_i(n) - n$, $n \geqslant 1$, are $\boldsymbol{P}_\infty$–martingales with mean zero. This implies that the statistic $\{R(n) - n\}_{n \geqslant 1}$ is also a $\boldsymbol{P}_\infty$–martingale with a zero mean. Therefore, inequality (17.2.4) follows from the optional stopping theorem, which yields $\boldsymbol{E}_\infty(R(\tau^*)) = \boldsymbol{E}_\infty(\tau^*)$, and the fact that $R(\tau^*) \geqslant A$. Note that the optional stopping theorem can be applied because

$$\lim_{n \to \infty} \int_{\{\tau^* > n\}} |R(n) - n| \, d\boldsymbol{P}_\infty = 0.$$

It follows from the fact that $0 \leqslant R(n) < A$ on the set $\tau^* > n$. ∎

It immediately follows from (17.2.4) that, by setting $A = T$, we guarantee the inequality $\boldsymbol{E}_\infty(\tau^*) \geqslant T$, that is, $A = T$ implies $\tau^*(T) \in \boldsymbol{\Delta}(T)$.

17.2.3 Asymptotic Performance for Low FAR

In this subsection, we study the behavior of the detection procedure τ^* for large values of A and T, that is, for small FAR. In order to obtain asymptotic expansions for moments of the detection delay, we assume that the normalized LLRs

$$\frac{1}{n}Z_i^\lambda(n + \lambda - 1) = \frac{1}{n}\sum_{k=\lambda}^{n+\lambda-1} \log \frac{p_{i,k}(X_{i,k}|X_{i,1},\ldots,X_{i,k-1})}{p_{0,k}(X_{i,k}|X_{i,1},\ldots,X_{i,k-1})}, \quad i = 1,\ldots,N,$$

converge *almost surely* (a.s.) as $n \to \infty$ to positive finite numbers J_i under $P_{\lambda,i}$. Furthermore, we impose the following conditions on the rate of convergence:

$$\sum_{n=1}^{\infty} n^{r-1} P_{\lambda,i} \left\{ |Z_i^{\lambda}(n + \lambda - 1) - nJ_i| > n\varepsilon \right\} < \infty \quad \text{for all } \varepsilon > 0, \quad (17.2.5)$$

where r is a positive real number. If (17.2.5) holds, we say that $n^{-1} Z_i^{\lambda}(n + \lambda - 1)$ *converges r-quickly* to J_i under $P_{\lambda,i}$ as $n \to \infty$ (Lai (1976)). For $r = 1$, (17.2.5) is the so-called complete convergence condition introduced by Hsu and Robbins (1947) in connection with the rates of convergence in the strong law of large numbers.

The following theorem establishes the asymptotic performance of the detection procedure $\tau^*(A)$ for large values of A. The proof is given in the Appendix. Hereafter we use the standard notation $X_A \sim Y_A$, which means that $X_A/Y_A \to 1$ as $A \to \infty$.

Theorem 17.2.1 *Let condition* (17.2.5) *hold for some* $r \geqslant 1$. *Then for all* $m \leqslant r$, $1 \leqslant \lambda < \infty$ *and* $i = 1, \ldots, N$,

$$E_{\lambda,i} \left\{ (\tau^*(A) - \lambda)^m | \tau^*(A) \geqslant \lambda \right\} \sim \left(\frac{\log NA}{J_i} \right)^m \quad \text{as } A \to \infty. \quad (17.2.6)$$

In addition, under some uniform conditions where (17.2.5) is strengthened to $\sup_{\lambda}$, first-order asymptotic optimality can be established with respect to "risks" $D_i^m(\tau^*)$ defined in (17.2.1). More precisely, under certain conditions, if $A = T$, then as $T \to \infty$,

$$\inf_{\tau \in \Delta(T)} D_i^m(\tau) \sim D_i^m(\tau^*) \sim \left(\frac{\log NT}{J_i} \right)^m, \quad i = 1, \ldots, N. \quad (17.2.7)$$

Note, however, that in a general non-i.i.d. case the supremum in (17.2.1) is attained for some unspecified point λ that might even depend on T and go to infinity when $T \to \infty$.

Now consider the i.i.d. case assuming that, if the change occurs in the ith population, then the observations $X_{i,1}, X_{i,2}, \ldots$ are i.i.d. before the change with the p.d.f. $p_0(x)$ and after the change with the p.d.f. $p_i(x)$ (with respect to a sigma-finite measure $\mu(x)$). For the sake of brevity, in the rest of this subsection, we omit the index λ when $\lambda = 1$. For instance, we simply write $Z_i(n)$, E_i, and P_i instead of $Z_i^1(n)$, $E_{1,i}$, and $P_{1,i}$, respectively.

The first important observation is that $D_i^m(\tau^*) = E_i(\tau^* - 1)^m$. Therefore, in further calculations related to the "risk" $D_i^m(\tau^*)$, we can concentrate on the evaluation of $E_i(\tau^* - 1)^m$, assuming that the change occurs at the point $\lambda = 1$.

Define

$$\nu_i(a) = \min\{n \geqslant 1 : Z_i(n) \geqslant a\} \quad (\nu_i(a) = \infty \text{ if no such } n \text{ exists}),$$

$$\gamma_i = \lim_{a \to \infty} \boldsymbol{E}_i \exp\{-[Z_i(\nu_i(a)) - a]\}, \quad \bar{\gamma}_N = \frac{1}{N}(\gamma_1 + \cdots + \gamma_N),$$

$$\bar{\varkappa}_i = \lim_{a \to \infty} \boldsymbol{E}_i\left[Z_i(\nu_i(a)) - a\right], \quad C_i = \boldsymbol{E}_i\left\{\log\left(1 + \sum_{k=1}^{\infty} e^{-Z_i(k)}\right)\right\},$$

$$\mathrm{I}_i = \int \log\left(\frac{p_i(x)}{p_0(x)}\right) p_i(x)\mu(dx), \quad i = 1, \ldots, N.$$

Note that I_i is the *Kullback-Leibler* (K-L) information number between the densities $p_i(x)$ and $p_0(x)$. In the i.i.d. case, this number plays the role of the number J_i that appeared in (17.2.5)–(17.2.7).

We will impose the following mild condition on the K-L information numbers I_i:

$$0 < \mathrm{I}_i < \infty \quad \text{for } i = 1, \ldots, N. \tag{17.2.8}$$

The above condition (finiteness) implies that $\boldsymbol{E}_i \exp\{-Z_i(1)\} = 1$, and hence, all moments of the negative part $Z_i^-(1) = -\min\{0, Z_i(1)\}$ of LLR's are finite, that is, $\boldsymbol{E}_i\{Z_i^-(1)\}^m < \infty$ for all $m > 0$. The latter property turns out to be crucial in establishing finiteness of moments of the stopping time τ^* and its asymptotic optimality with respect to moments of the detection delay. Note also that I_i's are positive whenever the p.d.f.'s $p_0(x)$ and $p_i(x)$ do not coincide almost everywhere, that is when $\mu\{x : p_0(x) \neq p_i(x)\} > 0$.

The following theorem, whose proof is given in the Appendix, establishes higher order asymptotic approximations to the average detection delay of the procedure τ^* and also its first-order asymptotic optimality.

Theorem 17.2.2 *Suppose that condition (17.2.8) holds.*

(i) Then, $A = T$ implies that $\boldsymbol{E}_\infty(\tau^*) \geqslant T$, and moreover,

$\lim_{T \to \infty}[\boldsymbol{E}_\infty \tau^*(T)/T]$ *is bounded;*

(ii) If $A = T$, then for all $m \geqslant 1$ and $i = 1, \ldots, N$

$$\inf_{\tau \in \Delta(T)} \mathrm{D}_i^m(\tau) \sim \mathrm{D}_i^m(\tau^*) \sim \left(\frac{\log T}{\mathrm{I}_i}\right)^m \quad \text{as } T \to \infty; \tag{17.2.9}$$

(iii) Also, assume that $Z_i(1)$ are non-arithmetic and $\boldsymbol{E}_i|Z_i(1)|^2 < \infty$. Then, as $A \to \infty$,

$$\frac{1}{\mathrm{FAR}} = \boldsymbol{E}_\infty \tau^*(A) = \frac{A}{\bar{\gamma}_N}(1 + o(1)) \tag{17.2.10}$$

and

$$E_i \tau^*(A) = \frac{1}{I_i} \left[\log(NA) + \overline{\varkappa}_i - C_i \right] + o(1), \qquad (17.2.11)$$

where $o(1) \to 0$ as $A \to \infty$.

It follows that the detection procedure considered minimizes any positive moment of the detection delay for asymptotically small FAR $(T \to \infty)$ whenever the K-L information numbers are finite. Note that finiteness of higher order moments of the LLR is not required. More importantly, if $A = T\tilde{\gamma}_N$, then $E_\infty(\tau^*) \sim T$, which allows us to get an almost exact constraint on the mean time to false alarm (or, equivalently, the FAR) rather than the conservative inequality (17.2.4). Furthermore, we can use (17.2.11) to obtain an asymptotically exact expression for the ADD defined in (17.2.2).

The constants $\overline{\varkappa}_i$ and γ_i come from renewal theory (see, for example, Siegmund (1985), Woodroofe (1982) or Ghosh et al. (1997)), and can be computed quite easily in many cases using the technique described in those three books as well as Tartakovsky (1991a). The following formulas are particularly useful:

$$\gamma_i = \frac{1}{I_i} \exp \left\{ - \sum_{n=1}^{\infty} n^{-1} [P_\infty(Z_i(n) > 0) + P_i(Z_i(n) \leqslant 0)] \right\},$$
$$\overline{\varkappa}_i = \frac{I_i^2 + \sigma_i^2}{2I_i} + \sum_{n=1}^{\infty} \frac{1}{n} E_i \left[\min \{0, Z_i(n)\} \right], \qquad (17.2.12)$$

where $\sigma_i^2 = E_i[Z_i(1) - I_i]^2$.

The constant C_i is not as straightforward to compute, but Monte Carlo techniques can be used to estimate it accurately in specific examples. Experimentation indicates that formulas (17.2.10) and (17.2.11), with γ_i and $\overline{\varkappa}_i$ coming from (17.2.12), give quite accurate approximations for the average run lengths even for moderate values of A.

Two other attractive candidates are the bank of CUSUM (Page's) tests and the bank of Shiryaev-Roberts tests (parallel implementation of CUSUM statistics and Shiryaev-Roberts statistics in channels) that are defined as follows:

$$\tau_h = \min \left\{ n \geqslant 1 : \max_{1 \leqslant i \leqslant N} U_i(n) \geqslant h \right\},$$
$$\tilde{\tau}_{\tilde{A}} = \min \left\{ n \geqslant 1 : \max_{1 \leqslant i \leqslant N} R_i(n) \geqslant \tilde{A} \right\}, \qquad (17.2.13)$$

where $U_i(n) = \max_{1 \leqslant \lambda \leqslant n} Z_i^\lambda(n)$ is the CUSUM statistic in the ith channel. These procedures have been studied in details in Tartakovsky (1988,1991a,b,

1992,1994), for i.i.d. models. It can be shown that $h = \log NT$ implies $E_\infty(\tau_h) \geqslant T$, and that $\tilde{A} = NT$ implies $E_\infty(\tilde{\tau}_{\tilde{A}}) \geqslant T$. Theorems 17.2.1 and 17.2.2 hold for τ_h and $\tilde{\tau}_{\tilde{A}}$ with corresponding modifications of the thresholds (for example, for the multi-channel CUSUM test, A is replaced with e^h). However, both procedures have a significant drawback compared to τ^* in the sense that there is no analog of (17.2.10) for these procedures. In fact, it is difficult to choose the thresholds h and $\tilde{A}$ so as to guarantee the approximate equalities $E_\infty(\tau_h) \approx T$ and $E_\infty(\tilde{\tau}_{\tilde{A}}) \approx T$.

17.2.4 Composite Hypotheses: Adaptive Detection Procedures

So far we have considered the case where both pre-change and post-change distributions were completely known. In practice, a more realistic situation is when the post-change distribution is known only partially — up to parameters θ. Assume that the observations follow a general model with a completely known baseline p.d.f. $p_{0,k}(X_{i,k} \mid \boldsymbol{X}_i^{k-1})$ but with $p_{i,k}(X_{i,k} \mid \boldsymbol{X}_i^{k-1}) = p_{i,k}(X_{i,k} \mid \boldsymbol{X}_i^{k-1}, \theta_i)$ being parameterized by $\theta_i \in \Theta_i$. Write $\boldsymbol{P}_{\lambda,i,\theta_i}$ and $\boldsymbol{E}_{\lambda,i,\theta_i}$ for the probability measure and the expectation respectively, when the change occurs in the ith channel at the time-point λ, and the parameter after the change is θ_i.

Let $\hat{\theta}_{i,n} = \hat{\theta}_i(\boldsymbol{X}_i^n)$ be an estimator of θ_i based on the n observations $\boldsymbol{X}_i^n$ from the ith channel. For any $k \geqslant 1$, define the adaptive versions of the partial LLRs

$$\widehat{Z}_i^k(k) = \log \frac{p_{i,k}(X_{i,k} \mid \boldsymbol{X}_i^{k-1}, \hat{\theta}_{i,k-1})}{p_{0,k}(X_{i,k} \mid \boldsymbol{X}_i^{k-1})},$$

which are obtained by replacing the unknown values of the parameters θ_i's in the LLR's $Z_i^k(k, \theta_i)$ with their estimates based on the previous $k - 1$ observations. We stress that the kth observation is not included in the estimate. Similarly, for $n \geqslant 1$, we define the adaptive versions of the statistics $\widehat{R}_i(n)$ and $\widehat{R}(n)$,

$$\widehat{R}_i(n) = \sum_{\lambda=1}^n \exp\left\{ \sum_{k=\lambda}^n \widehat{Z}_i^k(k) \right\}, \quad \widehat{R}(n) = N^{-1} \sum_{i=1}^N \widehat{R}_i(n). \tag{17.2.14}$$

The adaptive ("plug-in") detection procedure is defined as

$$\hat{\tau}(A) = \min\{n \geqslant 1 : \widehat{R}(n) \geqslant A\}. \tag{17.2.15}$$

The first important fact to note is that by replacing $R(n)$ with $\widehat{R}(n)$ in the proof of Lemma 17.2.1 and observing that $\widehat{R}(n) - n$ is a zero-mean martingale (with respect to $\boldsymbol{P}_\infty$), we immediately obtain: $\boldsymbol{E}_\infty \hat{\tau}(A) \geqslant A$.

It follows that $A = T$ implies $\boldsymbol{E}_\infty \hat{\tau}(A) \geqslant T$, and hence, the FAR is easily controlled.

Furthermore, if the quantities $n^{-1} \sum_{k=\lambda}^{n+\lambda-1} \widehat{Z}_i^k(k)$ converge r-quickly to the same positive numbers J_i's that appeared in Section 17.2.3, then an analog of Theorem 17.2.1 holds for the procedure $\hat{\tau}$. For i.i.d. data models, under certain conditions on the estimators $\hat{\theta}_{i,n}$, an analog of Theorem 17.2.2 holds as well. Specifically, as $A \to \infty$, for all $\theta_i \in \Theta_i$ and $i = 1, \ldots, N$

$$\boldsymbol{E}_{1,i,\theta_i} \hat{\tau}(A) = \frac{1}{I_i} \left[\log(NA) + \frac{1}{2} \log\log(NA) + \hat{C}_i \right] + o(1), \qquad (17.2.16)$$

where $\hat{C}_i$ is a constant. Comparing (17.2.11) with (17.2.16), we can see that an additional term appears which goes to infinity at the rate of double log of the threshold. This is an unavoidable penalty for the prior uncertainty with respect to the post-change parameter value. We also argue that the rate $(1/2) \log\log T$ cannot be improved, thus implying that the proposed adaptive procedure is asymptotically second-order optimal with respect to the average detection delay $D_{i,\theta_i}(\tau) = \sup_\lambda \boldsymbol{E}_{\lambda,i,\theta_i}(\tau - \lambda | \tau \geqslant \lambda)$ in the worst case scenario.

Choosing the estimators $\hat{\theta}_{i,n}$ is not a straightforward task: estimating the parameters in a naive way may affect the performance substantially. For example, if one tries to detect a change in the common shift parameter $\theta_i = \theta > 0$ of the i.i.d. Gaussian sequence, then the estimator $\hat{\theta}_{i,n} = n^{-1} S_{i,n}^+$, where $S_{i,n} = \sum_{k=1}^n X_{i,k}$, is not a good choice. This estimator works well only when the change occurs from the very beginning. For large λ, the performance degrades dramatically (see Dragalin (1996) and Tartakovsky (2003)). A good choice would be an estimator that "forgets" the past. Specifically, define $\hat{\theta}_{i,k-1}(\lambda) = (k-\lambda)^{-1}(\sum_{j=\lambda}^{k-1} X_{i,j})^+$ for $1 \leqslant \lambda < k$, $\hat{\theta}_{i,\lambda}(\lambda) = 0$ and $\widehat{Z}_i^k(k) = \hat{\theta}_{i,k-1}(\lambda) X_{i,k} - \hat{\theta}_{i,k-1}(\lambda)^2/2$. Then it can be shown that the performance of the corresponding adaptive procedure defined in (17.2.14) and (17.2.15) is almost identical to that of the mixture-based procedure. In particular, it is asymptotically optimal in the second-order sense.

In the context of the adaptive CUSUM procedure that is based on the statistic $\max_{1 \leqslant i \leqslant N} \widehat{U}_{i,n}$, where $\widehat{U}_{i,n} = \max_{1 \leqslant \lambda \leqslant n} \sum_{k=\lambda}^n \widehat{Z}_i^k(k)$ and $\widehat{Z}_i^k(k) = \tilde{\theta}_{i,k-1} X_{i,k} - \tilde{\theta}_{i,k-1}^2/2$, the following adaptive exponentially weighted estimators perform fairly well (see Dragalin (1996)). For $k = 1, 2, \ldots$, define $\tilde{\theta}_{i,k}$ by the recursion formula:

$$\tilde{\theta}_{i,k+1} = \begin{cases} \frac{\beta_k \tilde{\theta}_{i,k} + X_{i,k}}{1 + \beta_k} & \text{if } \beta_k \tilde{\theta}_{i,k} + X_{i,k} > 0 \text{ and } \widehat{U}_{i,k} > 0 \\ \theta_{i,0} & \text{otherwise} \end{cases}$$

with the initial condition $\tilde{\theta}_{i,1} = \theta_{i,0}$, where $\theta_{i,0}$ is a design positive number and β_k and $\widehat{U}_{i,k}$, $k = 1, 2, \ldots$ satisfy the recursions

$$\beta_{k+1} = (1 + \beta_k)\mathbb{1}_{\{\beta_k \tilde{\theta}_{i,k} + X_{i,k} > 0, \widehat{U}_{i,k} > 0\}}, \quad \beta_1 = 0,$$

$$\widehat{U}_{i,k+1} = \widehat{U}_{i,k}^+ + \tilde{\theta}_{i,k+1}(X_{i,k+1} - \tilde{\theta}_{i,k+1}/2), \quad \widehat{U}_{i,1} = 0.$$

Hereafter $\mathbb{1}_\Omega$ denotes the indicator of a set Ω. Therefore, the adaptive exponentially weighted estimate $\tilde{\theta}_{i,k}$ has a structure similar to a sample mean where the current sample size k is replaced with the "adaptive" number β_k, which is set to zero (renewed) whenever the statistic $\widehat{U}_{i,k}$ hits the zero level. This allows us to forget the observations that are not consistent with the "change" hypothesis. A reasonable choice for the initial condition $\theta_{i,0}$ is the minimum expected value of the change in the mean.

Note also that, unlike the estimates $\hat{\theta}_{i,k}(\lambda)$, the estimates $\tilde{\theta}_{i,k}$ do not depend on λ, which allows for the computation of the CUSUM statistics $\widehat{U}_{i,k}$ recursively.

17.3 DETECTION IN DISTRIBUTED SENSOR SYSTEMS

Distributed sensor systems capable of collecting, storing, and disseminating a variety of environmental data have the potential to enable the next revolution in information technology. An important application area for distributed sensors is in *environmental awareness* systems. Examples of applications include toxic agent detection, intruder detection for homes and businesses, child-care monitoring systems, automated airport surveillance and metal detection, detection of the onset of an epidemic, failure detection in manufacturing systems and large machines, and patient monitoring in hospitals and homes. A key component of these environmental awareness systems is change detection based on observations made by the sensors.

As described in Section 17.1, the distributed sensor problem could in general be a multichannel change-point detection problem. However, for simplifying the presentation, we assume a single-channel version of the problem, with the understanding that the multichannel generalization is straightforward.

Suppose there are L sensors in the system. At time n, an observation $X_{\ell,n}$ is made at sensor S_ℓ. Conditioned on the change-point λ, the observation sequences $\{X_{1,n}\}, \{X_{2,n}\}, \ldots, \{X_{L,n}\}$ are assumed to be mutually independent. Furthermore, throughout this section, we restrict our attention to the i.i.d. case where the observations in a particular sequence, say $\{X_{\ell,n}\}_{n \geqslant 1}$, are independent conditionally on λ, have a common p.d.f. $f_\ell^{(0)}$ before the change, and a common p.d.f. $f_\ell^{(1)}$ from the time of change. Note that we are assuming that all the sensors change distribution at the change-time λ. Let $\boldsymbol{P}_\lambda$ ($\boldsymbol{P}_\infty$) and $\boldsymbol{E}_\lambda$ ($\boldsymbol{E}_\infty$) stand for the probability measure and the corresponding expectation when the change occurs at time λ (does not occur, that is $\lambda = \infty$).

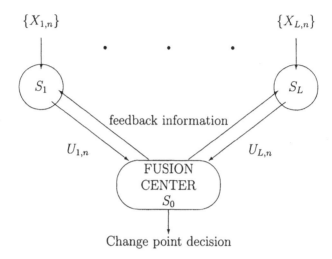

Figure 17.3.1: Block-Diagram of a Decentralized Multi-Sensor Change-Point Detection System.

A general block-diagram of the decentralized change-point detection system is shown in Figure 17.3.1. Based on the information available at S_ℓ at time n, a message $U_{\ell,n}$, belonging to a finite alphabet of size V_ℓ, is formed and sent to the fusion center. We will use the vector notation: $\boldsymbol{X}_n = (X_{1,n}, \ldots, X_{L,n})$ and $\boldsymbol{U}_n = (U_{1,n}, \ldots, U_{L,n})$. Based on the sequence of sensor messages, a decision about the change is made at the fusion center. The fusion center picks a time τ which is a *stopping time* on $\{\boldsymbol{U}_n\}_{n \geqslant 1}$, and a change is declared to have occurred at *tau*.

Various information structures are possible for the decentralized configuration depending on how feedback and local information are used at the sensors (Veeravalli (2001)). Consider the simplest information structure where the message $U_{\ell,n}$ formed by sensor S_ℓ at time n is a function of only its current observation $X_{\ell,n}$, i.e., $U_{n,\ell} = \psi_{\ell,n}(X_{\ell,n})$. Moreover, since for a particular ℓ, the sequence $\{X_{\ell,n}\}_{n \geqslant 1}$ is assumed to be i.i.d., it is natural to confine ourselves to *stationary* quantizers for which the quantizing functions $\psi_{\ell,n}$ do not depend on n, i.e. $\psi_{\ell,n} = \psi_\ell$ for all $n \geqslant 1$.

The quantizing functions $\{\psi_\ell; \ell = 1, \ldots, L\} = \boldsymbol{\psi}$, together with the fusion center stopping time τ, form a policy $\phi = (\tau, \boldsymbol{\psi})$. The goal is to choose the policy ϕ that minimizes $\mathrm{D}^m(\phi)$ defined by

$$\mathrm{D}^m(\phi) = \sup_{1 \leqslant \lambda < \infty} \boldsymbol{E}_\lambda\{(\tau - \lambda)^m \mid \tau \geqslant \lambda) \qquad (17.3.1)$$

for all $m > 0$, while maintaining the average time to false alarm, $\boldsymbol{E}_\infty(\tau)$, at a level greater than T.

Let H_λ be the hypothesis that the change occurs at time $\lambda \in \{1, 2, \dots\}$, and let H_∞ be the hypothesis that $\lambda = \infty$ (no change). Since the observations at each sensor S_ℓ, $\{X_{\ell,n}, n = 1, 2, \dots\}$, are i.i.d., the sensor outputs, $\{U_{\ell,n}, n = 1, 2, \dots\}$ will also be i.i.d. for stationary sensor quantizers. Let $g_\ell^{(j)}$ denote the distribution induced on $U_{\ell,n}$ when the observation $X_{\ell,n}$ is distributed as $f_\ell^{(j)}$, $j = 0, 1$.

Then, for *fixed* sensor quantizers, the LLR between the hypotheses H_λ and H_∞ at the sensor S_ℓ and at the fusion center are respectively given by

$$Z_\ell^\lambda(n) = \sum_{k=\lambda}^n \log \frac{g_\ell^{(1)}(U_{\ell,k})}{g_\ell^{(0)}(U_{\ell,k})} \quad \text{and} \quad Z^\lambda(n) = \sum_{\ell=1}^L Z_\ell^\lambda(n). \qquad (17.3.2)$$

For fixed sensor quantizers, the fusion center faces a standard change detection problem based on the vector observation sequence $\{U_k\}$. Hence we can define the Shiryaev-Roberts statistic $R(n)$ that obeys the recursion:

$$R(n) = (1 + R(n - 1))e^{Z^n(n)}, \quad R(0) = 0. \qquad (17.3.3)$$

Then the Shiryaev-Roberts detection procedure at the fusion center $\tau^* = \tau^*(A)$ is given by

$$\tau^*(A) = \min\{n \geqslant 1 : R(n) \geqslant A\} \quad (\tau^* = \infty \text{ if no such } n \text{ exists)}, \quad (17.3.4)$$

where A is a positive threshold which is selected so that $\mathrm{FAR}(\tau^*(A)) \leqslant 1/T$.

Let $\mathrm{I}(g_\ell^{(1)}, g_\ell^{(0)})$ denote the K-L information number between the densities $g_\ell^{(1)}$ and $g_\ell^{(0)}$. Assume that

$$0 < \sum_{\ell=1}^L \mathrm{I}(g_\ell^{(1)}, g_\ell^{(0)}) < \infty. \qquad (17.3.5)$$

Then an application of Theorem 17.2.2 (ii) (with one channel) gives us that the detection procedure $\tau^*(A)$ defined in (17.3.4), with $A = T$, is asymptotically optimal as $T \to \infty$ among all procedures with FAR no greater than $1/T$. To be specific, if $A = T$, then for all $m > 0$

$$\inf_{\tau \in \Delta(T)} \mathrm{D}^m(\tau) \sim \mathrm{D}^m(\tau^*) \sim \left(\frac{\log T}{\sum_{\ell=1}^L \mathrm{I}(g_\ell^{(1)}, g_\ell^{(0)})} \right)^m \quad \text{as } T \to \infty.$$

This result immediately reveals how to choose the sensor quantizers.

Corollary 17.3.1 *It is asymptotically optimum (as $T \to \infty$) for sensor S_ℓ to pick ψ_ℓ to maximize the K-L information number* $\mathrm{I}(g_\ell^{(1)}, g_\ell^{(0)})$.

Based on the results of Tsitsiklis (1993), it is easy to show that the optimal stationary quantizer $\psi_{\ell,\mathrm{opt}}$ is a *monotone likelihood ratio quantizer* (MLRQ), that is, there exist thresholds $\beta_{\ell,1}, \beta_{\ell,2} \ldots, \beta_{\ell,V_\ell-1}$ satisfying $0 = \beta_{\ell,0} \leqslant \beta_{\ell,1} \leqslant \beta_{\ell,2} \leqslant \cdots \leqslant \beta_{\ell,V_\ell-1} < \infty = \beta_{\ell,V_\ell}$ such that

$$\psi_{\ell,\mathrm{opt}}(X) = i \text{ only if } \beta_{\ell,i-1} < \frac{f_\ell^{(1)}(X)}{f_\ell^{(0)}(X)} \leqslant \beta_{\ell,i}, \quad i = 1, \ldots, V_\ell.$$

Thus, the asymptotically optimal policy ϕ_{opt} for a decentralized change detection problem consists of a stationary (in time) set of MLRQ's at the sensors followed by a Shiryaev-Roberts procedure based on $\{U_k\}$ at the fusion center (as described in (17.3.4)). In other words, if we denote by τ_{opt}^* the Shiryaev-Roberts stopping rule at the fusion center for the case where the sensor quantizers are chosen to be $\boldsymbol{\psi}_{\mathrm{opt}} = \{\psi_{\ell,\mathrm{opt}}\}$, then the asymptotically optimal policy $\phi_{\mathrm{opt}} = (\tau_{\mathrm{opt}}^*, \boldsymbol{\psi}_{\mathrm{opt}})$.

For each ℓ, let the p.d.f.'s induced on $U_{\ell,n}$ by the optimal MLRQ $\psi_{\ell,\mathrm{opt}}$ be given by $g_{\ell,\mathrm{opt}}^{(1)}$ and $g_{\ell,\mathrm{opt}}^{(0)}$. Then the effective K-L information number between the "change" and "no change" hypotheses at the fusion center is:

$$\mathrm{I}_{\mathrm{tot}} = \sum_{\ell=1}^{L} \mathrm{I}(g_{\ell,\mathrm{opt}}^{(1)}, g_{\ell,\mathrm{opt}}^{(0)}) \, .$$

Finally, we denote by $\boldsymbol{\Phi}_{\mathrm{st}}(T)$ the class of policies ϕ with all stationary quantizers and stopping rules at the fusion center such that $\tau \in \boldsymbol{\Delta}(T)$.

The asymptotic performance of the asymptotically optimal solution to the decentralized change detection problem described above is given in the following theorem, which follows directly from Theorem 17.2.2.

Theorem 17.3.1 *Suppose* $0 < \mathrm{I}_{\mathrm{tot}} < \infty$.

(i) *Then,* $A = T$ *implies that* $\boldsymbol{E}_\infty \tau_{\mathrm{opt}}^* \geqslant T$, *and moreover,*

$\lim_{T\to\infty}[\boldsymbol{E}_\infty \tau_{\mathrm{opt}}^*(T)/T]$ *is bounded;*

(ii) *If* $A = T$, *then for all* $m \geqslant 1$,

$$\inf_{\phi \in \boldsymbol{\Phi}_{\mathrm{st}}(T)} \mathrm{D}^m(\phi) \sim \mathrm{D}^m(\phi_{\mathrm{opt}}) \sim \left(\frac{\log T}{\mathrm{I}_{\mathrm{tot}}}\right)^m \quad \text{as } T \to \infty;$$

(iii) *In addition, assume that $Z^1(1)$ is non-arithmetic and that the second moment $E_1|Z^1(1)|^2$ is finite. Then, as $A \to \infty$,*

$$\text{FAR}(\tau^*(A)) = \frac{1}{E_\infty \tau^*_{\text{opt}}(A)} = \frac{\gamma}{A}(1 + o(1)), \qquad (17.3.6)$$

$$\text{ADD}(\phi_{\text{opt}}) = E_1 \tau^*_{\text{opt}}(A) - 1 = \frac{1}{I_{\text{tot}}}(\log A + \varkappa - C) - 1 + o(1). \quad (17.3.7)$$

The quantities γ, $\varkappa$, and C are defined in a similar manner as γ_i, $\varkappa_i$, and C_i in Theorem 17.2.2 if we replace $Z_i(n)$ by $Z(n) = \sum_{\ell=1}^{L} Z_\ell^1(n)$.

The above result not only allows us to choose the threshold A to meet the FAR constraint precisely, but also to calculate the corresponding ADD.

Remark 17.3.1 The condition that the LLR $Z^1(1)$ be non-arithmetic is imposed because one needs to consider certain discrete cases separately in the renewal theorem (Woodroofe (1982), Section 2.1). Since the data at the output of quantizers are discrete, it may happen that the LLR does not obey this condition. If $Z^1(1)$ is arithmetic with span $d > 0$, the results of Theorem 17.3.1 (iii) hold true as $\log A \to \infty$ through multiples of d (that is, $\log A = kd$, $k \to \infty$) and with respective modification of definition of $\varkappa$. However, even in most discrete cases the LLR remains non-arithmetic.

In Section 17.4.3, we will give an example of target detection where the sensor observations are Gaussian random variables with different statistics before and after the change. We will verify the asymptotic results given above via Monte-Carlo simulations.

17.4 APPLICATIONS AND EXPERIMENTAL RESULTS

17.4.1 Target Detection and Tracking in Surveillance Systems

Surveillance systems, such as those in ballistic and cruise missile defense, deal with the detection and tracking of moving targets. *Infrared Search and Track* (IRST) systems are one component of a multisensor suite which can meet the technical challenge of the timely detection/track/identification of small targets. The most challenging problem for an IRST system is the rapid detection of a maneuvering dim target against a heavily cluttered background. To illustrate the importance of this task, we remark that under certain conditions *a few seconds' decrease* in the time it takes to detect a sea/surface skimming cruise missile can lead to a significant increase in the *probability of raid annihilation*.

The problem of detecting moving targets is complicated by the fact that target tracks occur (appear or disappear) at unknown points in time. As a result, this problem can be naturally formulated as an abrupt change detection problem, which is the central theme of this chapter.

Assume that there is a mosaic of staring IR sensors that register the data $X_n = \{X_n(r_{ij}), i, j = 1, \ldots, M\}$, $n \geq 1$, which represent a sequence of 2D frames of intensities $X_n(r_{ij})$ at the pixels with coordinates r_{ij}. Typically, the observations are highly cluttered. Since clutter is usually hundreds or even thousands of times more intensive than sensor noise and the intensity of the target, the detection is impossible without clutter suppression. For this reason, we first apply a spatial-temporal clutter rejection and electronic scene stabilization algorithm (filter) which was developed by Tartakovsky and Blažek (2000). At the output of this filter, we observe the residuals

$$X_n(r) = \mathbb{1}_{\{\lambda \leqslant n\}} \theta S(r - y_n) + \xi_n(r),$$

where λ is an unknown moment of target appearance; $\mathbb{1}_\Omega$ is an indicator of the set Ω; θ is an unknown intensity of the target; $S(r)$ is a known *point spread function* (PSF) of the sensor; y_n is an unknown spatial location of a target on the plane at time n; $\xi_n(r)$ is the effective noise (residual clutter plus noise) having variance σ^2. It is worth mentioning that the PSF $S(r)$ has a very compact support—its effective spatial size is normally 2×2 or 3×3 pixels. Also, even after clutter removal, pixel SNR $q_0 = \theta^2/\sigma^2$ can be as small as 0 dB or even negative.[1]

The detection is complicated by the fact that the velocities and the positions of targets are unknown. In order to overcome this difficulty, we use a so-called *"track-before-detect"* (TBD) approach (Kligys et al. (1998), Petrov et al. (2004), Rozovskii and Petrov (1999)) which is embedded into the multichannel system that represents a bank of position and velocity filters. The idea of TBD is to perform "preliminary" tracking in an attempt to align successive frames according to typical patterns of the targets' dynamics. The second block in the developed system implements this idea. Specifically, we use switching multiple models for target dynamics y_n and a *bank of optimal nonlinear filters* (BONF) to estimate the position of the target at each point in time. The output of BONF is the unnormalized posterior density for y_n given previous data. Maximizing this density, we obtain the estimate $\hat{y}_n$ of the target location y_n. This estimate is then fed into the detection block along with the estimate of the target intensity $\hat{\theta}_n$. The BONF-based TBD algorithm has been developed by B. Rozovsky and A. Petrov, and its detailed description can be found in Petrov et al. (2004) and Rozovskii and Petrov (1999).

---

[1]SNR in decibels (dB) is defined as $10 \log_{10}(\theta^2/\sigma^2)$.

Now we are in a position to describe the detection algorithm. The algorithm has the form (17.2.14) and (17.2.15) where the statistic $\widehat{R}_i(n)$ corresponds to the ith position (spatial) channel in the bank of matched filters. Note that we need a bank of filters, since the estimates $\hat{y}_n$ of the target position are not perfect. As a result, there is almost always a mismatch between the actual target position and its estimated value. The required number of filters in the bank is, however, relatively small compared to the case where TBD is not performed. For example, we used $N = 9 - 32$ filters in the bank regardless of the size of the image. Compare this with $N = 2^{12}$ which would be required for the system without TBD comprised of the bank of 3D matched filters for the image size 128×128 and the effective target size 2×2 pixels. Equivalent noise is modeled by the zero-mean white Gaussian process, in which case the partial LLR for the ith position channel is defined by equation (17.4.1), with $\hat{\theta}_k$ being an estimate of the target intensity and δ_i being a two-dimensional shift that is measured in the number of pixels $(i = 0, \ldots, N - 1)$.

$$\widehat{Z}_i^k(k) = \frac{\hat{\theta}_{k-1}}{\sigma^2} \sum_{l,j} S(r_{lj} - \hat{y}_k - \delta_i) X_k(r_{lj}) - \frac{\hat{\theta}_{k-1}^2}{2\sigma^2} \sum_{l,j} S^2(r_{lj} - \hat{y}_k - \delta_i), \quad (17.4.1)$$

Thus, the results of Section 17.2.4 are applied by using the LLR's (17.4.1). To be precise, detections are declared and tracks are initiated when the statistic $\widehat{R}(n)$ exceeds the threshold A. The threshold is chosen so that the average frequency of false exceedances would be bounded above by a given level FAR $\leqslant \bar{F} = 1/T$. According to Lemma 17.2.1, $A = 1/\bar{F}$ guarantees this inequality.

Figure 17.4.1 illustrates the results of detection by the described adaptive change-point detection algorithm. In these experiments, we used the real cluttered and jittery IR background obtained from the NAVY SPAWAR Systems Center, San Diego, CA (staring shipboard IRST, LAPTEX field test). The picture in the upper half shows a typical data frame at the output of the clutter rejection filter. The effective size of the target is 3×3 pixels, and it is invisible to the naked eye, since the pixel SNR varies from frame to frame in the range −16 to 0 dB. For this reason, it is explicitly circled in the picture. The picture in the lower half depicts two statistics: $U_n(\hat{y}_n) = \log \widehat{R}(n)$ and Δ_n. the first one is sensitive to target appearance and the second one to target disappearance. The detection of tracks occurs when the adaptive statistic $U_n(\hat{y}_n)$ exceeds the threshold $\log A$ (the upper one) and track disappearance is declared when the statistic Δ_n exceeds the threshold C (the lower one). See Petrov et al. (2004) and Tartakovsky et al. (1999) for more details. The frame rate was 1 frame/second and the threshold A was chosen so as to guarantee that there would be no more than 1 false detection per minute (that is, per 60 frames). In the particular

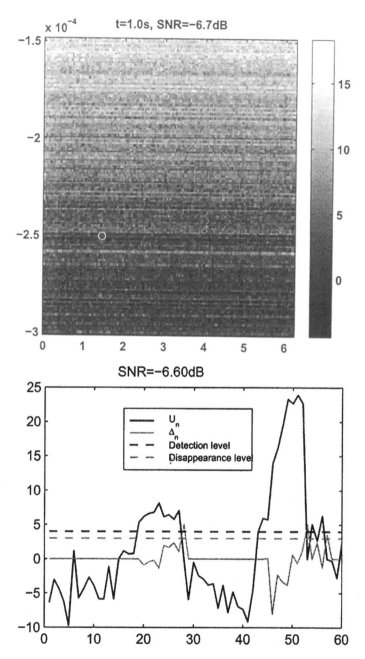

Figure 17.4.1: Detection of Track Appearance and Disappearance: Real IR Background at the Output of the De-Cluttering Filter (Top) and Target Detection (Bottom).

scenario shown in the figure, the algorithm detected the first target with the delay of about 20 seconds (20 frames). This number matches the first-order approximation given by Theorem 17.2.1 of section 17.2.4 (see (17.2.6)),

$$E_{i,\lambda,\theta}(\tau^* - \lambda | \tau^* \geqslant \lambda) \approx (2/q) \log(N/\bar{F}),$$

where $q = \frac{\theta^2}{\sigma^2} \sum_{l,j} S^2(r_{lj})$ is the cumulative SNR and N is the number of spatial (position) channels. Indeed, for $N = 9$ channels used in the experiments, $\bar{F} = 1/60$, and estimated cumulative SNR -1.14 dB, the average detection delay is ADD ≈ 16.4 according to this approximate formula.

The results obtained allow us to conclude that the adaptive algorithm developed above is able to detect even very low SNR targets with reasonable detection delays, and that the theoretical results of Section 17.2 are useful for performance evaluation.

17.4.2 Rapid Attack/Intrusion Detection

In computer networks, large-scale attacks in their final stages can readily be identified by observing very abrupt changes in the network traffic. In the early stage of an attack, however, these changes are hard to detect and difficult to distinguish from usual traffic fluctuations. Existing intrusion detection systems can be classified as either *Signature Detection Systems* or *Anomaly Detection Systems*. The latter systems compare the parameters of the observed traffic with "normal" (legitimate) network traffic. The attack is declared once a deviation from normal traffic is observed (Kent (2000)).

Our approach belongs to the class of Anomaly Detection Systems. The idea is based on the observation that an attack leads to relatively abrupt changes in the statistics of the traffic when compared to the traffic's "normal mode". Therefore, the problem of detecting an attack can be formulated and solved as a change-point detection problem: detect a change in the traffic model with minimal average delay, while controlling FAR. In this subsection, we briefly describe an efficient adaptive sequential method, recently developed in Blažek et al. (2001), for an early detection of intrusions from the class of *"Denial-of-Service* (DoS)" attacks.

In experiments, we used a Network Simulator NS-2[2] with a network consisting of 100 nodes configured into a transit-stub topology that is depicted in Figure 17.4.2. The network contained one transit domain, four transit nodes, and 12 stub domains with 96 nodes. Further details can be found in Blažek et al. (2001).

While monitoring network traffic, one can observe various kinds of features related to the headers and sizes of the received and transmitted pack-

---

[2]More information on the NS-2 can be found at http://www.isi.edu/nsnam/ns/

ets, the usage of system resources, service quality, and similar aspects associated with the utilization of the network. For example, in the transport layer, we observe the number of *Transmission Control Protocol* (TCP) packets categorized by size and/or type, the number of *User Datagram Protocol* (UDP) packets and their sizes, and the source and destination port for each packet. An intrusion leads to a change in the traffic intensity through

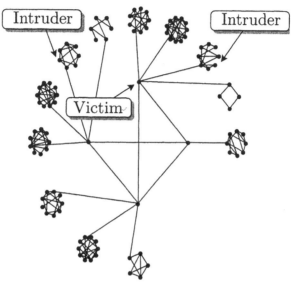

Figure 17.4.2: Transit-Stub Network Topology Used in Experiments.

changes in packet sizes. Therefore, the problem of detecting a DoS attack is regarded as the quickest change detection in the mean values of the numbers of packets. Let pt denote the type of the packet — ICMP, UDP, or TCP. In the experiments, we categorized the packets by their sizes into a number of size bins. Let $X_{pt}^{k,i}$ be the total number of packets of type pt with sizes in the ith bin received during the kth time interval, where $i = 1, \ldots, M$ (M being the number of bins). Therefore, the total monitoring system is a multichannel system with $N = 3M$ channels.

In network security applications, it is fairly difficult to build an exact statistical model. As a result, the post-change distributions $\boldsymbol{P}_{pt,i}$'s are usually not specified. For this reason, we used a nonparametric version of the CUSUM-type algorithm (17.2.13). Specifically, we performed simultaneous thresholding of the statistics $S_{k,i}^{pt}$ with reflection from the zero barrier that were functions of the numbers of packets $X_{pt}^{k,i}$ and their "historical" mean values $\mu_{i,k}$. If the statistic $\max_{pt,i}(S_{k,i}^{pt})$ exceeds a threshold h, then an alarm message is sent to the decision making engine. The statistic $S_{k,i}^{pt}$ represents a nonparametric adaptive version of the CUSUM statistic $U_i(k)$ defined at the end of Section 17.2.3. As shown in Figure 17.4.3 (top), information about the patterns of regular data flow is updated when the statistic $S_{k,i}^{pt}$ reaches the zero barrier. If the decision-making engine reports that a previously issued alarm message is a false alarm, then the information about regular data patterns and thresholds is updated accordingly (in particular, the pre-change mean value is re-estimated), and data monitoring starts all over again.

Under regular conditions, traffic consisted of approximately 5% ICMP packets, 20% UDP packets, and 75% TCP packets. After a 120-second period (measured using the simulator time) of regular traffic, we initiated an attack (TCP SYN Flooding, UDP Packet Storm or ICMP Ping Flooding; see Blažek et al. (2001)) and traffic rapidly increased, reaching 20% of all traffic.

As shown in Figure 17.4.3 (top), the sequential algorithm has detected the UDP DoS attack in its early stage. The plot at the bottom illustrates the operating characteristic (ADD versus FAR) obtained by simulations (dashed line) and by the first-order asymptotic formulas similar to (17.2.6) and (17.2.9) (solid line) for the UDP Packet Storm attack. The plot of the theoretical estimate of ADD versus $|\log \text{FAR}|$ is the straight line with the slope that can be computed from the asymptotic theory (see formulas (17.2.6) and (17.2.9)). It is seen that the experimental estimates of ADD are always bigger than the theoretical estimates. This is not surprising, since the first-order approximations used in calculations ignore excesses over the thresholds of the decision statistics. However, the accuracy of these approximations is high enough to be useful for preliminary estimates.

Therefore, as in the previous subsection, the results of the asymptotic theory can be used to predict the performance of the detection algorithm with a reasonable accuracy.

17.4.3 Decentralized Detection Example and Simulation Results

Consider the problem of detecting a slowly fluctuating target using L geographically separated sensors (for example, radars). The observations are corrupted by additive white Gaussian noise (sensor noise or residual clutter and noise) that is independent between each two sensors. The sensors pre-process the observations using a matched filter, matched to the signal corresponding to the target. The output of the matched filter at sensor S_ℓ at time n is given by:

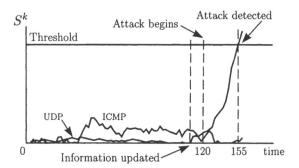

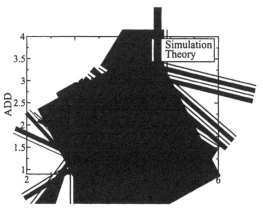

Figure 17.4.3: Intrusion Detection: One Particular Run of a UDP DoS Attack (Top) and Operating Characteristic of the Detection Algorithm (Bottom).

$$X_{\ell,n} = \begin{cases} \xi_{\ell,n} & \text{if } n < \lambda \\ \mu_\ell + \xi_{\ell,n} & \text{if } n \geqslant \lambda, \end{cases}$$

where λ is the time of appearance of the target, $\{\xi_{\ell,n}\}_{n\geqslant 1}$ is a sequence of i.i.d. zero-mean Gaussian random variables with variance σ_ℓ^2, and μ_ℓ is the mean value that is related to the average intensity of the target.

Therefore, the likelihood ratio at sensor S_ℓ is given by

$$Y_\ell(X_{\ell,n}) = \exp\left\{ (X_{\ell,n}\,\mu_\ell - \mu_\ell^2/2)/\sigma_\ell^2 \right\}, \quad n = 1, 2, \ldots.$$

Since $Y_\ell(x)$ is monotonically increasing in x, we can characterize the sensor quantizers in terms of thresholds on the observations, rather than on their likelihood ratios. To further simplify the example, we assume that the sensor messages are binary, that is, $V_\ell = 2$ for all ℓ. Then the quantizers reduce to binary tests that are characterized by a single threshold, that is,

$$U_{\ell,k} = \begin{cases} 1 & \text{if } X_{\ell,k} \geqslant \beta_\ell \\ 0 & \text{if } X_{\ell,k} < \beta_\ell \end{cases}.$$

The distribution induced on $U_{\ell,k}$ by this quantizer is given by:

$$g_\ell^{(j)}(0) = 1 - g_\ell^{(j)}(1) = \Phi\left(\frac{\beta_\ell - j\mu_\ell}{\sigma_\ell}\right) = q_\ell^{(j)}, \; j = 0, 1, \qquad (17.4.2)$$

where $\Phi(\cdot)$ is the distribution function of a standard Gaussian random variable. The optimal value of β_ℓ, i.e., the one that maximizes $I(g_\ell^{(1)}, g_\ell^{(0)})$, is easily found based on (17.4.2). Then we can compute the Shiryaev-Roberts statistic for the fusion center using (17.3.3). In particular, it is easy to show that

$$Z^n(n) = \sum_{\ell=1}^{L} \left\{ U_{\ell,n} \log \left[\frac{1 - q_{\ell,\text{opt}}^{(1)} \, q_{\ell,\text{opt}}^{(0)}}{1 - q_{\ell,\text{opt}}^{(0)} \, q_{\ell,\text{opt}}^{(1)}} \right] - \log \left[\frac{q_{\ell,\text{opt}}^{(0)}}{q_{\ell,\text{opt}}^{(1)}} \right] \right\}, \qquad (17.4.3)$$

where $q_{\ell,\text{opt}}^{(j)}$'s are the optimal values that correspond to (17.4.2) with the optimal threshold values $\beta_\ell = \beta_{\ell,\text{opt}} = \arg\max I(g_\ell^{(1)}, g_\ell^{(0)})$.

An example with five sensors having identically distributed observations is illustrated in Figure 17.4.4. The parameter values are $\mu_\ell = 0.4$ and $\sigma_\ell^2 = 1$. The K-L information number for the sensor observations is 0.08. The optimal threshold that maximizes the K-L information number at the output of the sensor is $\beta = 0.32$, and the corresponding maximum K-L number after quantization is 0.0509. We plot ADD versus FAR for the optimal decentralized detection policy and compare the performance with that of a centralized policy which has direct access to the observations at the radars. As we expect, for the centralized policy, the plot of ADD versus $-\log(\text{FAR})$ is a straight line with a slope that is approximately equal to $1/[5I(f^{(1)}, f^{(0)})] = 2.5$. For the optimal decentralized policy, the tradeoff curve between ADD and $-\log(\text{FAR})$ has a slope that is approximately equal to $1/I_{\text{tot}} \approx 3.93$, as expected from Theorem 17.3.1. The decentralized policy, of course, suffers a performance degradation relative to the centralized policy. However, the bandwidth requirements for communication with the fusion center are considerably smaller in a decentralized setting, especially with binary quantizers. Also shown in Figure 17.4.4 is the tradeoff curve for a centralized detection policy with a *single* sensor. As expected, the slope of ADD versus $-\log(\text{FAR})$ is five times greater than that in the five sensor centralized case. Furthermore, it can be seen that even if the

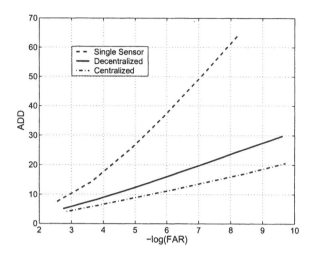

Figure 17.4.4: Operating Characteristics for Five Identically Distributed
Sensors.

sensor observations are quantized to one bit, the decentralized policy with
five sensors far outperforms the single-sensor centralized policy.

Based on (17.4.3), we may also compute higher order approximations
for FAR and ADD as given in Theorem 17.3.1 using formulas (17.2.12).
These higher-order approximations typically match the simulation results
very closely. Results of the detailed analysis will be presented elsewhere.

17.5 CONCLUDING REMARKS

For the problem of detecting abrupt changes in multichannel and distributed
systems, we have proposed easily implementable sequential procedures based
on multichannel versions of Shiryaev-Roberts and CUSUM-type statistics.
A major theoretical result is the asymptotic optimality property of these
procedures among all detection procedures with a guaranteed low rate of
false alarms for quite general non-i.i.d. models. As we have shown, the
regularity conditions can be relaxed considerably for the i.i.d. models. Further, we have derived lower bounds for the mean time to false alarm and
asymptotic expansions for the average detection delay. These bounds and
asymptotic expansions provide useful approximations for the operating characteristics and constitute the basis for the approximate design of detection
thresholds.

The usefulness of the asymptotic theory developed here has been verified by extensive simulations and, more importantly, by implementation using real data. In particular, the proposed detection algorithms have been implemented for rapid detection of moving dim targets against heavy clutter by IRST systems, as well as for real-time intrusion detection in high-speed computer networks. The evaluation of the performance of these algorithms in the existing testbeds illustrates their high efficiency. Based on these results, we can conclude that the technology developed here allows one to reliably detect even low-SNR targets while maintaining the desired frequency of false detections. In addition, it helps detect denial-of-service attacks in their early stages, well before the hostile traffic reaches its full potential.

APPENDIX

Appendix A1. Proof of Theorem 17.2.1

Write

$$M_i(A) = J_i^{-1} \log(NA), \quad \gamma_{\lambda,i}(\varepsilon, A) = P_{\lambda,i} \left\{ 0 \leqslant \tau^* - \lambda < (1 - \varepsilon) M_i(A) \right\}.$$

By Chebyshev's inequality, for every $\varepsilon > 0$

$$E_{\lambda,i}[(\tau^* - \lambda)^m \mathbb{1}_{\{\tau^* \geqslant \lambda\}}] \geqslant [\varepsilon M_i(A)]^m P_{\lambda,i} \left\{ \tau^* - \lambda \geqslant \varepsilon M_i(A) \right\}.$$

Obviously,

$$P_{\lambda,i} \left\{ \tau^* - \lambda \geqslant (1 - \varepsilon) M_i(A) \right\} = P_{\lambda,i} \left\{ \tau^* \geqslant \lambda \right\} - \gamma_{\lambda,i}(\varepsilon, A),$$

and $P_{\lambda,i} \left\{ \tau^* \geqslant \lambda \right\} = P_\infty \left\{ \tau^* \geqslant \lambda \right\}$, since the event $\left\{ \tau^* \geqslant \lambda \right\}$ belongs to the sigma algebra $\mathcal{F}_{\lambda-1}$. It follows that for every $0 < \varepsilon < 1$

$$E_{\lambda,i} \left[(\tau^* - \lambda)^m | \tau^* \geqslant \lambda \right] \geqslant [(1 - \varepsilon) M_i(A)]^m \left[1 - \frac{\gamma_{\lambda,i}(\varepsilon, A)}{P_\infty \left\{ \tau^* \geqslant \lambda \right\}} \right].$$

It can be shown that $P_\infty \left\{ \tau^* \geqslant \lambda \right\} \to 1$ as $A \to \infty$ for any fixed λ. Also, applying an argument similar to that used in the proof of Lemma 3.1 in Tartakovsky (2003), we conclude that for an arbitrary $\varepsilon \in (0, 1)$, $\gamma_{\lambda,i}(\varepsilon, A) \to 0$ as $A \to \infty$. Since ε is arbitrary, we obtain the following asymptotic lower bound

$$E_{\lambda,i} \left[(\tau^* - \lambda)^m | \tau^* \geqslant \lambda \right] \geqslant \left(\frac{\log NA}{J_i} \right)^m (1 + o(1)). \tag{A.1}$$

It suffices to show that this lower bound is, at the same time, asymptotically the upper bound. To this end, we introduce the sequence of auxiliary stopping times

$$\nu_i(\lambda) = \min \left\{ n \geqslant \lambda : Z_i^\lambda(n) \geqslant \log NA \right\}, \quad \lambda = 1, 2, \ldots \tag{A.2}$$

It follows from Theorem 4.2 in Dragalin et al. (1999) that for every $m \leqslant r$

$$E_{\lambda,i}[\nu_i(\lambda) - \lambda]^m \sim \left(\frac{\log NA}{J_i}\right)^m \quad \text{as } A \to \infty$$

whenever $(n - \lambda)^{-1}Z_i^\lambda(n)$ converges to J_i r–quickly (condition (17.2.5)). Since $\tau^*(A) \leqslant \nu_i(\lambda)$ on $\{\tau^*(A) \geqslant \lambda\}$ (see (A.5)–(A.7) below), we have

$$E_{\lambda,i}[(\tau^* - \lambda)^m|\tau^* \geqslant \lambda] \leqslant E_{\lambda,i}[(\nu_i(\lambda) - \lambda)^m|\tau^* \geqslant \lambda] = E_{\lambda,i}(\nu_i(\lambda) - \lambda)^m$$

$$= \left(\frac{\log NA}{J_i}\right)^m (1 + o(1)),$$

where the equality $E_{\lambda,i}[(\nu_i - \lambda)^m|\tau^* \geqslant \lambda] = E_{\lambda,i}(\nu_i - \lambda)^m$ follows from the fact that $\{\tau^* \geqslant \lambda\} \in \mathcal{F}_{\lambda-1}$ and $\nu_i(\lambda)$ does not depend on $\mathcal{F}_{\lambda-1}$.

Comparing this upper bound with (A.1) completes the proof. ∎

Appendix A2. Proof of Theorem 2

In order to prove (17.2.9), we again use an "upper-lower" bound technique.

We first observe that the stopping time τ^* does not exceed the one-sided stopping time $\nu_i(1) = \nu_A$ defined in (A.2), and hence, in order to derive an upper bound for $D_i^m(\tau^*)$ it is sufficient to find an asymptotic expansion for $E_i(\nu_A)^m$. Note that in the i.i.d. case the sequence of statistics $\{Z_i(n), n \geqslant 1\}$ is a random walk with mean $E_i Z_i(1) = I_i$, which is positive by condition (17.2.8).

The second important observation is that, again by condition (17.2.8) (finiteness), $E_i\{Z_i^-(1)\}^m < \infty$ for all $m > 0$, where

$$Z_i^-(1) = -\min\{0, Z_i(1)\}.$$

Indeed,

$$E_i e^{Z_i^-(1)} = E_i e^{-Z_i(1)} \mathbb{1}_{\{Z_i(1)<0\}} + E_i \mathbb{1}_{\{Z_i(1)\geqslant 0\}} \leqslant E_i e^{-Z_i(1)} + 1 = 2.$$

Therefore, Theorem III.8.1 of Gut (1988) can be applied to show that

$$E_i(\nu_A)^m = \left(\frac{\log NA}{I_i}\right)^m (1 + o(1)) \quad \text{as } A \to \infty \text{ for all } m \geqslant 1.$$

Setting $A = T$, we obtain that for all $m \geqslant 1$

$$D_i^m(\tau^*) \leqslant \left(\frac{\log NT}{I_i}\right)^m (1 + o(1)) \quad \text{as } T \to \infty. \tag{A.3}$$

On the other hand, by the strong law of large numbers

$$\lim_{n\to\infty} \boldsymbol{P}_i\left\{\max_{1\leqslant k\leqslant n} Z_i(k) \geqslant (1+\varepsilon)\mathrm{I}_i n\right\} = 0 \quad \text{for all } \varepsilon > 0,$$

and a slight modification of the proof of Theorem 1 of Lai (1998) applies, thereby showing that

$$\inf_{\tau\in\Delta(T)} D_i^1(\tau) \geqslant \frac{\log T}{\mathrm{I}_i}(1 + o(1)) \quad \text{as } T \to \infty.$$

Applying Jensen's inequality, we also deduce that for all $m \geqslant 1$

$$\inf_{\tau\in\Delta(T)} \mathrm{D}_i^m(\tau) \geqslant \left(\frac{\log T}{\mathrm{I}_i}\right)^m (1 + o(1)) \quad \text{as } T \to \infty,$$

which, along with the upper bound (A.3), proves (17.2.9).

To prove (17.2.10), we introduce the statistic $\Lambda(n) = N^{-1}\sum_{i=1}^N e^{Z_i(n)}$ and the auxiliary stopping time $\tau_A = \min\{n : \Lambda(n) \geqslant A\}$. By $\boldsymbol{E}$, we denote the expectation with respect to the measure $\boldsymbol{P} = N^{-1}\sum_{i=1}^N \boldsymbol{P}_i$. We have

$$\boldsymbol{P}_\infty\{\tau_A < \infty\} = \boldsymbol{E}\left[\frac{1}{\Lambda(\tau_A)}\mathbb{1}_{\{\tau_A<\infty\}}\right] = \frac{1}{A}\boldsymbol{E}\left[e^{-\rho_A}\mathbb{1}_{\{\tau_A<\infty\}}\right]$$

$$= \frac{1}{AN}\sum_{i=1}^N \boldsymbol{E}_i\left[e^{-\rho_A}\mathbb{1}_{\{\tau_A<\infty\}}\right],$$

where $\rho_A = \log\Lambda(\tau_A) - \log A$ is the excess of $\Lambda(n)$ over $\log A$ at the stopping time τ_A on $\{\tau_A < \infty\}$. Clearly,

$$\tau_A = \min\{n \geqslant 1 : Z_i(n) + Y_i(n) \geqslant \log(AN)\},$$

where

$$Y_i(n) = \log\left(1 + \sum_{j\neq i} e^{Z_j(n)-Z_i(n)}\right).$$

Since $Y_i(n) \to 0$ as $n \to \infty$ $\boldsymbol{P}_i$–a.s., the $Y_i(n)$, $n \geqslant 1$, are slowly changing under $\boldsymbol{P}_i$ (see Siegmund (1985), Woodroofe (1982) or Ghosh et al. (1997) for the definition of slowly changing sequences). Therefore, Theorem 4.1 in Woodroofe (1982) applies to show that $\rho_A \to \varkappa_i$ as $A \to \infty$ in $\boldsymbol{P}_i$– distribution, where $\varkappa_i = Z_i(\nu_i) - a$ is the overshoot in the one-sided test ν_i. Also, since $Z_i(n)/n \to \mathrm{I}_i$ $\boldsymbol{P}_i$–a.s., $\sup_n Z_i(n) = \infty$ with probability 1, which implies that $\boldsymbol{P}_i\{\tau_A < \infty\} = 1$. Therefore,

$$A\boldsymbol{P}_\infty\{\tau_A < \infty\} \xrightarrow[A\to\infty]{} \frac{1}{N}\sum_{i=1}^N \gamma_i = \bar{\gamma}_N. \tag{A.4}$$

Finally, generalizing the quite tedious argument used in Pollak (1987), we are able to show that

$$E_\infty R(\tau^*) = E_\infty \tau^* \sim \frac{1}{P_\infty \{\tau_A < \infty\}} \quad \text{as } A \to \infty,$$

which, along with (A.4), completes the proof of (17.2.10).

Consider the hypothesis $H_{i,1}$. Observe that the statistic $\log R(n)$ can be represented in the form

$$\log R(n) = Z_i(n) + V_{i,n} - \log N, \qquad (A.5)$$

where

$$V_{i,n} = \log \left(1 + \sum_{s=1}^{n-1} \exp\{-Z_i(s)\} \right.$$
$$\left. + \exp\{-Z_i(n)\} \sum_{\substack{j=1 \\ j \neq i}}^{N} \sum_{s=1}^{n} \exp\left\{ \sum_{k=s}^{n} Z_j^s(n) \right\} \right). \qquad (A.6)$$

Therefore, the stopping time $\tau^*(A)$ can be written as

$$\tau^*(A) = \min\{n \geq 1 : Z_i(n) + V_{i,n} \geq \log(AN)\}. \qquad (A.7)$$

The sequence $V_{i,n}$ is slowly changing under P_i and converges in P_i- distribution to a random variable

$$V_i = \log \left(1 + \sum_{k=1}^{\infty} e^{-Z_i(k)} \right)$$

with the expectation

$$E_i V_i = C_i = E_i \left\{ \log \left(1 + \sum_{k=1}^{\infty} e^{-Z_i(k)} \right) \right\}.$$

Thus, the proof of (17.2.11) can be completed by using Woodroofe's (1982) Theorem 4.5. To use this theorem, we have to check the conditions (4.13) and (4.16) in that monograph, which is straightforward but tedious. Note that condition (4.14) in Woodroofe (1982) holds trivially, since $V_{i,n}$ are nonnegative. Applying this theorem gives (17.2.11) ∎

ACKNOWLEDGMENTS

We wish to thank Dr. John Barnett and Dr. Steven Doss-Hammel of SPAWAR Systems Center, San Diego for useful comments and providing real IR data. We are also grateful to Dr. Boris Rozovsky and Dr. Vlad Repin for valuable discussions and help in the work. Application to IRST is our joint work with Rudolf Blažek (clutter rejection), Anton Petrov and Boris Rozovsky (track-before-detect). Application to network security is the joint work with Boris Rozovsky, Rudolf Blažek, and Hongjoong Kim. Finally, we are grateful to reviewers for valuable comments.

The research of A. Tartakovsky was supported in part by the U.S. ONR grants N00014-99-1-0068 and N00014-95-1-0229 and by the U.S. DARPA grant N66001-00-C-8044. The research of V.V. Veeravalli was supported in part by the U.S ONR grant N00014-97-1-0823 and by the U.S. NSF CA-REER/PECASE grant CCR-0049089.

REFERENCES

[1] Basseville, M. and Nikiforov, I.V. (1993). *Detection of Abrupt Changes: Theory and Applications*. Prentice Hall: Englewood Cliffs.

[2] Blažek, R., Kim, H., Rozovskii, B. and Tartakovsky, A. (2001). A novel approach to detection of "denial-of-service" attacks via adaptive sequential and batch-sequential change-point detection methods. In *Proc. IEEE Systems, Man, and Cybernetics Information Assurance Workshop*, West Point, New York.

[3] Dragalin, V.P. (1995). A multi-channel change point problem. In *Proc. 3rd Umea-Wuerzburg Conf. in Statistics* (E. von Collani and R. Goeb, eds.), 97-108. Wuerzburg Research Group on Quality Control, Wuerzburg University, Germany.

[4] Dragalin, V.P. (1996). Adaptive procedures for detecting a change in distribution. In *Proc. 4th Wuerzburg-Umea Conf. in Statistics* (E. von Collani, R. Goeb and G. Kiesmueller, eds.), 87-103. Wuerzburg Research Group on Quality Control, Wuerzburg University, Germany.

[5] Dragalin, V.P., Tartakovsky, A.G. and Veeravalli, V. (1999). Multihypothesis sequential probability ratio tests, I: Asymptotic optimality. *IEEE Trans. Inform. Theory*, **45**, 2448-2461.

[6] Ghosh, M., Mukhopadhyay, N. and Sen, P.K. (1997). *Sequential Estimation*. Wiley: New York.

[7] Gut, A. (1988). *Stopped Random Walks: Limit Theorems and Applications*. Springer-Verlag: New York.

[8] Hsu, P.L. and Robbins, H. (1947). Complete convergence and the law of large numbers. *Proc. Nat. Acad. Sci., U.S.A.*, **33**, 25–31.

[9] Kent, S. (2000). On the trial of intrusions into information systems. *IEEE Spectrum*, 52–56.

[10] Kligys, S., Rozovskii, B.L., and Tartakovsky, A.G. (1998). Detection algorithms and track before detect architecture based on nonlinear filtering for IRST. *Technical report CAMS-98.9.1*, Center for Applied Mathematical Sciences, University of Southern California, Los Angeles, U.S.A. http://www.usc.edu/dept/LAS/CAMS/usr/facmemb/tartakov

[11] Lai, T.L. (1976). On r-quick convergence and a conjecture of Strassen. *Ann. Probab.*, **4**, 612–627.

[12] Lai, T.L. (1995). Sequential changepoint detection in quality control and dynamical systems. *J. Roy. Statist. Soc., Ser. B*, **57**, No. 4, 613–658.

[13] Lai, T.L. (1998). Information bounds and quick detection of parameter changes in stochastic systems. *IEEE Trans. Inform. Theory*, **44**, 2917–2929.

[14] Lorden, G. (1971). Procedures for reacting to a change in distribution. *Ann. Math. Statist.*, **42**, 1987–1908.

[15] Nikiforov, I.V. (1995). A generalized change detection problem. *IEEE Trans. Inform. Theory*, **41**, 171–187.

[16] Petrov, A., Rozovskii, B.L. and Tartakovsky, A.G. (2004). Efficient nonlinear filtering methods for detection of dim targets by passive systems. In *Multitarget-Multisensor Tracking: Applications and Advances* (X.R. Li et al., eds.), **IV**. Artech House: Boston.

[17] Pollak, M. (1985). Optimal detection of a change in distribution. *Ann. Statist.*, **13**, 206–227.

[18] Pollak, M. (1987). Average run lengths of an optimal method of detecting a change in distribution. *Ann. Statist.*, **15**, 749–779.

[19] Rozovskii, B. and Petrov, A. (1999). Optimal nonlinear filtering for track-before-detect in IR image sequences. In *SPIE Proceedings: Signal and Data Processing of Small Targets* (O. Drummond, ed.), **3809**, 152–163.

[20] Shiryaev, A.N. (1963). On optimum methods in quickest detection problems. *Theor. Probab. Appl.*, **8**, 22–46.

[21] Shiryaev, A.N. (1978). *Optimal Stopping Rules*. Springer-Verlag: New York.

[22] Siegmund, D. (1985). *Sequential Analysis: Tests and Confidence Intervals*. Springer-Verlag: New York.

[23] Tartakovsky, A.G. (1988). Multi-alternative sequential detection and estimation of signals with random appearance times. *Statist. Control Problems*, **83**, 216–222.

[24] Tartakovsky, A.G. (1991a). *Sequential Methods in the Theory of Information Systems*. Radio i Svyaz': Moscow (in Russian).

[25] Tartakovsky, A.G. (1991b). Asymptotically optimal multi-alternative sequential detection of a disorder of information systems. In *Proc. IEEE Intern. Symp. Inform. Theory*, Budapest, 359.

[26] Tartakovsky, A.G. (1992). Efficiency of the generalized Neyman-Pearson test for detecting changes in a multichannel system. *Probl. Inform. Transmis.*, **28**, 341–350.

[27] Tartakovsky, A.G. (1994). Asymptotically minimax multialternative sequential rule for disorder detection. In *Statistics and Control of Random Processes: Proc. Steklov Institute of Mathematics*, **202**, Issue 4, 229–236. AMS: Providence.

[28] Tartakovsky, A.G. (2003). Extended asymptotic optimality of certain change-point detection procedures. *Preprint*, Center for Applied Mathematical Sciences, University of Southern California, Los Angeles.

[29] Tartakovsky, A.G. and Blažek, R. (2000). Effective adaptive spatial-temporal technique for clutter rejection in IRST. In *SPIE Proceedings: Signal and Data Processing of Small Targets* (O. Drummond, ed.), **4048**, 85–95.

[30] Tartakovsky, A.G., Kligys, S. and Petrov, A. (1999). Adaptive sequential algorithms for detecting targets in heavy IR clutter. In *SPIE Proceedings: Signal and Data Processing of Small Targets* (O. Drummond, ed.), **3809**, 119–130.

[31] Tsitsiklis, J.N. (1993). Extremal properties of likelihood-ratio quantizers. *IEEE Trans. Commun.*, **41**, 550–558.

[32] Veeravalli, V.V. (2001). Decentralized quickest change detection. *IEEE Trans. Inform. Theory*, **47**, 1657–1665.

[33] Woodroofe, M. (1982) *Nonlinear Renewal Theory in Sequential Analysis.* SIAM: Philadelphia.

Addresses for Communication:

ALEXANDER TARTAKOVSKY, Center for Applied Mathematical Sciences, University of Southern California, 1042 Downey Way, DRB-155, Los Angeles, CA 90089-1113, U.S.A. E-mail: tartakov@math.usc.edu
VENUGOPAL VEERAVALLI, ECE & CSL, University of Illinois at Urbana-Champaign, 1308 West Main Street, Urbana, IL 61801, U.S.A. E-mail: vvv@uiuc.edu

Chapter 18

Extension of Hochberg's Two-Stage Multiple Comparison Method

RAND R. WILCOX

University of Southern California, Los Angeles, U.S.A.

18.1 INTRODUCTION

Hochberg (1975) derived a two-stage multiple comparison procedure that addresses an issue of fundamental importance: How many additional observations must be sampled, if any, to achieve some desired level of precision when estimating a collection of linear contrasts. In particular, given that the length of a confidence interval should not exceed some specified value, how many additional observations are required to achieve this goal? A related problem is how to control the simultaneous probability coverage, or the familywise error rate (which is defined as the probability of at least one Type-I error among a collection of hypotheses being tested) in the event that additional observations can be acquired.

Extending a classic two-stage procedure introduced by Stein (1945), Hochberg assumed that the goal was to draw inferences about a set of linear contrasts based on the means of normal distributions with possibly unequal variances. Under normality, the method guaranteed that the length of the confidence intervals would not exceed some specified value and that the simultaneous probability coverage would be at least $1 - \alpha$. The goals in this paper are to comment on small-sample properties of Hochberg's method when sampling from non-normal distributions. An examination of these properties reveals some advantages over more conventional single-stage techniques. We also report some small-sample properties of an extension of Hochberg's method to trimmed means and illustrate the methods using data from a psychological study.

It is noted that a closely related method was derived by Tamhane (1977). One difference between the two methods is that Hochberg used the usual sample means when computing a confidence interval, while Tamhane used a generalized sample mean. It is not evident from a theoretical point of view how Tamhane's method might be properly generalized to trimmed means. Because one of the goals here is to describe an extension to trimmed means, Tamhane's method is not considered further. Another difference is that Tamhane's method yields confidence intervals having length $2m$ (where m is chosen by the investigator), while Hochberg's method produces variable-length confidence intervals where the lengths are guaranteed not to exceed $2m$.

Interest in trimmed means stems from the fact that its standard error can be substantially smaller than the standard error of a sample mean under arbitrarily small departures from normality toward a heavy-tailed distribution (see, for example, Staudte and Sheather (1990)). In practical terms, using a trimmed mean can result in much higher power and shorter confidence intervals. Single-stage procedures for trimmed means are known to provide reasonably accurate probability coverage over a much wider range of situations compared to well-known methods for means (see Wilcox (1997)). However, little has been done to extend methods based on trimmed means to two-stage sampling schemes. Wilcox and Keselman (2002) have examined a variety of approaches for estimating the power of a test aimed at comparing trimmed means, but many of the more intuitive methods have been found to be unsatisfactory. The power issue is not directly addressed here, but rather indirectly by attempting to control the length of a confidence interval. There are other robust estimators with excellent properties, but trimmed means are especially convenient for the situation at hand. In particular, even when the sample is from an asymmetric distribution, the influence function of a trimmed mean has a relatively simple form which makes it fairly straighforward to derive a reasonable extension of Hochberg's technique. Robust M-estimators, for example, have an influence function that takes on a very complicated form when dealing with an asymmetric distribution. R-estimators have excellent properties when one is sampling from symmetric distributions, but practical concerns arise when samples are from asymmetric distributions instead See, for example, Huber (1981) and Bickel and Lehmann (1975). Morgenthaler and Tukey (1991, p. 15) noted some concerns even with a symmetric distribution.

18.2 APPLICATION TO A SCHIZOPHRENIA STUDY

The question of interest was whether four groups differed in terms of skin resistance (measured in Ohms) following the presentation of a generalization stimulus. The four groups consisted of subjects with 1) no schizophrenic spectrum disorder, 2) schizotypal or paranoid personality disorder, 3) schizophrenia, predominantly negative symptoms, and 4) schizophrenia, predominantly positive symptoms. The data are reproduced in Table 18.2.1 and were supplied by S. Mednick. Pairwise comparisons based on means found no significant differences at the 0.05 level when using any of several multiple comparison procedures. A concern was that a boxplot revealed outliers in three of the four groups which inflated the corresponding sample variances, and this suggested that the power might be low due to relatively large standard deviations. Applying Hochberg's method with the goal of performing all pairwise comparisons using confidence intervals of length at most 0.6, the required sample sizes turned out to be 35, 10, 13, and 12. So, for example, the first group required an additional $35 - 10 = 25$ observations to ensure that the length of the confidence interval was at most 0.6.

Switching to 20% trimmed means, 1 and 6 additional observations are required from groups 1 and 2, respectively, and no additional observations are required from groups 3 and 4. Moreover, a significant difference between the population 20% trimmed means for groups 3 and 4 was found; the 0.95 confidence interval being $(-0.33, -0.14)$. In fairness, a single-stage procedure based on 20% trimmed means (Wilcox (1997), Section 6.2) also rejects the null hypothesis. However, often it is difficult to obtain second stage data, so any method that reduces the required number of observations beyond those available has practical value. In the present application, 0.6 is a relatively large length for the confidence intervals.

Table 18.2.1: Measures of Skin Resistance for Four Groups

(No Schiz.)	(Schizotypal)	(Schiz. Neg.)	(Schiz. Pos.)
0.49959	0.24792	0.25089	0.37667
0.23457	0.00000	0.00000	0.43561
0.26505	0.00000	0.00000	0.72968
0.27910	0.39062	0.00000	0.26285
0.00000	0.34841	0.11459	0.22526
0.00000	0.00000	0.79480	0.34903
0.00000	0.20690	0.17655	0.24482
0.14109	0.44428	0.00000	0.41096
0.00000	0.00000	0.15860	0.08679
1.34099	0.31802	0.00000	0.87532

If it is reduced to 0.4, then the required sample sizes are 78, 16, 28 and 27 when comparing means. Switching to 20% trimmed means, the required sample sizes are 23, 35, 10 and 10. So, groups 1, 3 and 4 require fewer observations in the second stage when comparing 20% trimmed means, but group 2 (which has no outliers according to a boxplot) requires more.

18.3 METHODOLOGY

Hochberg's (1975) method is applied as follows. Imagine that n observations are randomly sampled from each of J normal distributions and let s_j^2 be the usual sample variance associated with the jth group, $j = 1, \ldots, J$. The focus here is on all pairwise comparisons among the groups and in particular, the goal is to construct confidence intervals with simultaneous probability coverage at least $1 - \alpha$, and each confidence interval is to have length at most $2m$. Let h be the $1 - \alpha$ quantile of the range of J independent Student's t variates, each having $n - 1$ degrees of freedom. Tables of these quantiles can be found in Wilcox (1983). Let

$$d = (m/h)^2.$$

Then the total number of observations needed from the jth group is given by

$$N_j = \max\left\{n, \left\langle s_j^2/d \right\rangle + 1\right\}, \qquad (18.3.1)$$

where $\langle x \rangle$ is the value of x rounded down to the nearest integer.

The second stage of Hochberg's method consists of sampling an additional $N_j - n$ observations from the jth group and computing the sample mean, $\bar{X}_j$, based on all N_j values. For all pairwise comparisons, the confidence interval for $\mu_j - \mu_k$ is:

$$(\bar{X}_j - \bar{X}_k) \pm hb,$$

where

$$b = \max\left(s_j/\sqrt{N_j},\ s_k/\sqrt{N_k}\right).$$

Hochberg actually derived a more general method that would apply to a collection of linear contrasts, but the details are not crucial for the present purposes. Also, following Hochberg, the use of h is intended as an approximation to the augmented range (as defined in Miller (1966, Chapter 2)).

For a random sample $X_1, \ldots, X_n$, the γ-trimmed mean is

$$\bar{X}_t = (n - 2g)^{-1}(X_{(g+1)} + \cdots + X_{(n-g)}),$$

where $X_{(1)} \leq \cdots \leq X_{(n)}$ are the usual order statistics and $g = \langle \gamma n \rangle$, with $\langle \gamma n \rangle$ being the value of γn rounded down to the nearest integer. Various studies suggest that in terms of efficiency, $\gamma = 0.2$ is a good choice for general use. But, of course, exceptions will arise (for example, Wilcox (1997)). Let

$$\mu_t = \frac{1}{1 - 2\gamma} \int_{x_\gamma}^{x_{1-\gamma}} x dF(x),$$

be the population trimmed mean where x_γ and $x_{1-\gamma}$ are the γ and $1 - \gamma$ quantiles. Then

$$\bar{X}_t = \mu_t + \frac{1}{n} \sum_{i=1}^{n} IF_t(X_i),$$

plus a remainder term that goes to zero as n gets large at the usual rate of $1/\sqrt{n}$, where

$$(1 - 2\gamma)IF_t(X) = \begin{cases} x_\gamma - \mu_w, & \text{if } x < x_\gamma \\ X - \mu_w, & \text{if } x_\gamma \leq X \leq x_{1-\gamma} \\ x_{1-\gamma} - \mu_w, & \text{if } x > x_{1-\gamma}, \end{cases}$$

and

$$\mu_w = \int_{x_\gamma}^{x_{1-\gamma}} x dF(x) + \gamma(x_\gamma + x_{1-\gamma})$$

is the population Winsorized mean (for example, Huber (1981)). This suggests the following generalization of Hochberg's method to trimmed means.

The Winsorized values corresponding to $X_1, \ldots, X_n$ are

$$W_i = \begin{cases} X_{(g+1)}, & \text{if } X_i \leq X_{(g+1)} \\ X_i, & \text{if } X_{(g+1)} < X_i < X_{(n-g)} \\ X_{(n-g)}, & \text{if } X_i \geq X_{(n-g)}. \end{cases}$$

The Winsorized sample mean is

$$\bar{X}_w = \frac{1}{n} \sum_{i=1}^{n} W_i,$$

and the Winsorized sample variance is

$$s_w^2 = \frac{1}{n-1} \sum_{i=1}^{n} (W_i - \bar{X}_w)^2.$$

From Tukey and McLaughlin (1963), a good approximation of the distribution of

$$\sqrt{n}(1 - 2\gamma)s_w^{-1}(\bar{X}_t - \mu_t)$$

is a Student's t distribution with $n - 2g - 1$ degrees of freedom. In the present context, this suggests that the number of observations required from the jth group is approximately

$$N_j = \max\left\{ n, \left\langle s_{jw}^2(1 - 2\gamma)^{-2}d^{-1} \right\rangle + 1 \right\},$$

where s_{jw}^2 is the Winsorized sample variance from the jth group, and $d = (m/h)^2$. Now h is determined with degrees of freedom $\nu = n - 2g - 1$.

The second stage again consists of sampling an additional $N_j - n_j$ observations from the jth group. Let $\hat{\mu}_{jt}$ be the trimmed mean associated with the jth group based on all N_j observations. For all pairwise comparisons, the confidence interval for $\mu_{jt} - \mu_{kt}$, the difference between the population trimmed means corresponding to groups j and k, is

$$(\hat{\mu}_{jt} - \hat{\mu}_{kt}) \pm hb_t,$$

where

$$b_t = \max\left(\frac{s_{jw}}{(1 - 2\gamma)\sqrt{N_j}}, \frac{s_{kw}}{(1 - 2\gamma)\sqrt{N_k}} \right).$$

18.4 SIMULATION RESULTS

A serious concern with well-known single-stage methods for drawing inferences about means is that the actual probability coverage can be substantially smaller than the nominal level (Wilcox (1997)). This section briefly reports some simulation results on Hochberg's method, the main point being that the probability coverage is always greater than or equal to $1 - \alpha$ for the situations considered, while the same is not true among many single-stage techniques that have been examined. The *familywise error rate* (FWE) refers to the probability of at least one Type-I error. Simulations also suggest that over a broad range of situations, the extension to 20% trimmed means performs reasonably well in terms of guaranteeing a FWE less than or equal to α. The main concern, particularly when using means, is that FWE can be substantially smaller than the nominal level when sampling from non-normal distributions or from normal distributions with unequal variances.

First consider $J = 4$ with sampling from normal distributions. Theory indicates that FWE will be at most α when working with means, but how far below α can it be and how does the extension to trimmed means perform with small n's? Consider $n = 20$ with equal variances. Of course, the properties of Hochberg's method are related to m, the maximum allowable half-length of the confidence intervals. So, when carrying out simulations, the strategy was to consider increasingly smaller values of m until situations with $N_j > 500$ were encountered, and then the strategy was to increase m until typically $N_j = n$. That is, a range of m values was determined in the case where the expected number of additional observations was close to zero or fairly large.

Observations were generated from g-and-h distributions (Hoaglin (1985)) which includes normal distributions as a special case. More precisely, if Z has a standard normal distribution,

$$X = g^{-1}\{\exp(gZ) - 1\} \exp(hZ^2/2),$$

has a g-and-h distribution where g and h are parameters that determine the third and fourth moments. If $g = 0$, which corresponds to a symmetric distribution, this last equation is taken to be

$$X = Z \exp(hZ^2/2).$$

So, $g = h = 0$ corresponds to a normal distribution. As h increases, tail thickness increases as well.

After generating observations for each group, N_j was computed
based on some value of m chosen in a manner previously described,
an additional $N_j - n$ observations were sampled, and then it was determined whether one or more of the resulting confidence intervals would
reject the hypothesis of equal means (or trimmed means, whichever
was appropriate). This process was repeated 10,000 times and the proportion of times one or more rejections was observed (call it $\hat{\alpha}$), was
used to estimate FWE. To consider the effects of unequal variances,
observations in the jth group were multiplied by σ_j. Here, attention
was focused on $\sigma = (\sigma_1, \sigma_2, \sigma_3, \sigma_4) = (1, 2, 3, 4)$ with $\alpha = 0.05$. When
dealing with skewed distributions, distributions were first centered by
substracting the appropriate measure of location so that when multiplying the observations in the jth group by σ_j, the hypothesis of equal
measures of location remained true.

When sampling from normal distributions, these were some $\hat{\alpha}$-values:

σ	m	$\bar{X}$	$\bar{X}_t$
$(1,1,1,1)$	0.5	0.049	0.067
	1.0	0.029	0.032
$(1,2,3,4)$	2.0	0.020	0.033
	4.0	0.012	0.024

In the above table, $n = 20$ was used. So, when using 20% trimmed
means, $\hat{\alpha} = 0.067$. But with $n = 40$, it was found that $\hat{\alpha} = 0.051$.
The general pattern, which is reflected by these results, is that as m
increases, $\hat{\alpha}$ decreases and it can drop well below the nominal 0.05 level
in some cases, particularly when using means. For a skewed, light-tailed
distribution $(g = 0.5, h = 0)$, some of the $\hat{\alpha}$ values were as follows:

σ	m	$\bar{X}$	$\bar{X}_t$
$(1,1,1,1)$	1	0.015	0.027
	2	0.010	0.021
	4	0.009	0.022
$(1,2,3,4)$	3	0.034	0.030
	4	0.031	0.029
	6	0.030	0.029

For both types of heavy-tailed distributions considered (namely
$(g, h) = (0, 0.5)$ which is symmetric and heavy-tailed, and $(g, h) = (0.5, 0.5)$ which is skewed and heavy-tailed) the $\hat{\alpha}$ values were generally less than 0.01 when using means, regardless of the value of m, and

slightly higher than 0.01 when using 20% trimmed means instead. So, in general, it seems that the actual FWE will be less than or equal to the nominal level, it can be considerably smaller, and in some situations where 20% trimmed means are being used, it can be quite close to 0.05. For normal distributions with equal variances and $n = 20$, using trimmed means results in $\hat{\alpha} = 0.067$, but all indications are that in general, FWE will not exceed the nominal level when using the extension of Hochberg's method to trimmed means.

18.5 CONCLUDING REMARKS

Of course, comparing trimmed means is not the same as comparing means, and arguments can be made that at least in some situations, means are more meaningful. For example, when comparing investment strategies, surely the total amount earned is more important than any trimmed sum. But simultaneously, there are well-known situations where the mean is a highly misleading representation of what is typical and some type of a robust estimator is deemed more appropriate. Also, when sampling from approximately symmetric distributions, trimmed means can have substantially smaller standard errors resulting in substantially smaller samples in the 2nd stage of a two-stage procedure.

It is well-known that heteroscedastic single-stage procedures for comparing means can have actual Type-I error probabilities well above the nominal level (Wilcox (1997)). Note that Hochberg's method can be turned into a single-stage procedure by setting $m = \infty$. Apparently, an advantage of this single-stage procedure is that Type-I errors well above the nominal level can be avoided in situations where other methods are unsatisfactory. But a negative feature is that the actual probability of a Type-I error can drop well below the nominal level suggesting relatively low power.

REFERENCES

[1] Bickel, P.J. and Lehmann, E.L. (1975). Descriptive statistics for nonparametric models II. Location. *Ann. Statist.*, **3** , 1045–1069.

[2] Hoaglin, D.C. (1985). Summarizing shape numerically: The g-and-h distributions. In *Exploring data tables, trends, and shapes* (D. Hoaglin et al., eds.), 461–515. Wiley: New York.

[3] Hochberg, Y. (1975). Simultaneous inference under Behrens-Fisher conditions: A two sample approach. *Commun. Statist.*, **4**, 1109–1119.

[4] Huber, P. (1981). *Robust Statistics*. Wiley: New York.

[5] Miller, R.G.,Jr. (1966). *Simultaneous Statistical Inference*. McGraw Hill: New York.

[6] Morgenthaler, S. and Tukey, J.W. (1991). *Configural Polysampling*. Wiley: New York.

[7] Staudte, R.G. and Sheather, S.J. (1990). *Robust Estimation and Testing*. Wiley: New York.

[8] Stein, C. (1945). A two-sample test for a linear hypothesis whose power is independent of the variance. *Ann. Statist.*, **16**, 243–258.

[9] Tamhane, A. (1977). Multiple comparisons in model I one-way ANOVA with unequal variances. *Commun. Statist. Theory Meth.*, **6**, 15–32.

[10] Tukey, J.W. and McLaughlin D.H. (1963). Less vulnerable confidence and significance procedures for location based on a single sample: Trimming/Winsorization 1. *Sankhya, Ser. A*, **25**, 331–352.

[11] Wilcox, R.R. (1983). A table of percentage points of the range of independent t variables. *Technometrics*, **25**, 201-204.

[12] Wilcox, R.R. (1997). *Introduction to Robust Estimation and Hypothesis Testing*. Academic Press: San Diego.

[13] Wilcox, R.R. and Keselman, H.J. (2002). Power analysis when comparing trimmed means. *J. Mod. Appl. Statist. Meth.*, **1**, 24–31.

Address for communiation:

RAND R. WILCOX, Department of Psychology, University of Southern California, Los Angeles, CA 90089-1061, U.S.A. E-mail: rwilcox@usc.edu

Chapter 19

Sequential Testing in the Agricultural Sciences

LINDA J. YOUNG
University of Florida, Gainesville, U.S.A.

19.1 INTRODUCTION

In the United States, agricultural producers (farmers) have been able to increase the yield of crops, primarily through the use of hybridization, commercial fertilizers, and chemical pesticides (Young (2000)). When chemical pesticides were first developed, they were inexpensive and generally effective, leading to their extensive application. Some pests, particularly insects, began to develop resistance to the pesticides. New pesticides were developed as insects continued to show resistance. Concerns about the possible deleterious effects of pesticide use on human health and the environment were also raised. Many began to question whether pesticide use could be limited while still protecting the crops. See Young (2000) or Mulekar and Young (2003) for a fuller discussion of the issues.

Introduced in the 1960s, *integrated pest management* (IPM) strives for long-term prevention of pests (such as insects and weeds) and the damage they cause in a manner that minimizes chemical use. The agricultural system is viewed as a whole. A combination of techniques, including biological control, habitat manipulation, modification of cultural practices, and introduction of resistant varieties, is used to con-

trol pest populations. Careful monitoring of potential pests is conducted throughout a growing season. Chemicals are applied only if this
monitoring indicates that a pest population has exceeded the *economic
threshold*, that is, the population density or damage level at which the
loss in revenue is anticipated to exceed the cost of control. Pest control
materials are chosen and applied in a manner that minimizes risks to
human health, beneficial and nontarget organisms, and the environment.

Statistically, the challenge is to determine whether or not some insect or weed population has reached the economic threshold. Numerous sampling plans have been proposed for this purpose. Here we review the role of sequential hypothesis testing in this context. Oakland
(1950) and Morgan et al. (1951) presented early applications of Wald's
(1945,1947) *sequential probability ratio test* (SPRT) in the biological
sciences. After Waters (1955) provided details and examples of its application in forest insect studies, SPRT was used more often. In the
1960's, SPRT was widely applied by farmers in integrated pest management programs. Today, in IPM programs, SPRT and related sampling
plans form the foundation for determining whether or not a specified
region, usually a farmer's field, needs to be treated with pesticides so
that the crop's yield is protected from insects and/or weeds (Kogan and
Herzog (1980), Pedigo and Buntin (1994), Kuno (1991), Binns and Nyrop (1992)). These sequential testing methods are illustrated through
the development of sequential sampling plans for making treatment decisions in the context of controlling greenbugs (*Homoptera: Aphididae*)
on sorghum (*Sorghum bicolor* L.). Practical challenges and open issues
are discussed.

19.2   GREENBUGS ON SORGHUM

The greenbug was introduced into the United States around 1882. Initially, the damage caused by it was limited to grass and wheat. However, greenbugs have the potential to evolve into new biotypes and, in
1968, the first greenbug biotype adapted to sorghum causing economic
damage to this crop (Fernandes (1995)).

Although greenbugs continue to damage grass and wheat, they are
now the most important pest of sorghum in Nebraska and Kansas.
Both winged and wingless greenbug forms may be found. All wingless
greenbugs are female who give birth to living young, and these young

greenbugs are female as well. Under optimal conditions (75°F), greenbugs can begin producing young seven days after birth, and one female
can produce about 80 offspring over a period of 25 days. Thus, under
ideal conditions, large numbers of greenbugs can be produced in a short
time (Wright et al. (1994)).

Greenbugs suck sap with their needle-like mouthparts and inject a
toxic saliva into plants. Their feeding causes leaf discoloration and kills
plant tissues. The number of greenbugs that a plant can host before
experiencing economic damage increases as the plant matures. For example, when the plant is in the 0 to 1 leaf stage, an average of 10 to
25 greenbugs per plant is a threatening level, and an average of 25 to
50 greenbugs per plant is a treatment level. For plants in late whorl
through soft dough stages, averages of 700 and 1000 per plant represent
threatening and treatment levels, respectively (Brooks et al. (2001)).
Here, sampling plans are developed for the 0 to 1 leaf stage.

Preliminary data from which the sampling plans are to be developed were collected from four Nebraska sorghum fields in 1997 and six
fields in 1998. Within each field, a strip of grain sorghum that was
anticipated to be susceptible to greenbugs and another strip that was
expected to be resistant to greenbugs were planted. On a sampling occasion, two transects running diagonal to the sorghum rows were taken
within each strip, one on each end of the field. For each transect, the
number of greenbugs on each of fifteen plants was recorded (the distance between every pair of plants being twenty steps). The data from
the two transects within a strip in a field were combined to make a sample of size thirty. Sampling occasions were at approximately one-week
intervals. At times, especially early or late in the growing season or
after insecticide application, greenbug populations were extremely low.
Samples with two or fewer observed greenbugs on the thirty plants are
not included in this analysis. Table 19.2.1 gives, for each field, the
number of samples considered here.

19.3 THE SPRT

Ease of application is a major consideration in proposing any sampling
method for agriculture. Usually the individual responsible for collecting the sample has substantial training in biology, but little statistical
expertise (if any at all). In this setting, being able to complete all
computations prior to data collection is a definite advantage, if not a

requirement. This need has a major impact on the development and choice of sampling protocols.

Table 19.2.1: Number of Sampling Occasions for Fields in Study

1997		1998	
Field*	Sampling Occasions	Field*	Sampling Occasions
Fillmore	10	Saunders 1	12
Gage 1	13	Saunders 2	12
Gage 2	16	Saunders 3	12
Saunders	16	Gage 2	8
		Gage 3	10
		Saline 1	8

*Fields are identified by Nebraska county. Numbering within a county facilitates record keeping.

First, consider Wald's (1945,1947) sequential probability ratio test (SPRT). To develop an SPRT, the form of the population distribution must be known. The Poisson distribution is usually the first model considered for count data. However, this distribution rarely fits biological data well (Taylor et al. (1978), Young and Young (1998)). Typically, in population dynamics studies, the variance exceeds the mean, indicating over-dispersion. Although several distributions have the over-dispersion property, the negative binomial distribution is the most commonly applied in both insect control studies (Kogan and Herzog (1980), Taylor (1984), Kuno (1991), Binns and Nyrop (1992)) and weed control studies (Berti et al. (1992), Wiles et al. (1992), Johnson et al. (1995,1996), Mulugeta and Boerboom (1999), Wiles and Schweizer (1999)). Because the mean is of primary interest, the parameterization suggested by Anscombe (1949) is the one most frequently encountered in the ecological literature and is adopted here:

$$P(X = x) = \binom{k + x - 1}{x - 1} \left(\frac{k}{\mu + k}\right)^k \left(\frac{\mu}{\mu + k}\right)^x, x = 0, 1, 2, \ldots \quad (19.3.1)$$

where $\mu > 0$ is the mean and $k > 0$ is an aggregation parameter. The variance is a quadratic function of the mean: $\sigma^2 = \mu + \mu^2/k$.

For the greenbug data, the available samples of size 30 are small for evaluating the population distribution. In addition, a review of the data clearly indicates that the numbers were often rounded to the closest five or ten as the counts became large, further complicating evaluation of the population distribution. The sample variance consistently exceeded the sample mean. In addition to being a common model for over-dispersed count data, the negative binomial distribution has served as a model for the numbers of greenbugs on wheat and sorghum (Elliot and Kieckhefer (1986), Fernandes (1995)). Thus, it is assumed that the negative binomial model is adequate for these greenbug data.

Knowledge, or a precise estimate, of k is required to completely specify the sequential hypothesis tests based on a negative binomial distribution, and the value of k must be common to both hypotheses. SPRT is fairly robust to misspecification of k. If k is underestimated, the test is conservative. If the estimate of k is larger than its true value, the error rates are larger than specified. As k becomes larger, the effect of misspecification is less pronounced (Hubbard and Allen (1991)).

Sometimes k is not a constant. Most frequently k tends to increase with the mean. In spite of that, if k stays relatively constant for the hypothesized means and values in-between them, the assumption of a common k may be reasonable for the purpose of testing the hypotheses. If k varies in this region, a conservative test can be developed by using an estimate of k at the lower end of the range of these k values. Hefferman (1996) has suggested adjusting α and β so that a test based on a common k has the desired error rates if k varies with μ in a known manner or if the test is truncated. Taylor's power model, which uses the model $\sigma^2 = a\mu^b$, is a popular method of modeling the relationship between mean and variance (Taylor (1961,1965,1984)). In entomological applications, a fairly common practice is to fit the model, to estimate the variance for hypothesized values of the mean, and then to determine the k-value associated with that particular mean and variance (see Nyrop and Binns (1992)).

For the greenbug data, the method-of-moments estimator of k was first plotted against the sample mean. Although the estimates appeared fairly constant, they tended to be increasingly variable as the mean increased. Thus, estimation of a common k and evaluation of its adequacy for these data are considered next.

Several methods of estimating a common k have been proposed (Beall (1942), Anscombe (1949,1950), Bliss and Fisher (1953)). Although the maximum likelihood estimator has some nice optimality properties, it is not used here because it would be heavily affected by the rounding of some of the observed values mentioned earlier. Instead, the method suggested by Bliss and Owens (1958) is used. For the ith population sampled (i = 1, 2, ..., r), let n_i, s_i, and $\overline{X}_i$ be the sample

386   *Young*

size, estimated standard deviation, and sample mean, respectively. Define $x_i' = \overline{X}_i^2 - s_i^2/n_i$ and $y_i' = s_i^2 - \overline{X}_i$. If there is a k common to the r populations (call it k_c), the regression line of y' on x' will pass through the origin and have slope $1/k_c$. The precision of the estimate may be increased by weighting each population inversely to the variance; that is,

$$w_i = \frac{0.5(n_i - 1)k_c^4}{\mu_i^2 \left(\mu_i + k_c\right)^2 \left(k_c(k_c + 1) - (2k_c - 1)n_i^{-1} - 3n_i^{-2}\right)} \qquad (19.3.2)$$

Because the weight involves unknown parameters, including k_c, the estimation process is iterative. Using this weighted regression, a common k was estimated for the set of 117 samples and found to be 0.2. The estimated intercept was not significantly different from zero ($F = 1.1$, $df = 1,114$, $p = 0.2953$), and the estimated slope was significantly different from 0 ($F = 53.1$, $df = 1,114$, $p < 0.0001$). Thus, the remainder of this work is based on the assumption that a common k exists and is estimated to be 0.2.

Fields are sampled regularly during the growing season. If a decision is made not to treat a field for greenbugs, then it will not be checked again for a period of time, usually three to seven days. Thus, if a decision is made not to treat, it is important for the current infestation level to be low enough for it to be unlikely for the population to dramatically exceed the economic threshold before the field is evaluated again. For this reason, the null hypothesis of interest is that the mean number of greenbugs per plant is no more than a given safety level. For the 0 to 1 leaf stage, the lower limit of the treatment level is taken as the economic threshold, and the lower limit of the threatening level is taken as the safety level. Therefore, the hypotheses of interest are

$$H_0 : \quad \mu = 10(= \mu_0) \qquad (19.3.3)$$
$$H_1 : \quad \mu = 25(= \mu_1).$$

A Type-I error occurs if it is decided that the greenbug population exceeds the economic threshold, but it does not. In that case a field will be treated unnecessarily, resulting in increased expense and a possible deleterious effect on the environment. A Type-II error occurs if a decision is made not to treat the field, but in fact the population exceeds the economic threshold. In this case, the farmer will have a loss in yield that exceeds the cost of insecticide application, thereby reducing income. Both errors have significant negative impacts, and the farmer would like the probability of committing either error to be as small as possible. However, as the probability of committing an error decreases, the sampling effort required to make a decision increases. Thus, as a proposed plan is evaluated, the error rates may be adjusted to keep the sampling effort at an affordable level. Because both types of error are a concern here, the Type-I and Type-II error rates (α and β, respectively) will be initially specified to be $\alpha = \beta = 0.05$. After the properties of the proposed sampling plan based on these error rates are investigated, an adjustment may be needed.

For the SPRT, a sequence of random observations $\{X_1, X_2, X_3, ...\}$ is drawn from one of the two hypothesized distributions resulting in a sequence of likelihood ratios:

$$\lambda_n = \prod_{i=1}^{n} f_X(x_i; \mu_1) / \prod_{i=1}^{n} f_X(x_i; \mu_0).$$

A decision is made and the test terminated after the nth observation if $\lambda_n \geq A$ (accept H_1) or $\lambda_n \leq B$ (accept H_0), where A and B are constants depending on the predetermined levels α and β. A and B are approximated by

$$A \approx (1 - \beta)/\alpha \text{ and } B \approx \beta/(1 - \alpha).$$

Sampling continues as long as $B < \lambda_n < A$.

In this form, computations are required after each observation to determine whether the stopping criterion is met. However, by taking the natural logarithms of each quantity in the inequality, $b = \log(B) < \log(\lambda_n) < \log(A) = a$, the inequality can be rewritten as bounds on the cumulative sum of the observations. Let T_n denote the sum of the first n observations. Computational formulae for the bounds have been widely published (for example, Waters (1955), Fowler and Lynch (1987), Young and Young (1998)). Using those for the negative binomial distribution, the decision procedure may be stated as:

(1) terminate the test, accept the null hypothesis, and do not treat the field for greenbugs if $T_n \leq -248.8 + 15.3n$;

(2) terminate the test, accept the alternative hypothesis, and take steps to control greenbugs if $T_n \geq 248.8 + 15.3n$; and

(3) continue sampling if $-248.8 + 15.3n < T_n < 248.8 + 15.3n$, $n = 1, 2, \ldots$

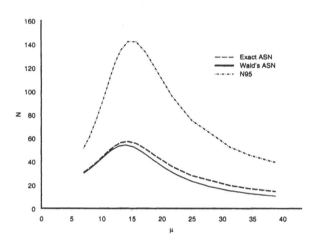

Figure 19.3.1: The Approximate and Exact ASN Functions Contrasted with the 95th Percentile of Sample Size Function for SPRT.

Wald (1945,1947) developed approximations of the *operating characteristic* (OC) and *average sample number* (ASN) functions for the SPRT. The accuracy of these approximations has been explored (for example, Morgan et al. (1951), Fowler and Lynch (1987)). In general, the OC function is approximated well enough for practical purposes. However, the approximation of the ASN function may be well below the true ASN function for parameter values intermediate to the hypothesized values. In addition, the distribution of the sample number tends to be highly skewed, resulting in some very large sample sizes. Corneliussen and Ladd (1970) developed an algorithm for rapid computation of the exact OC and ASN functions for a binomial distribution. Young (1994) extended this work to non-negative discrete distributions from the exponential family of distributions. Both approaches are incorporated into the ECOSTAT software (Young and Young (1998)). In addition, the 95th percentile of sample size is found so that practitioners can realistically assess whether enough resources are available to sample until a decision is made. For this example, the exact Type-I and

Type-II error rates of 0.06 and 0.04, respectively, are close to the nominal 0.05 values that are obtained through approximation. Although the differences are slightly larger for some sets of hypotheses, this is seldom of practical concern.

The approximate and exact ASN functions, as well as the 95th percentile of sample sizes for the hypotheses in (19.3.3) are shown in Figure 19.3.1. Here, the approximate ASN function only slightly underestimates the true ASN function. The much higher levels of the 95th percentile of sample sizes is a clear indication that the distribution of sample sizes is positively skewed. In general, sufficient resources should be available for sample sizes at the 95th percentile. As can be seen in Figure 19.3.1, the 95th percentile of sample size is less than 145 for all values of the mean. If a farmer is unable or unwilling to count the number of greenbugs on as many as 145 sorghum plants, the Type-I and/or Type-II error rates are increased, realizing the extra risk of error that this brings. Alternatively, if the farmer wants to devote more sampling efforts before making the management decision, then α and β can be decreased.

19.4   THE 2-SPRT

The SPRT is optimal in the sense that no other test of a simple null hypothesis versus a simple alternative with Type-I and Type-II error probabilities at most α and β, respectively, results in a smaller average sample size at the hypothesized values (Wald and Wolfowitz (1948)). However, since the mean may not assume one of the hypothesized values, the behavior of the ASN function over the full parameter space is of interest. As seen in Figure 19.3.1, the ASN function peaks at a point intermediate to the two hypothesized values, and the peak may represent a substantially larger average sample size than that at either hypothesized value. Kiefer and Weiss (1957) addressed this issue when they posed the problem of finding a test that minimizes the maximum average sample number (the minimax expected sample size) while controlling the Type-I and Type-II error probabilities. Lorden (1976) proposed the 2-SPRT as a means of minimizing the average sample number at a specified value of the parameter to within $o(1)$ as the error probabilities go to zero. Huffman (1983) extended this work by showing how to construct 2-SPRT's for testing in one-dimensional exponential families so as to attain the minimax expected sample size to within $o(\sqrt{E(N)})$.

Although Dragalin and Novikov (1987) refined Huffman's results to achieve the minimax expected sample size to within $O(1)$, Huffman's results are applied here. Consider the hypotheses $H_0 : \mu = \mu_0$ versus $H_1 : \mu = \mu_1(> \mu_0)$. Let μ^* be a value intermediate to μ_0 and μ_1 for which the ASN is to be minimized. Define a third hypothesis $H_2 : \mu = \mu^*$. A one-sided hypothesis of H_2 against H_0 is conducted for possible rejection of H_0. Simultaneously, another one-sided SPRT of H_2 against H_1 is conducted for the possible rejection of H_1. The decision boundaries for this test are two converging lines that produce a triangular continuation region. First, consider the 2-SPRT for a given μ^*. Then, how to choose μ^* so that the ASN is minimized asymptotically is discussed.

Let $X_1, X_2, X_3, \ldots$ be a random sample from a density belonging to the Koopman-Darmois family; that is, $f(x; \theta) = exp\{a(x) + \theta x - b(\theta)\}$ where $a(x)$ is a function of x alone and $b(\theta)$ is a function of θ alone. For the negative binomial distribution, $\theta = \log(\mu/(\mu + k))$, and the original set of hypotheses must be restated in terms of computational hypotheses involving θ to develop a test. The desired Type-I and Type-II error probabilities are specified to be α and β, respectively. Sampling continues until

(1) $T_n \geq h_1 n + S_1$ (accept H_1); or

(2) $T_n \leq h_0 n + S_0$ (accept H_0).

It remains to determine h_0, h_1, S_0, and S_1.

The Kullback-Leibler information numbers are:

$$I_i(\theta) = (\theta - \theta_i)b'(\theta) - [b(\theta) - b(\theta_i)], i = 0, 1.$$

Define $a_i(\theta') = (\theta' - \theta_i)/I(\theta'), i = 0, 1$. Since two SPRT's are to be conducted simultaneously, two likelihood ratios need to be formed. Let us define $\lambda_{0n} = \prod_{i=1}^{n} f_X(x_i; \theta_0) / \prod_{i=1}^{n} f_X(x_i; \theta^*)$ and $\lambda_{1n} = \prod_{i=1}^{n} f_X(x_i; \theta^*) / \prod_{i=1}^{n} f_X(x_i; \theta_1)$. The 2-SPRT takes the form:

(1) terminate the test, reject H_0, and accept H_1 if $\lambda_{0n} \leq A$;

(2) terminate the test, reject H_1, and accept H_0 if $\lambda_{1n} \geq B$; and

(3) otherwise, continue sampling.

The quantities A and B are chosen so that the desired error rates α and β are attained when both one-sided tests are conducted simultaneously. Ignoring the excess over the boundary, Huffman (1983) recommended for practical uses that

$$A(\theta) = \frac{a_0(\theta) - a_1(\theta)}{a_0(\theta)}\alpha, \tag{19.4.1}$$

and

$$B(\theta) = \frac{a_1(\theta) - a_0(\theta)}{a_1(\theta)}\beta. \tag{19.4.2}$$

The form of the distribution permits the quantities in the decision rule to be computed as follow: $h_0 = \log\{B(\theta^*)/(1 - A(\theta^*))\}/(\theta_1 - \theta^*)$, $h_1 = \log\{(1 - B(\theta^*))/A(\theta^*)\}/(\theta^* - \theta_0)$, $S_0 = (b(\theta_1) - b(\theta^*))/(\theta_1 - \theta^*)$, and $S_1 = (b(\theta^*) - b(\theta_0))/(\theta^* - \theta_0)$. This gives a solution to the modified Kiefer-Weiss problem. The maximum sample size for the SPRT occurs at the point of intersection of the two decision boundaries: $M = (S_1 - S_0)/(h_0 - h_1)$.

When the hypotheses concerning the binomial parameter p are symmetric about 0.5, the maximum average sample number occurs at the midpoint between p_0 and p_1, $p^* = 0.5$. However, the maximum of the ASN function does not occur at the midpoint between θ_0 and θ_1 when the distribution is skewed at that point. Hence, attention is now focused on determining the θ^* for which the ASN is a maximum. By constructing the 2-SPRT at that point, an asymptotic solution to the Kiefer-Weiss problem is obtained.

To obtain θ^*, θ' is first determined such that

$$\log(A(\theta'))^{-1}/I_1(\theta') = \log(B(\theta'))^{-1}/I_0(\theta'). \tag{19.4.3}$$

Let n' be the common value of the two sides in equation (19.4.3). Denote $a_i(\theta')$ as a_i'. Next find r' such that $\Phi(r') = a_i'/(a_1' - a_0')$ where Φ is the distribution function of a standard normal random variable. Also, for $\theta = \theta'$, we have $\sigma' = \sqrt{Var_{\theta'}(X)}$. The point θ^* for which the ASN is to be minimized may be expressed as $\theta^* = \theta' + r'/(\sigma'\sqrt{n'})$. The adjusted error rates based on $H_2 : \theta = \theta^*$ are computed by evaluating $A(\theta^*)$ and $B(\theta^*)$ as in equations (19.4.1) and (19.4.2). Finally the values needed to construct the decision boundaries for Huffman's extension of the 2-SPRT are as in the equations for h_0, h_1, S_0, and S_1 given above, with the computed value of θ^* and the corresponding

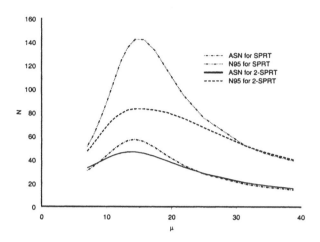

Figure 19.4.1: Comparison of the ASN and 95th Percentile of Sample Size Functions for the SPRT and the 2-SPRT.

corrected error probabilities. As mentioned earlier, this provides an asymptotic solution to the Kiefer-Weiss problem.

Consider again the greenbug data. For the hypotheses in (19.3.3), $\alpha = \beta = 0.05$, and $\widehat{k}_c = 0.2$, μ^* is 15.0. The decision rule is then:

(1) terminate the test, accept the null hypothesis, and do not treat the field if $T_n \leq -393.2 + 19.1n$; or

(2) terminate the test, accept the alternative hypothesis, and treat to control greenbugs if $T_n \geq 379.9 + 12.2n$; or

(3) continue sampling if $-393.2 + 19.1n < T_n < 379.9 + 12.2n$.

The maximum sample size is 110. The Type-I and Type-II errors are 0.06 and 0.04 for SPRT, and they are 0.03 and 0.07 for 2-SPRT. The ASN functions and the 95th percentile of sample sizes for SPRT and 2-SPRT are shown in Figure 19.4.1. Notice that, for 2-SPRT, slightly larger sample sizes are needed at the hypothesized values when compared to SPRT, but substantially smaller ones are needed at values in-between the hypothesized ones.

Mulekar et al. (1993) illustrated an application of 2-SPRT to insect populations. However, it continues to have only limited application compared to SPRT, probably because the computations involved are more complex. Young and Young (1998) have incorporated the 2-SPRT computations in the ECOSTAT software, and it is hoped that this will lead to its wider use.

19.5 BINOMIAL SAMPLING BASED ON THE NEGATIVE BINOMIAL

One of the practical challenges in collecting count data is to make accurate counts (number of greenbugs on a sorghum plant in this case). This is especially challenging when counts in excess of a thousand occur, which is not uncommon in heavily affected fields. Therefore, alternatives to counting every insect have long been sought. One of the earliest and most popular alternatives is presence/absence sampling. Two approaches to converting hypotheses about the mean to hypotheses about the proportion of infested plants are considered. The first is based on a negative binomial distribution. The other is founded on an ecological model.

Binns and Bostanian (1988) first suggested using the properties of a negative binomial distribution to establish hypotheses on the proportion p_T of the population with more than T individuals. See Wilson and Room (1983) for a discussion of presence-absence sampling without regard to a threshold T. If $T = 0$, this is simply presence/absence sampling, and the proportion p_T is computed as $p_T = 1 - (k/(\mu + k))^k$.

Consider again the greenbug data. Using $T = 0$, the properties of a negative binomial distribution and $k_c = 0.2$, the hypotheses in (19.3.3) can be written as

$$H_0: \quad p_T = 0.54 \qquad (19.5.1)$$
$$H_1: \quad p_T = 0.62.$$

These hypotheses can be tested using either SPRT or 2-SPRT. The decision rule for SPRT is:

(1) terminate the test, accept the null hypothesis, and do not treat if $T_n \leq -8.94 + 0.58n$;

(2) terminate the test, accept the alternative hypothesis, and treat the field if $T_n \geq 8.94 + 0.58n$; and

(3) continue sampling if $-8.94 + 0.58n < T_n < 8.94 + 0.58n$.

The decision rule for 2-SPRT is:

(1) terminate the test, accept the null hypothesis, and do not treat if $T_n \leq -13.90 + 0.60n$;

(2) terminate the test, accept the alternative hypothesis, and treat the field if $T_n \geq 14.08 + 0.56n$; and

(3) continue sampling if $-13.90 + 0.60n < T_n < 14.08 + 0.56n$.

Here, both SPRT and 2-SPRT have Type-I and Type-II error probabilities at the nominal 0.05 level. The functions for the ASN's and the 95th percentiles of sample sizes are shown in Figure 19.5.1. Notice that the ASN and the 95th percentile of sample size functions for 2-SPRT are substantially below those for SPRT for parameter values in-between the hypothesized ones. At the hypothesized values of the parameters, and at values more extreme than those, both functions are slightly higher for 2-SPRT than for SPRT. Comparing Figures 19.3.1 and 19.5.1, it is seen that larger sample sizes are required for presence/ absence sampling than when actual counts are recorded. This is to be expected since some information is lost when only presence/absence is recorded. Yet, when using presence/absence sampling instead of recording counts, the time needed to record an observation is less, and may be substantially less in heavily affected fields. An agriculturalist must decide which is a better approach.

Binns and Bostanian (1990) and Nyrop and Binns (1992) studied the behavior of these binomial tests when using Taylor's power model to obtain estimates of k. They found that imprecise knowledge of k can have a significant impact on both the OC and the ASN functions. To investigate the effect of imprecise knowledge about k here, a 95% confidence interval was set on k_c, and the behavior of the tests for values of k at the limits of this interval was explored. A 95% confidence interval on k_c derived from the regression that provided the estimate of 0.2 is (0.19, 0.32). As shown in Figures 19.5.2 and 19.5.3, the OC and ASN functions are affected by imprecise knowledge of k. The effect

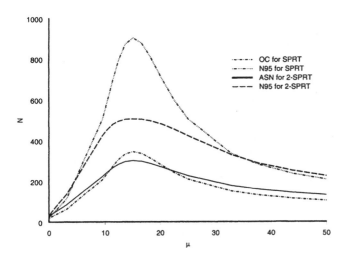

Figure 19.5.1: Comparison of the ASN and 95th Percentile of Sample Size Functions for the SPRT and the 2-SPRT Based on Binomial Probabilities.

on the 95th percentile of sample size is not shown here but is similar to that on the ASN function. If k is overestimated, the hypotheses in (19.5.1) correspond to smaller means than those in (19.3.3) causing the OC, ASN, and 95th percentile of sample size functions to shift to the left. For example, if $k = 0.32$, the upper limit of the 95% confidence interval, then the hypotheses in (19.5.1) are equivalent to $H_0 : \mu = 3.3$ versus $H_1 : \mu = 6.6$. Thus the OC and ASN functions associated with $k = 0.32$ in Figures 19.5.2 and 19.5.3 are shifted to the left of those for $k = 0.2$, which was used to set the hypotheses in (19.5.1) based on those in (19.3.3). Similarly, if k is overestimated, the OC, ASN, and 95th percentile of sample size functions shift to the right.

Binns and Bostanian (1990) proved that as T increases, the OC curves for minimum and maximum values of k will converge. If T is increased further, the curves will again begin to diverge. Although not proven by Binns and Bostanian, the same is true for the ASN function (Nyrop and Binns (1992)). Increasing the threshold T leads to increased sampling costs, and once again, an agriculturalist must decide whether this is worthwhile.

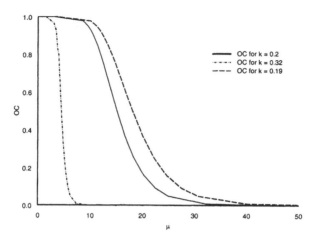

Figure 19.5.2: The OC Function for the 2-SPRT Based on Precise
Knowledge, Underestimation, and Overestimation of k.

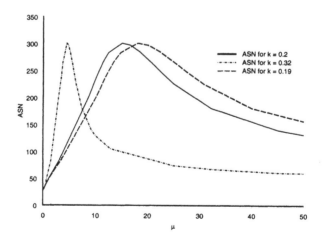

Figure 19.5.3: The ASN Function for the 2-SPRT Based on Precise
Knowledge, Underestimation, and Overestimation of k.

19.6   BINOMIAL SAMPLING BASED ON AN ECOLOGICAL MODEL

Gerrard and Chiang (1970) first proposed an ecological model relating the probability of observing a zero to the observed sample mean:

$$\log(-\log((1 - p_T))) = \gamma + \delta \log(m) + \epsilon \qquad (19.6.1)$$

where p_T is the proportion of sampling units with more than T individuals, m is the sample mean, γ and δ are estimated through regression, and ϵ is the error-term in the model. Linear regression is used to fit the model to the data (Nachman (1984), Nyrop et al. (1989), Binns and Bostanian (1990), Giles et al. (2000)). Model (19.6.1) has led to the development of sequential sampling plans based on a dichotomous response variable arising from count data. This will be illustrated for greenbugs in sorghum.

For the greenbug data, it was decided to use a tally threshold of zero corresponding to presence/absence sampling. A plot of $\log(-\log(1 - p_T))$ against $\log(m)$ indicated a strong linear relationship and the variability appeared to increase with $\log(m)$. To obtain a more accurate estimate for the range of means that is of interest, $\log(m)$ values in excess of four (m values in excess of 54.6) were omitted from the regression. The estimated model is

$$\log(-\log(1 - \hat{p}_T)) = -2.043 + 0.628 \log(m). \qquad (19.6.2)$$

The R^2 from this regression is 0.93. The standard error of the estimated slope is 0.031.

Using the hypotheses in (19.3.3) with the estimated linear relationship in equation (19.6.2) gives the set of hypotheses:

$$H_0 : \quad p_T = 0.42 \qquad (19.6.3)$$
$$H_1 : \quad p_T = 0.62.$$

It is interesting to compare the hypotheses in (19.6.3) to those in (19.5.1). Although H_1 is the same in both cases, H_0 differs. Now, either a binomial SPRT or 2-SPRT can be conducted. The OC, ASN, and 95th percentile of sample size functions can be computed as before. However, these do not account for the additional variability that is introduced through the modeling process.

Like the other sampling plans developed thus far, the performance in this case is influenced by binomial variation in the sample observations. However, this sampling plan is also affected by the variability in the modeled $p_T - m$ relationship (Schaalje et al. (1991), Nyrop and Binns (1992)). For each mean, there is a population of $\log(-\log(1 - p_T))$ values. Assume that these values are normally distributed about the mean of the $\log(-\log(1 - p_T))$ values and that this mean lies on the population regression line. Now, the variance of a predicted value of $\log(-\log(1 - p_T))$ for a specified m is:

$$\sigma^2_{\log(-\log(1-p_T))} = \sigma^2 \left(1 + \frac{1}{N} + \frac{(\log(m) - \overline{\log(m)})^2}{\sum(\log(m) - \overline{\log(m)})^2} \right) \qquad (19.6.4)$$

where σ^2 is the variance of the $\log(-\log(1 - p_T))$ about the regression line at a value of m, N is the total number of observations used to fit the line, and $\overline{\log(m)}$ is the average of the observed $\log(m)$ values. To assess the impact of the variability associated with fitting the regression model on the sampling plan, values of $\log(-\log(1 - p_T))$ at a series of distances from the fitted regression line are considered. For a specified $\log(m)$, these values are of the form $\log(-\log(1 - p_T)) \pm z s_{\log(-\log(1-p_T))}$ for $z = 0$, 0.125, 0.375, 0.625, 0.875, 1.125, 1.375, 1.625, 1.875, 2.125, 2.375, 2.625. The exact OC, ASN, and 95th percentile of sample size values for the 2-SPRT developed from this model are computed for each value of $\log(-\log(1 - p_T))$. Then the weighted averages of the functions are computed to obtain estimates of the expected OC, ASN and 95th percentile of sample size functions. The weights are the probabilities associated with the normal density for z-values between the midpoints of the computational z's. In addition, values associated with $\log(-\log(1 - p_T))$ values 1.625 standard deviations above and below the regression line are considered separately to represent extreme values. The results are shown in Figures 19.6.1 to 19.6.3. Notice that the expected OC, ASN, and 95th percentile of sample size functions differ markedly from those computed assuming only binomial variation. The expected Type-I and Type-II error probabilities are 0.18 and 0.17, respectively, and the extremes are even further from the nominal 0.05 values. Binns and Nyrop (1992) observed a similar behavior when applying the SPRT based on the model in equation (19.6.1).

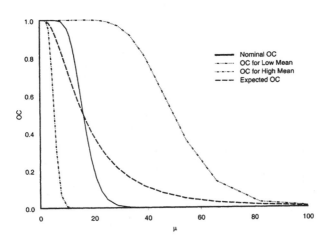

Figure 19.6.1: Comparison of the Nominal, Expected, Extreme Lower, and Extreme Upper Values of the OC Function for the 2-SPRT Involving the Hypotheses in (19.5.1).

19.7 CONCLUDING REMARKS

Wald's sequential probability ratio test has been widely implemented in integrated pest management. The 2-SPRT has been introduced to practitioners, but its application has been limited, probably due to computational complexity. Availability of software that facilitates these computations may speed its adoption. Emphasis here has been on the use of sequential tests in determining whether or not control of insects is needed. Although insect control decisions are the most common application of sequential methods in agriculture, some sequential methods have also been considered in making management decisions about the control of weeds (Johnson et al. (1995)).

The form of the population distribution must be assumed known in order to implement either the SPRT or the 2-SPRT. The negative binomial distribution often fits biological data well, perhaps because it can arise from several different biological models (Boswell and Patil (1970)). However, the form of the population distribution may change with the population mean. This has led some to question the use of probability models (see Taylor (1984)) and others to try to model the

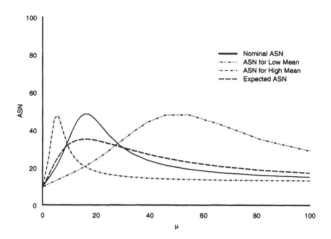

Figure 19.6.2: Comparison of the Nominal, Expected, Extreme Lower, and Extreme Upper Values of the ASN Function for the 2-SPRT Involving the Hypotheses in (19.5.1).

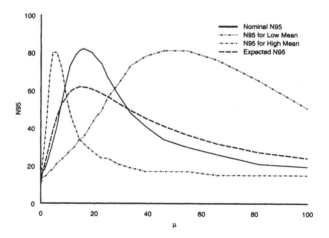

Figure 19.6.3: Comparison of the Nominal, Expected, Extreme Lower, and Extreme Upper Values of the 95th Percentile of Sample Size Function for the 2-SPRT Involving the Hypotheses in (19.5.1).

change in distribution with population mean (Binns (1986), Hefferman (1996)). Although the effect of imprecise knowledge about k on the properties of SPRT and 2-SPRT has been investigated, that of a departure from the assumed form of a negative binomial distribution has not.

One of the assumptions underlying sequential hypothesis testing is that the observations are independently and identically distributed. Possible spatial correlation of insect and weed populations has been investigated. Some pest populations have exhibited spatial dependence; others have not (Setzer (1985), Borth and Huber (1987), Liebhold and Elkinton (1989), Schotzko and O'Keeffe (1989), Liebhold et al. (1993), Cardina et al. (1995), Heisel et al. (1996), Johnson et al. (1996), Ellsbury et al. (1998), González-Andújar et al. (2001), Wright et al. (2002), Wyse-Pester et al. (2002)). The effect spatial dependence may have on current sampling programs has received some limited attention (Hoffmann et al. (1996)), but more investigation is warranted.

Attempts to reduce the total sampling effort have led to presence/absence sampling and binomial sequential tests. If the form of the population distribution is used to derive the hypotheses to be tested, the tests are highly dependent on precise modeling of the relationship between the mean μ and the proportion of infested plants p. If the population distribution is assumed to be negative binomial, the hypothesized means are effectively increased (decreased) if k is under-(over-) estimated.

Sequential tests based on modeled relationships between the proportion infested and the sample mean, such as the ecological model in equation (19.6.1), have been introduced. Scientists generally prefer these because no distributional assumptions are needed. As illustrated here, these model-based tests do not always have good statistical properties. The use of linear regression to estimate model parameters is questionable because $0 < p_T < 1$. Candy (2002) suggested fitting model (19.6.1) to data using a generalized linear model with the binomial response variable m, conditional on n, a complementary $\log - \log$ link function, and Williams's method III to account for over-dispersion. Comparisons of the OC functions based on linear and generalized linear models indicated that the linear model led to substantial underestimation of the probability of correct decisions and overestimation of the probability of incorrect decisions. The generalized linear model has not been used for model (19.6.1) in developing sequential sampling plans where n is variable. The effect of measurement-errors on model (19.6.1) should also be considered. Some work on this has been done (Schaalje and Butts (1992)), but only for fixed sample sizes. More research is needed in this area.

All of the work cited here focuses on testing a set of hypotheses for a specified species. However, two pests may simultaneously attack a crop, leading farmers to sample both at the same time. Binns and Nyrop (1992) suggested that this could be handled in one of three ways:

(1) Sampling plans could be constructed for the pest group instead of for each species (Hoy et al. (1983)).

(2) A sampling plan could be devised for the pest with the most variable sampling distribution, and this plan applied to all species.

(3) Dynamic sampling procedures based on estimated species composition and sampling distribution models for each species could be coded on a hand-held computer.

Further statistical development in these areas could be helpful.

Sometimes the growth of a pest population can be controlled if enough of its natural enemies, (for example, beneficial insects), are present (Young and Willson (1984)). This too leads to a need to simultaneously monitor more than one species. In addition to estimating the populations of each, use of the ratio of pests to natural enemies has been proposed (Hutchison et al. (1988)). Time-sequential sampling (Pedigo and van Shaik (1984)) provides a possible framework for sampling processes that are influenced by natural enemies (Binns and Nyrop (1992)). Here again the need for further statistical development is evident.

Determining whether or not chemical control of a pest is needed is an important decision, both economically and environmentally. Extensive efforts have been made to develop plans that are both statistically sound and feasible for a farmer to implement. Procedures requiring extensive field computations are unlikely to be implemented. In this setting, opportunities abound for statisticians to provide improved methods.

ACKNOWLEDGMENTS

Z.B. Mayo, Entomology Department, University of Nebraska, generously shared his greenbug data, which served as the foundation for the example in this work. Dr. Robert J. Wright, Entomology Department, University of Nebraska, provided helpful guidance on current practices in greenbug control.

REFERENCES

[1] Anscombe, F.J. (1949). The statistical analysis of insect counts based on the negative binomial distribution. *Biometrics*, **5**, 165-173.

[2] Anscombe, F.J. (1950). Sampling theory of the negative binomial and logarithmic series distributions. *Biometrika*, **37**, 358-382.

[3] Beall, G. (1942). The transformation of data from entomological field experiments so that the analysis of variance becomes applicable. *Biometrika*, **32**, 243-262.

[4] Berti, A., Zanin, G., Baldoni, G., Grignani, C., Mazzoncinni, M., Montemurro, P., Tei, F., Vazzana and Viggiani, P. (1992). Frequency distribution of weed counts and applicability of a sequential sampling method to integrated weed management. *Weed Res.*, **32**, 39-44.

[5] Binns, M.R. (1986). Behavioral dynamics and the negative binomial distribution. *Oikos*, **47**, 315-318.

[6] Binns, M.R. and Bostanian, N.J. (1988). Binomial and censored sampling in estimation and decision making for the negative binomial distribution. *Biometrics*, **44**, 473-483.

[7] Binns, M.R. and Bostanian, N.J. (1990). Robustness in empirically based binomial decision rules for integrated pest management. *J. Econ. Entomol.*, **83**, 420-427.

[8] Binns, M.R. and Nyrop, J.P. (1992). Sampling insect populations for the purpose of IPM decision making. *Annu. Rev. Entomol.*, **37**, 427-453.

[9] Bliss, C.I. and Fisher, R.A. (1953). Fitting the negative binomial distribution to biological data and note on the efficient fitting of the negative binomial. *Biometrics*, **9**, 176-200.

[10] Bliss, C.I. and Owens, A.R.G. (1958). Negative binomial distributions with a common *k*. *Biometrika*, **45**, 37-58.

[11] Borth, P.W. and Huber, R.T. (1987). Modelling pink bollworm establishment and dispersion in cotton with the Kriging technique. *Proc. Beltwide Cotton Prod. Res. Conf.*, 267-274. National Cotton Council of America: Memphis.

[12] Boswell, M.T. and Patil, G.P. (1970). Chance mechanisms generating the negative binomial distributions. In *Random Counts in Scientific Work* (G.P. Patil, ed.), **1**, 3-22. Pennsylvania State University Press: University Park.

[13] Brooks, H.L., Higgins, R.A. and Sloderbeck, P.E. (2001). *Sorghum Insect Management 2001*. Kansas State University: Manhattan.

[14] Candy, S.G. (2002). Empirical binomial sampling plans: Model calibration and testing using Williams' method III for generalized linear models with overdispersion. *J. Agric. Biol. Environ. Statist.*, **7**, 373-388.

[15] Cardina, J., Sparrow, D.H. and McCoy, E.L. (1995). Analysis of spatial distribution of common lambsquarters (*Chenopodium album*) in no-till soybean (*Glycine max*). *Weed Sci.*, **43**, 258-268.

[16] Corneliussen, A.H. and Ladd, D.W. (1970). On sequential tests of the binomial distribution. *Technometrics*, **12**, 635-646.

[17] Dragalin, V.P. and Novikov, A.A. (1987). Asymptotic solution of the Kiefer-Weiss problem for processes with independent increments. *Teor. Veroyatnost. I Primenen.*, **32**, 95-122 (in Russian).

[18] Elliot, N.C. and Kieckhefer, R.W. (1985). Cereal aphid populations in winter wheat: Spatial distributions and sampling with fixed levels of precision. *Environ. Entomol.*, **15**, 954-958.

[19] Ellsbury, M.M., Woodson, W.D., Clay, S.A., Malo, D., Schumacher, J., Clay, D.E. and Carlson, C.G. (1998). Geostatistical characterization of the spatial distribution of adult corn rootworm emergence. *Environ. Entomol.*, **27**, 910-917.

[20] Fernandes, O.A. (1995). Population Dynamics and Spatial Distribution of Lysiphlebus testacelpes (Cresson) (Hymenoptera: Braconidae) and its Host, the Greenbug, Schizaphis graminum Rondani (Homoptera: Aphididae): Studies for the Development of an Augmentation Program. *Ph.D. dissertation*, University of Nebraska-Lincoln.

[21] Fowler, G.W. and Lynch, A.M. (1987). Sampling plans in insect pest management based on Wald's sequential probability ratio test. *Environ. Entomol.*, **16**, 345-354.

[22] Gerrard, D.J. and Chiang, H.C. (1970). Density estimation of corn rootworm egg populations based upon frequency of occurrence. *Ecology*, **51**, 237-245.

[23] Giles, K.L., Royer, T.A., Elliott, N.C. and Kindler, S.D. (2000). Development and validation of a binomial sequential sampling plan for the greenbug (Homoptera: Aphididae) infesting winter wheat in the southern plains. *J. Econ. Entomol.*, **5**, 1522-1530.

[24] González-Andújar, J.L., Martinez-Cob, A., López-Granados, F. and García-Torres, L. (2001). Spatial distribution and mapping of crenate broomrape infestations in continuous broad bean cropping. *Weed Sci.*, **49**, 773-779.

[25] Hefferman, P.M. (1996). Improved sequential probability ratio tests for negative binomial populations. *Biometrics*, **52**, 152-157.

[26] Heisel, T., Andreasen, C. and Arsbøll, A.K. (1996). Annual weed distributions can be mapped with kriging. *Weed Res.*, **36**, 325-337.

[27] Hoffmann, M.P., Nyrop, J.P., Kirkwyland, Riggs, D.M., Gilrein, D.O. and Moyer, D.D. (1996). Sequential sampling plan for scheduling control of Lepidopteran pests of fresh market corn. *J. Econ. Entomol.*, **89**, 386-395.

[28] Hoy, C.W., Hennison, C., Shelton, A.M. and Andaloro, J.T. (1983). Variable intensity sampling: A new technique for decision making in cabage pest management. *J. Econ. Entomol.*, **76**, 139-143.

[29] Hubbard, D.J. and Allen, O.B. (1991). Robustness of the SPRT for a negative binomial to misspecification of the dispersion parameter. *Biometrics*, **47**, 419-427.

[30] Huffman, M.D. (1983). An efficient approximate solution to the Kiefer-Weiss problem. *Ann. Statist.*, **11**, 306-316.

[31] Hutchison, W.D., Poswal, M.A., Berberet, R.C. and Cuperus, G.W. (1988). Implications of the stochastic nature of Kuno's and Green's fixed-precision stop lines: Sampling plans for the pea aphid (Homoptera: Aphididae) in alfalfa as an example. *J. Econ. Entomol.*, **81**, 749-758.

[32] Johnson, G.A., Mortensen, D.A. and Gotway, C.A. (1996). Spatial and temporal analysis of weed seedling populations using geostatistics. *Weed Sci.*, **44**, 704-710.

[33] Johnson, G.A., Mortenssen, D.A., Young, L.J. and Martin, A.R. (1995). The stability of weed seedling population models and parameters in eastern Nebraska corn (*Zea Mays*) and Soybean (*Glycine max*) fields. *Weed Sci.*, **4**, 604-611.

[34] Kiefer, J. and Weiss, L. (1957). Some properties of generalized sequential probability ratio test. *Ann. Math. Statist.*, **28**, 57-75.

[35] Kogan, M. and Herzog, D.C. (1980). *Sampling Methods in Soybean Entomology*. Springer-Verlag: New York.

[36] Kuno, E. (1991). Sampling and analysis of insect populations. *Annu. Rev. Entomol.*, **36**, 285-304.

[37] Liebhold, A.M. and Elkinton, J.S. (1980). Characterizing spatial patterns of gypsy moth regional defoliation. *Forest Sci.*, **35**, 557-568.

[38] Liebhold, A.M., Rossi, M.R. and Kemp, W.P. (1993). Geostatistics and geographic information systems in applied insect ecology. *Annu. Rev. Entomol.*, **38**, 303-327.

[39] Lorden, G. (1976). 2-SPRT and the modified Kiefer-Weiss problem of minimizing the expected sample size. *Ann. Statist.*, **4**, 281-291.

[40] Morgan, M.E.P., MacLeod, P., Anderson, E.O. and Bliss, C.I. (1951). A sequential procedure for grading milk by microscopic counts. *Conn. (Storrs) Agric. Exp. Sta. Bul.*, **276**, 1-35.

[41] Mulekar, M.S. and Young, L.J. (2003). Sequential estimation in the agricultural sciences. In *Applied Sequential Methodologies* (N. Mukhopadhyay et al., eds.), 291-316, Marcel Dekker: New York.

[42] Mulekar, M.S., Young, L.J. and Young, J.H. (1993). Introduction of 2-SPRT for testing insect population densities. *Environ. Entomol.* **22**, 346-351.

[43] Mulugeta, D. and Boerboom, C.M. (1999). Seasonal abundance and spatial pattern of *Setaria faberi, Chenopodium albu,* and *Abutilon theophrasti* in reduced-tillage soybeans. *Weed Sci.*, **47**, 95-106.

[44] Nachman, G. (1984). Estimats of mean population density and spatial distribution of *Tetranychus urticas* (Acarina: Tetranychidae) and *Phytosiulus persimilis* (Acarina: Phytoseiidae) based upon the proportion of empty sampling units. *J. Appl. Ecol.*, **21**, 903-913.

[45] Nyrop, J.P., Agnello, A.M., Kovach, J. and Reissig, W.H. (1989). Binomial sequential classification of sampling plans for European red mite (Acari: Tetranychidae) with special reference to performance criteria. *J. Econ. Entomol.*, **82**, 482-490.

[46] Nyrop, J.P. and Binns, M.R. (1992). Algorithms for computing operating characteristic and average sample number functions for sequential sampling plans based on binomial count models and revised plans for European red mite (Acari: Tetranychidae) on apple. *J. Econ. Entomol.*, **65**, 1253-1273.

[47] Oakland, G.B. (1950). An application of sequential analysis to whitefish sampling. *Biometrics*, **6**, 59-67.

[48] Pedigo, L.P. and Buntin, G.D. (1994). *Handbook of Sampling Methods for Arthropods in Agriculture*, edited volume. CRC Press: Boca Raton.

[49] Pedigo, L.P. and van Schaik, J.W. (1984). Time-sequential sampling: A new use of the sequential probability ratio test for pest management decisions. *Bul. Entomol. Soc. Amer.*, **30**, 32-36.

[50] Schaalje, G.B. and Butts, R.A. (1992). Binomial sampling for predicting density of Russian wheat aphid (Homoptera: Aphididae) on winter wheat in the fall using a measurement error model. *J. Econ. Entomol.* **84**, 140-147.

[51] Schaalje, G.B., Butts, R.A. and Lysyk, T.J. (1991). Simulation studies of binomial sampling: a new variance estimator and density predictor, with special reference to the Russian wheat aphid (Homoptera: Aphididae). *J. Econ. Entomol.*, **84**, 140-147.

[52] Schotzko, D.J. and O'Keeffe, L.E. (1989). Geostatistical description of the spatial distribution of *Lygus hesperus* (Heteroptera: Miridae) in lentils. *J. Econ. Entomol.*, **82**, 1277-1288.

[53] Setzer, R.W. (1985). Spatio-temporal patterns of mortality in *Pemphigus populitransversus* and *P. populicaulis* on cottonwoods. *Oecologia*, **67**, 310-321.

[54] Taylor, L.R. (1961). Aggregation, variance and the mean. *Nature*, **189**, 732-735.

[55] Taylor, L.R. (1965). A natural law for the spatial disposition of insects. In *Proc. 12th International Congress on Entomology*, 396-397.

[56] Taylor, L.R. (1984). Assessing and interpreting the spatial distributions of insect populations. *Annu. Rev. Entomol.*, **29**, 321-357.

[57] Taylor, L.R., Woiwood, I.P. and Perry, J.N. (1978). The density-dependence of spatial behavior and the rarity of randomness. *J. Animal Ecol.*, **47**, 383-406.

[58] Wald, A. (1945). Sequential tests of statistical hypotheses. *Ann. Math. Statist.*, **16**, 426-482.

[59] Wald, A. (1947). *Sequential Analysis*. Wiley: New York.

[60] Wald, A. and Wolfowitz, J. (1948). Optimum character of the sequential probability ratio test. *Ann. Math. Statist.*, **19**, 326-339.

[61] Waters, W.E. (1955). Sequential sampling in forest insect surveys. *Forest Sci.*, **1**, 68-79.

[62] Wiles, L.J., Oliver, G.W., York, A.C., Gold, H.J. and Wilkerson, G.G. (1992). Spatial distribution of broadleaf weeds in North Carolina soybean (*Glycine max*) fields. *Weed Sci.*, **50**, 554-557.

[63] Wiles, L.J. and Schweizer, E.E. (1999). The cost of counting and identifying weed seeds and seedlings. *Weed Sci.*, **47**, 667-673.

[64] Wilson, L.T. and Room, L.T. (1983). Clumping patterns of fruit and arthropods in cotton, with implications for binomial sampling. *Environ. Entomol.*, **12**, 50-54.

[65] Wright, R., Danielson, S. and Mayo, Z.B. (1994). *Management of Greenbugs in Sorghum*. NebGuide G87-838-A. Cooperative Extension, Institute of Agriculture and Natural Resources, University of Nebraska, Lincoln.

[66] Wright, R.J., DeVries, T.A., Young, L.J., Jarvi, K.J. and Seymour, R.C. (2002). Geostatistical analysis of the small-scale distribution of European corn borer (Lepidoptera: Crambidae) larve and damage to whorl stage corn. *Environ. Entomol.*, **31**, 160-167.

[67] Wyse-Pester, D.Y., Wiles, L.J. and Westra, P. (2002). Infestation and spatial dependence of weed seedling and mature weed populations in corn. *Weed Sci.*, **50**, 54-63.

[68] Young, J.H. and Willson, L.J. (1984). A model to predict damage reduction to flower buds or fruit by *Heliothis spp.* in the absence or presence of two Coleoptera predators. *The Southwestern Entomol.*, **9**, 33-38.

[69] Young, L.J. (1994). Computation of some exact properties of Wald's SPRT when sampling from a class of discrete distributions. *Biometrical J.*, **36**, 627-635.

[70] Young, L.J. (2000). Production agriculture versus the environment: The role of statistics. *Great Plains Res.*, **10**, 89-106.

[71] Young, L.J. and Young, J.H. (1998). *Statistical Ecology: A Population Perspective*. Kluwer Academic: Boston.

Address for communication:

LINDA J. YOUNG, Department of Biostatistics, University of Florida, Gainesville, FL 32610-0212, U.S.A. E-mail: lyoung@biostat.ufl.edu

Chapter 20

Bayesian Sequential Procedures For Ordering Genes

SHELEMYAHU ZACKS
Binghamton University, Binghamton, U.S.A.

ANDRÉ ROGATKO
Fox Chase Cancer Center, Philadelphia, U.S.A.

20.1 INTRODUCTION: THE GENE ORDERING PROBLEM

One of the most important problems in genome mapping is that of ordering a set of genes, which are known to belong to the same chromosome. This is a very difficult task considering the fact that there are over 100,000 genes to order on twenty-three pairs of chromosomes in humans. The construction of the human genetic map was boosted by DNA sequencing (*restriction fragment-length polymorphism* (RFLP)) and new techniques in molecular genetics. For a good introduction, the reader is referred to the recent papers of Goradia and Lange (1990) and Karlin and Macken (1991). See also Boehnke et al. (1989), Buetow et al. (1991) and Donis-Keller et al. (1987).

The problem of gene ordering is complicated because generally, the order can only be inferred by observing possible recombinants from the genotypes of heterozygous individuals. Crossovers at the meiosis stage are random phenomena which occur with small probabilities. These crossover probabilities depend on the distance between the loci of genes on the chromosomes.

The likelihood functions depend on crossover probabilities. There are two possible models. Models of *independence* and models of *interference*. Interference models express the probabilities of recombination in terms of a genetic map distance w, $0 < w < \infty$. Such models have more biological appeal than the independence model (Ott (1988)). There are many models expressing recombination probabilities as functions of w. Such functions are described by Haldane (1919), Kosambi (1944) and Pascoe and Morton (1987). In the present chapter we restrict attention to the independence model. Similar results can be obtained for the interference models.

Gene orders are often compared by calculating the likelihoods at the maximum likelihood values of the parameters, under various gene orders. One then selects the order whose likelihood exceeds that of other orders by a large preset ratio (such as 100:1). Lathrop et al. (1987) wrote that the "interpretation of the likelihood statistics is hampered by the lack of a distribution theory that could lead to the calculation of significance levels". The standard asymptotic theory of likelihood ratio statistics does not apply to the comparison of orders.

It is well-known that fixed-sample-size procedures may not yield a definite decision concerning the correct order, especially if the number of observations is not sufficiently large. Sequential procedures, on the other hand, always yield a definite decision since one does not stop sampling before a definite decision is reached. However, if a stopping rule tends to terminate sampling too soon, then the associated error probabilities might be too high. Goradia and Lange (1990) studied sequential procedures for ordering three genes by sperm typing. They also provided a search algorithm for ordering more than three genes. Zacks (1994) studied theoretical approximations for the distributions of the stopping times recommended by Goradia and Lange.

It should be clear that, if there are g genes to order, the number of hypotheses to test is $g!/2$. This becomes an astronomical number very fast. We start therefore with the problem of ordering $g = 3$ genes and discuss an algorithm for ordering a large number of genes. The present article is based on articles by Rogatko and Zacks (1993), Babb et al. (1998) and Rogatko et al. (1999). Applications of the theory to genetic risk analysis were developed by Rogatko (1995), Rogatko et al. (1995) and by Rebbeck et al. (1997).

20.2 BASIC MODEL OF INDEPENDENCE

20.2.1 Orders, Gametes and Recombination

Suppose that A, B and C are labels of three gene loci on a given chromosome, and a, b, c are the corresponding loci on the paired chromosome. We distinguish between three possible orders of these loci: BAC, ABC or ACB (correspondingly bac, abc, or acb). The question is which order is the correct one. Let R_1 and R_2 be the chromosomal regions between the three loci. During meiosis, recombination of chromosomal regions might occur with small probabilities. A gamete is an unordered set of markers linked with the genes. Thus, if the original order was ABC, four types of gametes might be observed, according to the recombination patterns in R_1 and R_2, namely:

$$\{A, b, c\}, \{A, B, c\}, \{A, b, C\}, \text{ or } \{A, B, C\}.$$

The first one is due to recombination in R_1 and not in R_2. The second one is due to recombination in R_2 and not in R_1. The third shows recombinations in both R_1 and R_2, and $\{A, B, C\}$ shows recombination in neither R_1 nor R_2. However, if the order is BAC, a recombination in R_1 and not in R_2 yields the gamete $\{a, B, c\}$. The gamete $\{A, b, C\}$ is equivalent to $\{a, B, c\}$. If we observe this gamete, what can be said about the correct order? Obviously, to be able to draw reliable inference regarding the correct order, we need a large sample of gametes on which, recombination patterns can be observed unambiguously.

20.2.2 The General Case

Consider the case where g genes are to be ordered, $g > 3$. In such a situation, there are $g - 1$ regions (namely, $R_1, \cdots, R_{g-1}$) between the loci of these genes. We can think of 2^{g-1} possible crossover events. Let $(i_1, \cdots, i_{g-1})$, $i_j = 0, 1$, represent an event in which, $i_j = 1$ if there is a crossover in region R_j, and $i_j = 0$ if there is none ($j = 1, \cdots, g - 1$). Let $\nu = \sum_{j=1}^{g-1} i_j 2^{j-1}$ be the index of such a recombination event, $\nu = 0, \cdots, 2^{g-1} - 1$. In the case of $g = 3$, we have argued in the previous section that there are $2^{g-1} = 4$ types of observable gametes $\{A, B, C\}$, $\{A, b, C\}$, $\{A, B, c\}$ and $\{A, b, c\}$. In general, there are $m = 2^{g-1}$ observable types $\{A_1, A_2 \vee a_2, A_3 \vee a_3, \cdots, A_g \vee a_g\}$, where the symbol $\vee$ denotes the logical "or" operation ($i = 2, \cdots, g$). We say that a gamete is of type μ, $\mu = \sum_{j=1}^{g-1} l_j 2^{j-1}$, if the first locus is A_1 and the jth locus is either a_j if $l_j = 1$ or A_j if $l_j = 0$. If, on the other hand, the first locus is a_1, then the jth locus is A_j if $l_j = 1$ or it is a_j if $l_j = 0$ ($j = 2, \cdots, g - 1$). If we have a sample of n gametes, we denote by N_μ the number of gametes of type μ. Notice that N_0 denotes the number of gametes without any crossover.

In the case of g genes there are $k = g!/2$ possible orders of those genes, excluding the symmetrically equivalent orders. In general, if $(A_{i_1}, \cdots, A_{i_g})$ is a distinguishable order of g genes, let $\sigma(1, 2, 3, 4, \cdots, g) = (i_1, \cdots, i_g)$ be a corresponding order permutation. Let $\lambda(\tau, \nu)$ be the type of observable loci under order index τ and recombination index ν. In the following table we present the observable types of loci under certain different orders and different values of ν, for the case of $g = 4$. In this table we find, for example, that $\lambda(1, 3) = 3$ and $\lambda(11, 4) = 1$.

Table 20.2.1: Type Indices Mapping $\lambda(\tau, \nu)$

Recomb. Index				Gene order ($\tau = 0, 1, \ldots, 11$)					
ν	i_1	i_2	i_3	BACD		BADC	$\cdots$		DACB
0	0	0	0	(0) ABCD	(0)	ABCD		(0)	ABCD
1	1	0	0	(1) Abcd	(1)	Abcd		(4)	abcD
2	0	1	0	(2) ABcd	(2)	ABcd		(7)	AbcD
3	1	1	0	(3) abCD	(3)	abCD		(3)	abCD
4	0	0	1	(4) ABCd	(6)	ABcD		(1)	AbCD
5	1	0	1	(5) AbcD	(7)	abCd		(5)	abcD
6	0	1	1	(6) ABcD	(4)	ABCd		(6)	ABcD
7	1	1	1	(7) abCd	(5)	abcD		(2)	abCD

Finally, if p_ν ($\nu = 0, \cdots, 2^{g-1} - 1$) denotes the probability of the νth kind of recombination, the likelihood function of order index τ (when the n gametes are independent) is multinomial:

$$L(\tau, \mathbf{p}; \mathbf{N}, n) = C(n, \mathbf{N}) \prod_{\nu=0}^{2^{g-1}} p_\nu^N \lambda(\tau, \nu) \qquad (20.2.1)$$

where $C(n, N)$ is the multinomial coefficient.

In *models of independence* we assume that recombination in different regions are independent events. Let θ_i be the probability of recombination in region R_i, $(i = 1, \ldots, g - 1)$. Thus, if $\nu = \sum_{j=1}^{g-1} i_j 2^{j-1}$, then

$$p_\nu = \prod_{j=1}^{g-1} \theta_j^{i_j} (1 - \theta_j)^{1-i_j}. \tag{20.2.2}$$

Notice also that $0 < \theta_i \leq 0.5$ $(i = 1, \ldots, g)$, since crossovers can only take place between the two inner strands of the paired duplicated homologous chromosomes.

For example, under the independence assumption, the likelihood function of $(\tau, \theta_1, \theta_2)$ when $g = 3$ is:

$$L(\tau, \theta_1, \theta_2) = C(n, N) \sum_{j=0}^{2} I\{\tau = j\}$$
$$\times \theta_1^{J_1(j)} (1 - \theta_1)^{n - J_1(j)} \theta_2^{J_2(j)} (1 - \theta_2)^{n - J_2(j)}, \tag{20.2.3}$$

where the values $0, 1, 2$ of τ uniquely identify the orders BAC, ABC and ACB,

$$J_1(j) = \begin{cases} N_1 + N_3, & \text{if } j = 0 \text{ or } 1, \\ N_2 + N_3, & \text{if } j = 2, \end{cases} \tag{20.2.4}$$

and

$$J_2(j) = \begin{cases} N_2 + N_3, & \text{if } j = 0, \\ N_1 + N_2, & \text{if } j = 1 \text{ or } 2. \end{cases} \tag{20.2.5}$$

Moreover, it should be clear that

$$C(n, N) = \frac{n!}{N_1! N_2! N_3! (n - N_1 - N_2 - N_3)!}. \tag{20.2.6}$$

Models of interference express recombinations in terms of a genetic map distance w, $0 < w < \infty$, which is the mean number of exchanges in a given region. There are many models relating the probabilities p_ν with w. For example, if $f(w)$ is a mapping function and $g = 3$, we have:

$$p_1 = (f(w_1) - f(w_2) + f(w_1 + w_2))/2,$$
$$p_2 = (-f(w_1) + f(w_2) + f(w_1 + w_2))/2,$$
$$p_3 = (f(w_1) + f(w_2) - f(w_1 + w_2))/2.$$

Thus, for models of interference, the likelihood function can be expressed in terms of τ and w_1, w_2.

20.3   BAYESIAN TESTING

20.3.1   Posterior Order Probabilities

Generally, under a Bayesian model, if $L(\tau, \mathbf{p}; \mathbf{N}, n)$ is the likelihood function (20.2.1), the marginal predictive likelihood of τ is given by:

$$\tilde{L}(\tau; \mathbf{N}, n) = \iiint_{\Omega} L(\tau, \mathbf{p}; \mathbf{N}, n) h(\mathbf{p}; \tau) d\mathbf{p}, \qquad (20.3.1)$$

where

$$\Omega = \{(p_0, \ldots, p_{2^g-1}) : 0 < p_i, \sum_{i=0}^{2^g-1} p_i = 1\}, \qquad (20.3.2)$$

and $h(\mathbf{p}; \tau)$ is the *prior* density of $\mathbf{p} = (p_0, \ldots, p_{2^g-1})$. Furthermore, if π_τ denotes the prior probability of the order indexed by τ, the posterior probability of τ can be written as:

$$\pi(\tau \mid \mathbf{N}, n) = \frac{\tilde{L}(\tau; \mathbf{N}, n)\pi_\tau}{\sum_{j=0}^{k} \tilde{L}(j; \mathbf{N}, n)\pi_j}. \qquad (20.3.3)$$

A very sensitive issue related to the Bayesian analysis is the appropriate choice of a joint prior distribution for τ and $\boldsymbol{\theta}$. One way to avoid this problem and simplify computations, especially when the number of genes is larger than 3, is to replace $\tilde{L}(\tau; N, n)$ by $L^*(\tau; \hat{\boldsymbol{\theta}})$, where $\hat{\boldsymbol{\theta}} = (\hat{\theta}_1, \cdots, \hat{\theta}_{g-1})$ consists of the maximum likelihood estimators of $\theta_1, \ldots, \theta_g$. See Rogatko (1990). This approximation is effective in large samples. Then the posterior probabilities (20.3.3) can be approximated by

$$\pi(\tau; \mathbf{N}, n) \approx \left[1 + \sum_{j \neq \tau} \frac{\pi_j}{\pi_\tau} \frac{L^*(j, \hat{\theta}_1, \hat{\theta}_2)}{L^*(\tau, \hat{\theta}_1, \hat{\theta}_2)} \right]^{-1} \qquad (20.3.4)$$

Later, we present simulation results concerning the ordering of 4 genes. In this simulation study, the computations were performed by using the Jeffreys' prior and the large sample approximation (20.3.4). We see that the results are almost identical. We note that by using the large sample approximation, one can determine the posterior probabilities of different order hypotheses with the help of existing computer software for genetic map construction, such as CRIMAP (Green et al. (1990)).

20.3.2   Bayesian Sequential Testing

We are concerned with $k = g!/2$ order hypotheses: $H_1, H_2, \ldots, H_k$. Suppose that the loss function for accepting $H_{\tau'}$ when H_τ is true is

$$L(\tau, \tau') = CI\{\tau \neq \tau'\}, \tag{20.3.5}$$

where $C > 0$ is the penalty for accepting the wrong decision. Let

$$\pi_n^{(k)}(\mathbf{N}) = (\pi(1 \mid N, n), \ldots, \pi(k \mid N, n))$$

be the vector of posterior probabilities (20.3.3). The posterior risk after n observations reduces to

$$R(\tau; \pi_n^{(k)}(\mathbf{N})) = C \sum_{\tau' \neq \tau} \pi(\tau' \mid \mathbf{N}, n). \tag{20.3.6}$$

It follows immediately that the order which minimizes the posterior risk, given $(\mathbf{N}, n)$, is

$$\hat{\tau}_n = \text{arg. max } \pi(\tau \mid \mathbf{N}, n). \tag{20.3.7}$$

It was shown in Rogatko and Zacks (1993), and Babb et al. (1998) that if

$$\pi(\hat{\tau}_n \mid \mathbf{N}, n) \geq \pi^* \tag{20.3.8}$$

for some π^* close to 1, then it is optimal to stop sampling. After stopping, the hypothesis to accept is $H_{\hat{\tau}_n}$. We call this stopping rule the BSR rule. The optimal order at stopping is denoted by $\hat{\tau}_B$. The stopping time according to BSR is denoted by n_B.

Another Bayesian stopping rule can be based on predictive likelihood ratios. We present this rule and investigate its properties since it is applied in a somewhat simplified form in various studies on gene ordering (Donis-Keller et al. (1987) and Buetow et al. (1991)). Let $\Lambda_{ij}(\mathbf{N}, n)$ denote the predictive likelihood ratio (the Bayes factor) of $\tau = i$ relative to $\tau = j$, that is,

$$\Lambda_{ij}(\mathbf{N}, n) = \frac{\tilde{L}(i; \mathbf{N}, n)}{\tilde{L}(j; \mathbf{N}, n)}. \tag{20.3.9}$$

If for some τ^0, $\min_{\tau \neq \tau^0} \Lambda_{\tau^0 \tau}(\mathbf{N}, n) \geq \lambda^*$ for some critical value λ^*, then the decision $\hat{\tau} = \tau^0$ is made. The corresponding stopping rule is:

BLR: *Stop sampling at the first $n \geq n_0$, such that*

$$\max_{i=0,\cdots,k-1} \min_{j \neq i} \Lambda_{ij}(\mathbf{N}, n) \geq \lambda^*.$$

We call this rule a *maximin* likelihood ratio rule, and denote by S_{BLR} the stopping variable associated with BLR. We denote the Bayes decision based on this maximin rule by $\hat{\tau}_{BLR}$, and the corresponding stopping time by n_{BLR}. Notice that n_{BLR} is the first time that $\min_{j \neq \tau} \Lambda_{\tau,j}(N, n)$ $\geq \lambda^*$, for some τ.

In Table 20.3.1 we present the values of

$$\hat{\tau}_B \text{ and } \pi(\hat{\tau}_B \mid \mathbf{N}, n) = \max_{\tau=0,1,2} \pi(\tau \mid \mathbf{N}, n),$$

along with the values of

$$\hat{\tau}_{BLR} \text{ and } \Lambda^*(\mathbf{N}, n) = \max_{\tau=0,1,2} \min_{j \neq \tau} \Lambda_{\tau j}.$$

The data were simulated under the order $\tau = 0$, with $\theta_1 = \theta_2 = 0.2$. The prior probabilities of the three orders were assumed equal, and truncated Beta(0.5,0.5) distributions on $(0, 0.5) \times (0, 0.5)$ were employed as prior distributions for (θ_1, θ_2). We see that with $\pi^* = 0.95$ and $\lambda^* = 19$, BSR and BLR stop at $n = 46$, with $\hat{\tau}_B = \hat{\tau}_{BLR} = 0$.

In Table 20.3.2 we present the joint empirical frequency distribution of $\hat{\tau}_B$ and $\hat{\tau}_{BLR}$ as obtained from a sequence of 10,000 independent simulation runs, with $\theta_1 = \theta_2 = 0.2$, $\tau = 0$, $\pi_0 = \pi_1 = \pi_2 = 1/3$, $\pi^* = 0.95$ and $\lambda^* = 19$.

In Table 20.3.2, we see that the error probability of $\hat{\tau}_B$ under BSR is nearly 1.5%, whereas that of $\hat{\tau}_{BLR}$ under BLR is about 2.5%. In Table 20.3.3, we present the stopping times associated with these stopping rules, estimated from the simulation runs with $\theta_1 = \theta_2 = \theta$. Notice that BLR tends to stop slightly before BSR. This explains why the error probability of BLR is higher than that of BSR.

If τ^* is the correct order index, let P_{τ^*} denote the probability measure which, for each n (≥ 1) assigns $\mathbf{N}$ the following probability distribution:

$$P_{\tau^*}(N_1 = \nu_1, \cdots, N_m = \nu_m) = L(\tau^*; \nu, n) \qquad (20.3.10)$$

where $\nu = (\nu_1, \cdots, \nu_m)$, $\nu_i \geq 0$, $i = 1, \cdots, m$ and also $\sum_{i=1}^{m} \nu_i = n$.

Table 20.3.1: Decision Statistics $\pi(\tau_1^0 \mid \mathbf{N}, n)$, $\Lambda^*(\mathbf{N}, n)$
for BSR and BLR: $\pi^* = 0.95$ and $\lambda^* = 19$

n	N_1	N_2	N_3	$\pi(\tau_1^0 \mid \mathbf{N}, n)$	$\Lambda^*(\mathbf{N}, n)$	$\hat{\tau}_B$	$\hat{\tau}_{BLR}$
10	0	1	1	0.7435	5.7967	–	–
20	2	3	1	0.7414	16.4166	–	–
30	2	3	2	0.4595	5.6670	–	–
40	6	4	2	0.9193	11.9047	–	–
41	6	4	2	0.9241	12.6828	–	–
42	6	4	2	0.9284	13.4862	–	–
43	7	4	2	0.9154	10.9203	–	–
44	8	4	2	0.9000	9.0255	–	–
45	9	4	2	0.8835	7.5877	–	–
46	9	5	2	0.9509	19.4232	0	0
47	9	5	2	0.9548	21.1933	0	0
48	9	5	2	0.9584	23.0738	0	0
49	9	5	2	0.9615	25.0670	0	0
50	9	5	2	0.9644	27.1754	0	0

Table 20.3.2: Empirical Frequency Distribution
of $\hat{\tau}_B$ and $\hat{\tau}_{BLR}$

	$\hat{\tau}_{BLR}$			Total
$\hat{\tau}_B$	0	1	2	
0	9762	41	40	9843
1	1	72	0	73
2	0	0	84	84
Total	9763	113	124	10000

Let δ_{ij} denote the (predictive) probability of accepting the order hypothesis $\{\tau = i\}$ when the correct order is $\{\tau = j\}$. For each $i, j = 1, \cdots, k$, we have:

$$\delta_{ij} = \sum_{n \geq n_0} \sum_{\mathbf{N}} I\{n_s = n, \mathbf{N} \in D_i^{(n)}\} L(j; \mathbf{N}, n), \qquad (20.3.11)$$

where

$$D_i^{(n)} = \{\mathbf{N}; \pi(i \mid \mathbf{N}, n) \geq \pi^*\}, \ i = 1, \cdots, k. \qquad (20.3.12)$$

Table 20.3.3: Statistics of Stopping Variables
as Obtained from Simulation Runs

Procedure	θ	Mean	Median	Std. Dev.
	0.20	39.5	35	17.9
BSR	0.15	36.1	32	16.1
	0.10	37.8	33	17.8
	0.05	55.8	52	27.0
	0.01	153.5	122	108.2
	0.20	36.6	32	18.3
BLR	0.15	33.8	30	16.5
	0.10	35.8	31	18.0
	0.05	50.4	43	27.2
	0.01	150.6	121	110.7

Notice that, since $\pi^* > 0.5$, $D_1^{(n)}, \cdots, D_k^{(n)}$ are disjoint. Moreover, $\pi(i \mid \mathbf{N}, n) \geq \pi^*$ if and only if,

$$\sum_{j \neq i} \pi(j) L(j; \mathbf{N}, n) \leq \frac{1 - \pi^*}{\pi^*} \pi(i) L(i; \mathbf{N}, n). \tag{20.3.13}$$

Theorem 20.3.1 *If* $\lambda^* = \pi^*/(1 - \pi^*)$ *and* $\pi(i) = 1/k$, $i = 1, \ldots, k$, *then* $n_{BLR} \leq n_B$ *with probability one.*

Proof: According to (20.3.13), we have:

$$\sum_{j \neq \hat{\tau}_B} \Lambda_{j\hat{\tau}_B}(\mathbf{N}, n_B) \leq \frac{1 - \pi^*}{\pi^*} \tag{20.3.14}$$

or, in other words,

$$\max_{j \neq \hat{\tau}_B} \Lambda_{j, \hat{\tau}_B}(\mathbf{N}, n_B) \leq \frac{1 - \pi^*}{\pi^*}.$$

This is equivalent to claiming that

$$\min_{j \neq \hat{\tau}_B} \Lambda_{\hat{\tau}_B, j}(\mathbf{N}, n_B) \geq \frac{\pi^*}{1 - \pi^*} = \lambda^*. \tag{20.3.15}$$

Hence, $n_{BLR} \leq n_B$ with probability one. ∎

Theorem 20.3.2 *The predicted average error probability* $\bar{\epsilon}_\pi$ *associated with the BSR stopping rule is smaller than or equal to* $1 - \pi^*$.

Proof: According to (20.3.12), for every $i \neq j$, we have:

$$\delta_{ij} \leq \sum_{n \geq n^0} \sum_N I\{n_s = n, \mathbf{N} \in D_i^{(n)}\} \left[\frac{1 - \pi^*}{\pi^*} \frac{\pi(i)}{\pi(j)} \right.$$

$$\left. \text{x } L(i; \mathbf{N}, n) - \sum_{l \neq i \neq j} \frac{\pi(l)}{\pi(j)} L(l; \mathbf{N}, n) \right] \tag{20.3.16}$$

$$= \frac{1 - \pi^*}{\pi^*} \frac{\pi(i)}{\pi(j)} (1 - \epsilon_i) - \sum_{l \neq i \neq j} \frac{\pi(l)}{\pi(j)} \delta_{il}.$$

Summing the two sides of (20.3.16) over i ($i \neq j$) and multiplying by $\pi(j)$, we obtain:

$$\pi(j)\epsilon_j \leq \frac{1 - \pi^*}{\pi^*} \sum_{i \neq j} \pi(i)(1 - \epsilon_i) - \sum_{i \neq j} \sum_{l \neq i \neq j} \pi(l)\delta_{il}. \tag{20.3.17}$$

Hence, by summing both sides of (20.3.17) over j, one obtains

$$\bar{\epsilon}_\pi \leq \frac{1 - \pi^*}{\pi^*} \sum_j \sum_{i \neq j} \pi(i)(i - \epsilon_i) - \sum_j \sum_{i \neq j} \sum_{l \neq i \neq j} \pi(l)\delta_{il}. \tag{20.3.18}$$

The first term on the right hand side of (20.3.18) yields

$$\frac{1 - \pi^*}{\pi^*} \sum_j \sum_{i \neq j} \pi(i)(1 - \epsilon_i) = \frac{1 - \pi^*}{\pi^*} \sum_j (1 - \bar{\epsilon}_\pi - \pi(j)(1 - \epsilon_1))$$

$$= \frac{1 - \pi^*}{\pi^*} (k - 1)(1 - \bar{\epsilon}_\pi). \tag{20.3.19}$$

The second term on the right hand side of (20.3.18) is

$$-\sum_j \sum_{l \neq j} \sum_{l \neq i \neq j} \pi(l)\delta_{il} = -\sum_j \sum_{l \neq j} \pi(l) \sum_{i \neq l \neq j} \delta_{il}$$

$$= -\sum_j \sum_{l \neq j} \pi(l)(\epsilon_l - \delta_{jl}) = -\sum_j (\bar{\epsilon}_\pi - \pi(j)\epsilon_j) + \sum_j \sum_{l \neq j} \pi(l)\delta_{jl}$$

$$= -(k - 1)\bar{\epsilon}_\pi + \bar{\epsilon}_\pi = -(k - 2)\bar{\epsilon}_\pi \tag{20.3.20}$$

Indeed, $\sum_j \sum_{l \neq j} \pi(l)\delta_{jl} = \sum_l \pi(l) \sum_{j \neq l} \delta_{jl} = \sum_l \pi(l)\epsilon_l = \bar{\epsilon}_\pi$. Substituting (20.3.19) and (20.3.20) in (20.3.18) we obtain:

$$\bar{\epsilon}_\pi \leq \frac{1 - \pi^*}{\pi^*} (k - 1)(1 - \bar{\epsilon}_\pi) - (k - 2)\bar{\epsilon}_\pi \tag{20.3.21}$$

or, in other words,

$$\bar{\epsilon}_\pi \le \frac{1 - \pi^*}{\pi^*}(1 - \bar{\epsilon}_\pi) \qquad (20.3.22)$$

which is equivalent to $\bar{\epsilon}_\pi \le 1 - \pi^*$. ∎

The above theorem can also be proven by referring to Bechhofer et al. (1968, p. 20). The BSR has the Wald structure and $E_\tau\{n_B\} < \infty$ for all τ, where n_B denotes the stopping variable associated the BSR. Bechhofer et al. (1968) also provided formulae from which $E_\tau\{n_B\}$ can be approximated.

Next, let S be any stopping variable such that

$$P_\pi\{S < \infty\} = \sum_\tau \pi_\tau P_\tau\{S < \infty\} = 1.$$

Let $P_\pi(CD \mid S)$ denote the predictive probability of correct decision under S, and $\bar{\alpha}_\pi(S)$ denote the corresponding predictive error probability.

Theorem 20.3.3 *For any stopping variable S such that $P_\pi(S < \infty) = 1$, we have:*

$$\bar{\alpha}_\pi(S) > (1 - \pi^*)P_\pi\{S < n_B\}. \qquad (20.3.23)$$

Proof: On the set $\{S = n, S < n_B\}$ we have $\pi(\hat{\tau}_n \mid \mathbf{N}, n) < 1 - \beta$. Hence,

$$
\begin{aligned}
\bar{\alpha}_\pi(S) & = 1 - E_\pi\{\pi(\hat{\tau}_S \mid \mathbf{N}, S)I\{n_B \le S\}\} \\
& \quad - E_\pi\{\pi(\hat{\tau}_S \mid \mathbf{N}, S)I\{n_B > S\}\} \\
& > 1 - E_\pi[I\{n_B \le S\}] - \pi^* E_\pi[I\{n_B > S\}] \quad (20.3.24) \\
& = (1 - \pi^*)P_\pi\{n_B > S\}. \quad ∎
\end{aligned}
$$

Thus, if S is such that $P_\pi\{n_B > S\} = 1$ then $\bar{\alpha}_\pi(S) > 1 - \pi^*$ or $P_\pi(CD \mid S) < P_\pi(CD \mid n_B)$. In particular, we get $P_\pi(CD \mid n_{BLR}) < P_\pi(CD \mid n_B)$.

We have proven in this section that if all prior probabilities π_i are the same, then n_{BLR} is always smaller or equal to n_B. That is, BSR is a (somewhat) more conservative test procedure. Then, we have proven that the average predictive error probability $\bar{\alpha}_\pi(n_B)$ of n_B does not exceed $1 - \pi^*$. Finally, we have proven that the probability of correct ordering of BLR is smaller than that of BSR, that is, $\bar{\alpha}_\pi(n_{BLR}) > 1 - \pi^*$. These results were empirically validated in Table 20.3.2 and Table 20.3.3.

20.3.3   Bayesian Testing with Fixed Sample Sizes

In some applications, a statistician often relies upon available datasets for which the sample size is fixed. Babb et al. (1998) investigated testing multihypotheses under fixed sample sizes. Let n be the fixed sample size and $C_n = \left(\cup_{i=1}^{k} D_i^{(n)} \right)^c$ be the set on which a decision cannot be made. C_n is actually a continuation set in a sequential setup. Thus, the *fixed-sample Bayesian rule* (FSBR) is:

For $\pi^* > 0.5$, if $\mathbf{N} \in D_i^{(n)}$ the decision $\{\tau = i\}$ is taken (we accept H_i).

If $N \in C_n$, no decision is made.

Since $D_i^{(n)}$ $(i = 1, \ldots, k)$ are disjoint sets, the predictive probability of reaching a decision, when H_i is true, is given by

$$P_i(n) = \sum_{i=1}^{k} \sum_{\{\mathbf{N} \in D_j^{(n)}\}} \bar{L}(i; N, n). \qquad (20.3.25)$$

Notice that, for each i,

$$\lim_{n \to \infty} P_i(n) = 1. \qquad (20.3.26)$$

Thus, for sufficiently large n, $P_\pi(n) > 0$, where

$$P_\pi(n) = \sum_{i=1}^{k} \pi_i P_i(n). \qquad (20.3.27)$$

Let $\alpha_i(n)$ be the predictive probability of a wrong decision, given n, and

$$\bar{\alpha}_\pi(n) = \sum_{i=1}^{k} \pi_i \alpha_i(n). \qquad (20.3.28)$$

Denote the conditional predictive probability by $\tilde{\alpha}_\pi(n)$, that is,

$$\tilde{\alpha}_\pi(n) = \frac{\bar{\alpha}_\pi(n)}{P_\pi(n)}, \qquad (20.3.29)$$

provided that $P_\pi(n) > 0$.

Theorem 20.3.4 *For each fixed sample size n, such that $P_\pi(n) > 0$, we have $\tilde{\alpha}_\pi(n) \leq 1 - \pi^*$.*

For a proof of this theorem, see Babb et al. (1998). It is also proven there that, with any testing procedure ϕ and for a given n, if the inequality $\tilde{\alpha}_\pi(n, \phi) \le \tilde{\alpha}_\pi(n)$ holds, then $P_\pi(CD \mid n, FSBR) > P_\pi(CD \mid n, \phi)$. Here, $\tilde{\alpha}_\pi(n, \phi)$ denotes the predictive error probability of ϕ. This is the optimality of the FSBR under fixed sample sizes.

20.4   STEPWISE ORDERING

Since for $g = 10$ genes there are $(10!)/2 = 1,814,400$ hypotheses to test, it is impossible to evaluate simultaneously the posterior probabilities of all g genes, when g is large. Rogatko and Zacks (1993) proposed a stepwise Bayesian algorithm.

20.4.1   Stepwise Search For Maximal Posterior (SSMAP)

Step O:
  For given g genes (with g linked markers), consider the set $\mathcal{S}_3$ of all $t = \begin{pmatrix} g \\ 3 \end{pmatrix}$ triplets. Let $K, K = 1, \ldots, t$, be an index of a triplet, and Ω_3 be the set of three corresponding orders. Let $\pi^*(K)$ be the maximal *posterior probability* over Ω_3 and

$$K^* = \arg\max_{K \in \Omega_3} \pi(K).$$

Furthermore, let $K_3^0 = \arg\max_{K \in \mathcal{S}_3} \pi^*(K)$, and $\pi_3^0 = \pi(K_3^0)$. We choose the triplet corresponding to K_3^0 as the initial map, provided that $\pi_3^0 > \pi^*$. This condition is always satisfied if sampling of gametes is sequential and BSR is applied.

Step M ($M = 1, \ldots, g - 3$) :
  Let R_M be an index set of genes not yet mapped. For $j \in R_M$, there are $3 + M$ order hypotheses to test. Let $K^{(M,j)}$ be the order having maximal posterior probability among $3 + M$ hypotheses. Let $\pi(K^{(M,j)})$ be the corresponding posterior probability. Let $J_M \in R_M$ be the index maximizing $\pi(K^{(M,j)})$, that is,

$$\psi^{(M)} = \pi(K^{(M,J_M)}) = \max_{j \in R_M} \pi(K^{(M,j)}).$$

Table 20.4.1: Single-Stage Procedure
for Ordering Genes

| | | | | One Stage | | |
| | | | | $\hat{\tau} = 0$ | Stopping Times | |
θ_1	θ_2	θ_3	θ_4	(60,5)	Mean	Std. Dev.
0.10	0.10	0.10	0.10	9841	44.0	17.4
0.05	0.10	0.10	0.10	9804	59.8	31.9
0.10	0.05	0.10	0.10	9856	46.7	20.8
0.10	0.10	0.05	0.10	9818	46.5	20.5
0.10	0.10	0.10	0.05	9806	59.7	31.7
0.05	0.05	0.10	0.10	9850	56.6	25.9
0.10	0.05	0.05	0.10	9880	47.6	21.6
0.10	0.10	0.05	0.05	9819	56.4	26.0
0.05	0.10	0.05	0.10	9811	61.7	32.8
0.10	0.05	0.10	0.05	9821	61.5	32.7
0.05	0.10	0.10	0.05	9781	73.0	37.9
0.01	0.05	0.10	0.20	9666	159.4	117.0
0.20	0.10	0.05	0.01	9662	157.5	116.1
0.20	0.20	0.20	0.20	9884	51.1	17.8
0.30	0.30	0.30	0.30	9924	99.1	33.6

Table 20.4.1 (contd.): Two-Stage Procedure
for Ordering Genes

| θ_1 | θ_2 | θ_3 | θ_4 | Two Stages | | | |
| | | | | $\hat{\tau} = 0$ | | Stopping Times | |
				(12,4)	(5,5)	Mean	Std. Dev.
0.10	0.10	0.10	0.10	9852	9800	46.3	18.7
0.05	0.10	0.10	0.10	9805	9761	61.2	32.2
0.10	0.05	0.10	0.10	9879	9814	48.1	20.8
0.10	0.10	0.05	0.10	9804	9768	59.6	31.5
0.10	0.10	0.10	0.05	9817	9668	60.5	31.5
0.05	0.05	0.10	0.10	9855	9812	56.7	25.7
0.10	0.05	0.05	0.10	9859	9817	55.4	25.4
0.10	0.10	0.05	0.05	9824	9703	66.9	33.4
0.05	0.10	0.05	0.10	9771	9750	71.7	37.6
0.10	0.05	0.10	0.05	9886	9757	61.5	32.7
0.05	0.10	0.10	0.05	9804	9687	72.9	37.5
0.01	0.05	0.10	0.20	9671	9647	156.3	117.7
0.20	0.10	0.05	.01	9818	9501	155.4	115.3
0.20	0.20	0.20	0.20	9868	9821	51.1	19.0
0.30	0.30	0.30	0.30	9920	9882	95.0	34.9

Table 20.4.1 (contd.): Three-Stage Procedure
for Ordering Genes

| | | | | Three Stages | | | | |
| | | | | $\hat{\tau} = 0$ | | | Stopping Times | |
θ_1	θ_2	θ_3	θ_4	(3,3)	(4,4)	(5,5)	Mean	Std. Dev.
0.10	0.10	0.10	0.10	9877	9792	9743	47.9	19.5
0.05	0.10	0.10	0.10	9868	9806	9762	60.9	31.6
0.10	0.05	0.10	0.10	9844	9786	9734	60.7	32.8
0.10	0.10	0.05	0.10	9863	9709	9676	60.1	30.8
0.10	0.10	0.10	0.05	9866	9775	9631	61.7	31.5
0.05	0.05	0.10	0.10	9822	9770	9735	61.1	26.2
0.10	0.05	0.05	0.10	9842	9731	9700	65.5	32.7
0.10	0.10	0.05	0.05	9850	9713	9597	67.2	33.0
0.05	0.10	0.05	0.10	9847	9715	9690	70.8	37.0
0.10	0.05	0.10	0.05	9825	9788	9684	71.3	37.1
0.05	0.10	0.10	0.05	9827	9776	9659	72.6	37.0
0.01	0.05	0.10	0.20	9664	9640	9616	152.8	117.9
0.20	0.10	0.05	0.01	9722	9613	9302	156.7	114.8
0.20	0.20	0.20	0.20	9850	9787	9730	50.8	19.6
0.30	0.30	0.30	0.30	9864	9821	9776	91.3	34.6

At the end of the Mth stage, the gene with index J_M is embedded in the map, in the order $K^{(M,J_M)}$. The posterior probability of correct ordering, after M stages is

$$PCO^{(M)} = \pi_3^0 \prod_{j=1}^{M} \psi^{(j)}. \qquad (20.4.1)$$

The mapping is continued as long as $PCO^{(M)} \geq \pi^*$. The number of genes that can be ordered by SSMAP depends on the sample size n, the recombinant probabilities $\theta_1, \ldots, \theta_{g-1}$, and other factors. We remark that if the procedure is sequential so that in each stage, observations are added until $\psi^{(M)} \geq \pi^*$, then after M stages, $PCO^{(M)} \geq \pi^{*1+M}$. Thus, if we wish to map $g = 33$ genes by sequential sampling, then we need $\pi^* = 0.998$ in each stage to achieve $PCO^{(31)} = 0.95$. See Goradia and Lange (1990) for an example of sequential gene ordering.

The reader can find interesting results from simulation runs in Rogatko et al. (1999) which described the characteristics of SSMAP under fixed-sample-size designs.

In Table 20.4.1, we first present simulation results on ordering 5 genes in one stage (60 hypotheses). Then we provide simulation results for a two-stage procedure with four genes in the first stage (12 hypotheses) and 5 hypotheses in the second stage. Finally we provide the three-stage results that follow the SSMAP. We see that in all cases, the estimated probability of correct ordering turns out to be larger than $\pi^* = 0.95$.

20.5   APPLICATIONS IN GENETIC COUNSELING

The genetic risk of a disease, when markers are linked to the disease gene, is a function of inheritance, pedigree configuration, and genetic distances between the markers and the disease gene, among other factors. All these factors depend on parameters that have to be estimated from data, including probabilities of recombination in chromosomal regions and the order of the linked markers with respect to the disease gene. The question is how to estimate the genetic risk, that is, the probability that a person is a "carrier". Moreover, point estimates are insufficient. We have to provide *credible intervals* for the parameters.

Rogatko (1995) developed the theory and methodology for risk prediction with linked markers. The theory hinges on:

(i) developing a conditional density (p.d.f.) of the risk function, given an order between the disease gene, D, and linked markers, M and N;

(ii) determining the posterior probabilities of the three orders DMN, DNM and MDN.

If $g(r \mid DMN)$, $g(r \mid DNM)$ and $g(r \mid MDN)$ are the conditional risk densities given the gene orders, and $\pi(DMN)$, $\pi(DNM)$ and $\pi(MDN)$ are the corresponding posterior probabilities of these orders, then the risk density is:

$$g_\pi(r) = \pi(DMN)g(r \mid DMN) + \pi(DNM)g(r \mid DNM)$$
$$+ \pi(MDN)g(r \mid MDN). \qquad (20.5.1)$$

Since the risk depends only on the probabilities of recombination with the closest linked marker, we have $g(r \mid DMN) = g(r \mid DM)$, and $g(r \mid DNM) = g(r \mid DN)$.

Let θ_1 and θ_2 be the recombination probabilities in the region between D and M and that between D and N, respectively. The risk, as a function of (θ_1, θ_2), is obtained by an analysis of pedigree maps and other genetic information. Let $R = f(\theta_1, \theta_2)$. Notice that $f^{-1}(r) = \{(\theta_1, \theta_2) : f(\theta_1, \theta_2) = r\}$.

Let $\psi(\theta_1, \theta_2; \mathcal{D})$ be the posterior density of (θ_1, θ_2), given the data $\mathcal{D}$. The posterior distribution of R, given $\mathcal{D}$, is:

$$P\{R \le r \mid \mathcal{D}\} = P\{f_0(\theta_1, \theta_2) \le r \mid \mathcal{D}\}$$
$$= \iint\limits_{\{(\theta_1, \theta_2): f_0(\theta_1, \theta_2) \le r\}} \psi(\theta_1, \theta_2; \mathcal{D}) d\theta_1 d\theta_2. \qquad (20.5.2)$$

The risk function $f_0(\theta_1, \theta_2)$ depends on the order configuration, θ. For DMN or DNM it is a function of one variable, θ_1 or θ_2, only. One will find a fully worked-out example in Rogatko (1995). In that example, for DMN or DNM, we have

$$g(r \mid DM) = \frac{d}{dr} P\{R \le r \mid \mathcal{D}, DM\}$$
$$= \frac{l_{DM}\left(\dfrac{1 - \sqrt{2r - 1}}{2}\right)}{2\sqrt{2r - 1}}, \quad 0.5 \le r \le 1,$$

where
$$l_{DM}(\theta) = \frac{\theta^{N_1+N_3}(1-\theta)^{n-N_1-N_3}}{B_{0.5}(N_1+N_3+1, n-N_1-N_3+1)}$$
is the normalized likelihood, given $\mathcal{D} = \{\mathbf{N}, n\}$. Similarly, we can write
$$l_{DN}(\theta) = \frac{\theta^{N_2+N_3}(1-\theta)^{n-N_2-N_3}}{B_{0.5}(N_2+N_3+1, n-N_2-N_3+1)}.$$

The function $g(r \mid MDN)$ is much more complicated (Rogatko (1995)). The point is that, after determining the risk distribution $G_\pi(r) = \int_0^r g_\pi(r)dr$, the quantiles of this distribution yield *risk prediction intervals*. For more applications, see Rogatko et al. (1995) and Rebbeck et al. (1997).

REFERENCES

[1] Babb, J., Rogatko, A. and Zacks, S. (1998). Bayes sequential and fixed sample testing of multihypotheses. In *Asymptotic Methods in Probability and Statistics* (B. Szyszkowicz, ed.), 801-809. Elsevier Sciences: Amsterdam.

[2] Bechhofer, R.E., Kiefer, J. and Sobel, M. (1968). *Sequential Identification and Ranking Procedures*. University of Chicago Press: Chicago.

[3] Boehnke, M., Arnheim, N., Li, H. and Collins, F.S. (1989). Fine-structure genetic mapping of human chromosomes using the polymerase chain reaction on single sperm. Experimental design considerations. *Amer. J. Hum. Genet.*, **45**, 21-32.

[4] Buetow, K.H., Shiang, R., Yang, P., Nakamura, Y., Lathrop, G.M., White, R., Wasmuth, J.J., Wood, S., Berdahl, L.D., Leysens, N.J., Ritty, T.M., Wise, M.E. and Murray, J.C. (1991). A detailed multipoint map of human chromosome 4 provides evidence for linkage heterogeneity and position-specific recombination rates. *Amer. J. Hum. Genet.*, **48**, 911-925.

[5] Donis-Keller, H., Green, P., Helms, C., Cartinhour, S., Weiffenbach, B., Stephens, K., Keith, T.P., Bowden, D.W., Smith, D.R., Lander, E.S., Botstein, D., Akots, G., Rediker, K.S., Gravius, T., Brown,

V.A., Rising, M.B., Parker, L.C., Powers, J.A., Watt, D.E., Kauffman, E.R., Bricker, A., Phipps, P., Muller-Kahle, H., Fulton, T.R., Ng, S., Schumm, J.W., Braman, J.C., Knowlton, R.G., Barker, D.F., Crooks, S.M., Lincoln, S.E., Daly, M.J. and Abrahamson, J. (1987). A genetic linkage map of the human genome. *Cell*, **51**, 319-337.

[6] Goradia, T.M. and Lange, K. (1990). Multilocus ordering strategies based on sperm typing. *Ann. Hum. Genet.*, **54**, 49-77.

[7] Green, P., Falls, K. and Crooks, S. (1990). *CRI-MAP Documentation* (version 2.4).

[8] Haldane, J.B.S. (1919). The combination of linkage values, and the calculation of distance between the loci of linked factors. *J. Genet.*, **8**, 299-309.

[9] Johnson, N.L. and Kotz, S. (1983). *Encyclopedia of Statistical Sciences* (edited volume), **4**. Wiley: New York.

[10] Karlin, S. and Macken, C. (1991). Inhomogeneity of DNA sequence data. *J. Amer. Statist. Assoc.*, **86**, 27-35.

[11] Kosambi, D.D. (1944). The estimation of map distances from recombination values. *Ann. Eugen.*, **12**, 172-175.

[12] Lathrop, G.M., Chotai, J., Ott, J. and Lalouel, J.M. (1987). Tests of gene order from three-locus linkage data. *Ann. Hum. Genet.*, **51**, 235-249.

[13] Morton, N.E. (1955). Sequential tests for the detection of linkage. *Amer. J. Hum. Genet.*, **7**, 277-318.

[14] Ott, J. (1988). *Analysis of Human Genetic Linkage*. Johns Hopkins University Press: Baltimore.

[15] Ott, J. and Lathrop, G.M. (1987). Goodness-of-fit tests for locus order in three point mapping. *Genet. Epidemiol.*, **4**, 51-57.

[16] Pascoe, L. and Morton, N.E. (1987). The use of map functions in multipoint mapping. *Amer. J. Hum. Genet.*, **40**, 174-183.

[17] Rebbeck, T.R., Jordan, H.A., Schnur, H.E. and Rogatko, A. (1997). Utility of linked markers in genetic counseling: Estimation of carrier risks in X-linked ocular albinism. *Amer. J. Med. Genet.*, **70**, 58-66.

[18] Rogatko, A. (1990). Statistical inference in the gene order problem: generalized likelihood and Bayesian approaches. *Paper presented at the 7th Genetic Analysis Workshop*, Dayton, Ohio.

[19] Rogatko, A. (1995). Risk prediction with linked markers: Theory. *Amer. J. Med. Genet.*, **59**, 14-23.

[20] Rogatko, A., Babb, J., Jordan, H. and Zacks, S. (1999). Consructing miotic maps with known error probability. *Genet. Epidemiol.*, **16**, 274-289.

[21] Rogatko, A., Rebbeck, T. and Zacks, S. (1995). Risk prediction with linked markers: Pedigree analysis. *Amer. J. Med. Genet.*, **59**, 24-32.

[22] Rogatko, A. and Zacks, S. (1993). Ordering genes: Controlling the decision error probabilities. *Amer. J. Hum. Genet.*, **52**, 947-957.

[23] Zacks, S. (1994). The time until the first two order statistics of independent Poisson processes differ by a certain amount. *Commun. Statist.–Stoch. Models*, **10**, 853-866.

Addresses for communication:

SHELEMYAHU ZACKS, Department of Mathematical Sciences, Binghamton University, Binghamton, NY 13902-6000, U.S.A. E-mail: shelly @math.binghamton.edu
ANDRÉ ROGATKO, Department of Biostatistics, Fox Chase Cancer Center, Philadelphia, U.S.A. E-mail: A_Rogatko@fccc.edu

Author Index

A

Abraham, D.A. 2-6,16-17
Abrahamson, J. 411,417, 431
Agnello, A.M. 397,407
Akots, G. 411,417,431
Alexandrou, V.A. 108,121
Allen, O.B. 385,406
Allsopp, P.G. 298,305,314
Andaloro, J.T. 402,405
Andersen, J.S. 70,84
Andersen, P.K. 210,214
Anderson, E.O. 382,388,407
Anderson, T.W. 125,136,139
Andreasen, C. 401,405
Andrews, A.F. 70,83
Anscombe, F.J. 42,44-46,48,99,152,
    167,257,287,289,302,313,
    384-385,403
Aoshima, M. 26,33,273,289
Aras, G. 127,129-130,132-134,139
Armitage, P. 197,211,214,232,321,336
Arnheim, N. 411,430
Arnold, B.C. 21-22,34
Arpaia, S. 308,314
Arsbøll, A.K. 401,405
Astrachen, E. 36,51

B

Babb, J. 412,417,423-424,428,430,432
Backofen, R. 54,66
Bai, Z.D. 83
Balakrishnan, N. 105,119
Baldoni, G. 384,403
Bandyopadhyay, U. 78-80,83
Barker, D.F. 411,417,431
Barker, W.C. 36,48
Barón, M. 53-54,58,60-62,65,67
Bar-Shalom, Y. 2,17,220,223,238,
    244-245
Bartlett, R.H. 70,83
Basseville, M. 2,5,17,54,60,65,223, 225,
    244,339,367

Baumgartner, J. 294,313
Beall, G. 385,403
Bechhofer, R.E. 422,430
Begg, C.B. 78,83
Berberet, R.C. 402,406
Berdahl, L.D. 411,430
Berti, A 384,403
Berry, K.J. 36,48
BHAT 194-197,199,201,209-210,212-214
Bibby, J.M. 88,102
Bickel, P.J. 98,101,373,379
Bilias, Y. 202,214
Binns, M.R. 310,312,316,382,384-385,
    393-395,397-398,401-403,407
Bishop, T.A. 257
Biswas, A. 71,74,78-81,83
Blackburn, P.R. 210,217
Blažek, R. 354,357,367,369
Bliss, C.I. 303,313,382,385,388,403-404,
    407
Boehnke M. 411,430
Boerboom, C.M. 384,407
Bonato, 0. 294,313
Bonney, G.E. 75,84
Bogush, A.J. 106,119
Borth, P.W. 401,404
Bostanian, N.J. 393-395,397,403
Boswell, M.T. 301,313,399,404
Botstein, D. 411,417,431
Boutsikas, M.V. 108,119
Bowden, D.W. 411,417,431
Bradley, J.V. 37,48
Bradley, J.R.,Jr. 294,305,317
Braman, J.C. 411,417,431
Bratley, P. 267,290
Breiman, L. 86,88,101
Breslow, N. 213
Bricker, A. 411,417,431
Bridges, C.B. 54,65
Brillinger, D.R. 36,51
Brook, D. 5,17
Brooks, H.L. 383,404
Brown, A.W.A. 296,314
Brown, V.A. 411,417,431
Bryngelson, J.D. 53,58,67
Buetow, K.H. 411,417,430

Subject Index

E

T - #0022 - 111024 - C0 - 229/152/28 - PB - 9780367394561 - Gloss Lamination